THE PROTON:
APPLICATIONS TO ORGANIC CHEMISTRY

This is Volume 46 of
ORGANIC CHEMISTRY
A series of monographs
Editor: HARRY H. WASSERMAN

A complete list of the books in this series appears at the end of the volume.

THE PROTON:
APPLICATIONS TO
ORGANIC CHEMISTRY

Ross Stewart

Department of Chemistry
University of British Columbia
Vancouver, British Columbia
Canada

1985

ACADEMIC PRESS, INC.

(Harcourt Brace Jovanovich, Publishers)

Orlando   San Diego   New York   London
Toronto   Montreal   Sydney   Tokyo

ACADEMIC PRESS, INC.
Orlando, Florida 32887

United Kingdom Edition published by
ACADEMIC PRESS INC. (LONDON) LTD.
24–28 Oval Road, London NW1 7DX

LIBRARY OF CONGRESS CATALOGING IN PUBLICATION DATA

Stewart, Ross,   Date
    The proton: applications to organic chemistry.

    Includes index.
    1. Chemistry, Physical organic.   2. Protons.
I. Title.
QD476.S74   1985   547.1'3   85-757
ISBN 0-12-670370-1   (alk. paper)

PRINTED IN THE UNITED STATES OF AMERICA

85 86 87 88   9 8 7 6 5 4 3 2 1

Contents

4
PROTON TRANSFER, HYDROGEN ATOM TRANSFER, AND HYDRIDE TRANSFER

5
ALTERNATIVE SITES OF PROTONATION AND DEPROTONATION

6
ACIDITY AND BASICITY OF UNSTABLE AND METASTABLE ORGANIC SPECIES

7
ACTIVATION OF ORGANIC MOLECULES BY ACIDS AND BASES

Preface

The proton occupies a central place in organic chemistry, which the present monograph attempts to describe. It is a quarter of a century since the appearance of R. P. Bell's important work, *The Proton in Chemistry,* a book that dealt with a number of aspects of the subject drawn from the chemical sciences as a whole. The present work differs from Bell's in being concerned specifically with organic chemistry and, as befits this branch of the subject, being somewhat less rigorous in its treatment.

An introductory chapter is followed by two chapters that treat in fairly exhaustive fashion the strengths of neutral organic acids and neutral organic bases. Subsequent chapters deal with the mode of transfer of hydrogen in its three forms (H^+, $H\bullet$, and H^-), with alternative sites of protonation or deprotonation of organic compounds, with the acid–base chemistry of unstable and metastable species (including free radicals and excited states), and finally, with the activation induced in organic molecules by proton addition or removal and the catalytic effects that ensue.

The book is intended to be of use to practicing organic chemists of whatever stripe, and I hope that for a number of years it will be the first book to be pulled from the shelf when a question arises about the acid–base chemistry of some organic compound. Should the answer not be found herein, there is a reasonable likelihood that the reader will at least find a helpful reference to the chemical literature. In this connection an attempt has been made to include as many recent literature references as possible.

I am grateful to the following friends and colleagues who reviewed parts of the manuscript: Drs. C. A. Bunton, P. Y. Bruice, R. A. Cox, J. T. Edward, A. J. Kresge, C. Sandorfy, R. Srinivasan, and K. Yates. Dr. J. Peter Guthrie, who reviewed the entire manuscript, deserves special thanks for his very helpful criticisms and suggestions. For anything dubious that might remain I take full responsibility. I am also grateful to the Izaak Walton Killam Memorial Fund for Advanced Studies for the award of a Senior Fellowship that enabled me to devote an uninterrupted year to the completion of the work. Finally, thanks are due to my wife, whose help in a much earlier work (ref. *18,* Chapter 4) was attributed in the preface to "a charming lady," an acknowledgment that was accurate but considered to be open to misinterpretation. At any rate the same charming lady helped prepare the manuscript and index of the present work.

1

Introduction

The proton is a unique chemical species, being a bare nucleus. As a consequence it has no independent existence in the condensed state and is invariably found bound by a pair of electrons to another atom. It is proton *transfer* that is at the heart of the concept of acidity, which is the cornerstone on which much of modern chemical science has been built. Measurements of the extent of proton transfer to and from such compounds as carboxylic acids, phenols, sulfonic acids, and amines have provided a vast body of data from which much of the theory of organic reactivity has emerged. Inductive effects, linear free-energy relationships, and acid and base catalysis are all subjects that owe their development to the availability of such data.

Acids and bases have been defined in a number of ways over the years, but we shall follow herein the Brønsted definition, in which an acid is a species that gives up a proton and a base is a substance that accepts a proton (*1–4*). Alternative definitions of acidity and basicity have been well described elsewhere (*5–7*) and need not concern us, inasmuch as the present work's focus is explicitly on the proton and the chemistry associated with the transfer of this unique entity.

I. Water and the Reference State

We live in what has been called an aquocentric environment, and it is only natural that water should have become the standard medium for the quantitative determination of acid and base strengths. Thus, infinite dilution in water at 25°C is generally regarded as the standard state for acid–base equilibria. Apart from its ubiquity water has a number of other advantages over most alternative solvent systems. It has a high dielectric constant, thus reducing the effect of ion pairing on acid–base equilibria, and it can be easily obtained in a highly pure form. Above all, the teaching of chemical equilibria is still based to a very great extent on acid–base reactions in water, and one suspects that there are few practicing chemists who do not regard aqueous pH 7 as being somehow *the* point of neutrality, even though their work may normally involve reactions in organic media or in the solid state or in the gas phase. Further reasons for aqueous systems having become the focus of acid–base studies have been elegantly set forth by Arnett and Scorrano (*8*).

On the other hand, there are some disadvantages to referring all questions of acid and base strength to the aqueous state. The principal objections that can be raised to such an approach are the following. First, there is extensive solvation by water molecules of the components of an acid–base equilibrium, particularly the ionic components, making the system less satisfactory for examining fundamental properties than gas phase studies. Second, many organic compounds whose acidity or basicity is of interest are not particularly soluble in water. Third, the autoprotolysis constant of water, though fairly small, is not nearly small enough to encompass the range of acidities that would enable us to measure directly acid strengths as low as those of alkanes and as high as those of sulfonic acids. Similarly, the base strengths of very weak bases such as aldehydes or nitriles are well beyond the reach of direct measurement in water. [The paucity of very strong neutral bases means that water is able to accommodate measurements of virtually all of the stronger neutral bases that do exist; the strongest reported neutral base, apart from some cryptands that react very slowly, is the diaminonaphthalene **1**, whose pK_{BH^+} is estimated to be near 16 (9), just beyond the range of measurements in water.]

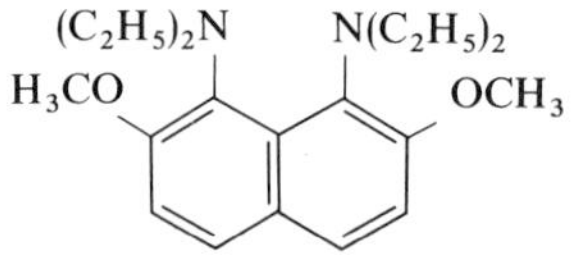

$$(C_2H_5)_2N \quad N(C_2H_5)_2$$
$$H_3CO \quad\quad\quad OCH_3$$

1

How critical are these three points? They will be addressed in Section III, but first it is important that the terms and symbols used to express acid and base strengths be clarified.

II. Acid–Base Equilibria: Terms and Symbols

It has become common practice to use the symbol pK_a to express the strengths of both acids and bases. In the latter case, of course, the quantity referred to is actually the strength of the conjugate acid of the base. Thus, the equilibrium constant for the dissociation of anilinium ion in water at 25°C is 2.5×10^{-5} [Eq. (1-1)], the negative logarithm of this quantity, 4.60, being

$$C_6H_5NH_3^+ \rightleftharpoons H^+ + C_6H_5NH_2 \quad K = 2.5 \times 10^{-5} \quad (1\text{-}1)$$

sometimes referred to as the "pK_a of aniline." However, aniline is also an acid in its own right, although a very weak one in water, and it is the pK for the equilibrium shown in Eq. (1-2) that would more aptly be so described (*10*).

$$C_6H_5NH_2 \rightleftharpoons H^+ + C_6H_5NH^- \qquad K \approx 10^{-28} \qquad (1\text{-}2)$$

The strength in water of a base such as aniline can always be related to the reaction shown in Eq. (1-3), with K_b and pK_b being the appropriate symbols

$$C_6H_5NH_2 + H_2O \rightleftharpoons C_6H_5NH_3^+ + HO^- \qquad K = 4.0 \times 10^{-10} \qquad (1\text{-}3)$$

to use in this case. In aqueous solution at 25°C the pK_a and pK_b for weak bases and their conjugate acids are related by Eq. (1-4). [For moderately

$$pK_a + pK_b = 14.0 \qquad (1\text{-}4)$$

strong acids this relationship may not hold rigorously (*11*).]

Why, in the case of bases, have K_b and pK_b been largely superseded by the terms K_a and pK_a, ambiguous though the latter terms sometimes can be? The advantage of writing all acid–base reactions in the direction of proton loss and then using K_a and pK_a to describe the equilibria as written is that one has a unified scale running from the most powerful acids, whether they be neutral species such as trifluoromethanesulfonic acid or protonated species such as the conjugate acids of nitroalkanes, through to the weakest acids, the alkanes. There are two ways of avoiding ambiguity when using the unified approach whereby all acid–base reactions are written in the direction of proton loss. One is always to refer to the acid or conjugate acid by its appropriate name or formula. Thus, it would always be "the pK of anilinium ion" or "the pK of $C_6H_5NH_3^+$" that is given as 4.60 and never "the pK of aniline" that is so designated. There is little difficulty in following this stricture with regard to chemical formulas; problems arise, however, with regard to names for the conjugate acids of ketones, esters, nitro compounds, and others. One can always attach the prefix "conjugate acid of" to the name of the neutral compound, but this practice can weary both the writer and the reader if followed rigorously.

The second way to avoid ambiguity, and the one that will be followed herein, is to use the terms K_{HA} and pK_{HA} for the acid dissociation of the species in question and the terms K_{BH^+} and pK_{BH^+} for its conjugate acid. Thus, we shall write "the pK_{HA} of aniline is 28" and the "pK_{BH^+} of aniline is 4.60." [The term "pK_{HA} (or pK_a) of anilinium ion" causes no difficulty, but the need for such a term will only rarely be encountered herein.] Throughout this work the terms pK_{HA} and pK_{BH^+} will usually be used, since this practice allows both weakly acidic and weakly basic organic compounds to be referred to by their usual names and formulas.

There remains the problem of molecules containing more than one acidic or basic site, for example, polycarboxylic acids or polyamino compounds. The traditional terms pK_1, pK_2, and so on seldom give rise to ambiguity in the case of polyprotic acids since it is usually perfectly clear which processes are being referred to; doubts are more likely to arise with polybasic compounds, and in these cases reference to the appropriate chemical equation can be made. When it is necessary to refer to the basicity of a polycarboxylic acid, the term pK_{BH^+} (plus the name or formula of the neutral compound) is available, and when the acidity of a polyamino compound is in question, the term pK_{HA} (plus name or formula) can be used, just as for monofunctional compounds.

In the case of unsymmetrical polyprotic molecules, microscopic equilibrium processes become relevant. In these cases the microscopic dissociation in question should be unambiguously identified by, for example, a chemical equation; the unvarnished symbol K (or pK) is really all that is needed if an equation is used, since the reaction to which the equilibrium constant refers has been specified.

In summary, pK_{HA} refers to the acid strength of the species whose name or formula is given (usually, but not necessarily, a neutral molecule); pK_{BH^+} refers to the acid strength of the conjugate acid of the species whose name or formula is given (again, usually a neutral molecule). For proton dissociation from polyprotic species, pK_1, pK_2, and so on will be used without explanation when there is no possibility of ambiguity; when this possibility exists or when microscopic dissociations are being considered, the reaction in question will be suitably identified (e.g., by means of an equation).

With regard to amphiprotic molecules, such as glycine and other simple amino acids, there are four microscopic acid dissociation processes that must be considered; they and their associated pK values are shown in Eqs. (1-5) to (1-8) (*12*).

$$H_3\overset{+}{N}CH_2CO_2H \rightleftharpoons H^+ + H_3\overset{+}{N}CH_2CO_2^- \qquad pK = 2.35 \qquad (1\text{-}5)$$

$$H_3\overset{+}{N}CH_2CO_2H \rightleftharpoons H^+ + H_2NCH_2CO_2H \qquad pK \approx 7.6 \qquad (1\text{-}6)$$

$$H_3\overset{+}{N}CH_2CO_2^- \rightleftharpoons H^+ + H_2NCH_2CO_2^- \qquad pK = 9.78 \qquad (1\text{-}7)$$

$$H_2NCH_2CO_2H \rightleftharpoons H^+ + H_2NCH_2CO_2^- \qquad pK \approx 4.5 \qquad (1\text{-}8)$$

The quantities pK_1 and pK_2, usually used for the experimentally determined dissociation constants of (protonated) glycine, are identical, within experimental error, to the pK values shown with Eqs. (1-5) and (1-7). The pK_1 value is actually a composite constant for reactions (1-5) and (1-6), with reaction (1-5) being the overwhelmingly dominant quantity, as can be seen from the size of the respective dissociation constants. [The estimate of 7.6 for the pK of reaction (1-6) comes from the value of pK_{BH^+} of glycine ethyl ester.] The pK_2 value of glycine includes the reactions shown in Eqs. (1-7) and (1-8), but because almost all glycine of zero net charge is in the dipolar form, the pK for the dissociation of the ammonio group [Eq. (1-7)] is virtually identical with the experimental pK_2. A complete analysis of the relationships between microscopic and macroscopic dissociation constants is given in Chapter 5.

III. Limitations on Water as a Medium for Acid–Base Measurements

It was pointed out in Section I that there are three disadvantages to using water as the standard state for acidity measurements. First, rather strong solvation effects will generally be present in aqueous solution; second, many compounds of interest have only slight solubility in water; and third, the size of the autoprotolysis constant of water sets a limit on the range of pK's that can be measured directly in this medium.

With respect to the first point, it does indeed seem reasonable to regard the fundamental properties of a compound as those of the molecule in isolation, that is, in the infinitely dilute gas phase, since molecular interactions in the liquid and solid states modify the intrinsic properties that molecules in isolation possess. Whether such a state is the most relevant starting point for discussing chemical reactivity is another matter altogether. Most chemistry is conducted in the condensed phase, where molecules are affected by the presence of neighboring matter, even when the molecules comprising such matter are completely unreactive and nonpolar.

Thus, a condensed medium, by virtue of its polarizability, will provide a substantial degree of interaction with solute molecules, even though the medium might have an extremely low dielectric constant. The difference in this respect between a medium such as liquid hexane, with a dielectric constant of 1.9, and a vacuum, with its dielectric constant of 1.0, is very much greater than a simple comparison of the two numbers would suggest.

Partly for these reasons the condensed state (and in particular the aqueous state) will be used in this work as the reference state for questions of acid and base strength. That is not to say that the important work done since the mid-1970s on proton transfer in the gas phase by Kebarle, McIver, Taft, Boehme, DePuy, McDonald, and others (*13–18*) is irrelevant to our concerns; far from it. It tells us a great deal about the fundamental properties of acids and bases, and reference to such work will be made where it seems appropriate to do so.

With regard to the limited solubility in water of many organic compounds, it turns out that this factor is not of particular consequence as far as extremely weak acids and bases are concerned. Most determinations of the strengths of such compounds in water require the use of mixed solvents in any case, which generally dissolve organic compounds to a greater extent than does water.

This solubilizing effect is observed not only with aqueous dimethyl sulfoxide (DMSO) and like systems that are often used in pK_{HA} measurements of very weak acids, but also with aqueous sulfuric acid and like systems that are often used in pK_{BH^+} measurements of very weak bases. For example, nitrobenzene is ~ 10 times as soluble in 80 wt % sulfuric acid as it is in water (19), with the increase not being caused by protonation of the substrate, which is in fact negligible at this acidity. In those other cases where substrate solubility is insufficient, water can be replaced by ethanol (20–24), dioxane (25), or other solvent (26) and corrections made in order to refer the pK_{BH^+} values so obtained back to the standard state of water (26).

There remains the matter of the magnitude of the autoprotolysis constant of water. Dimethyl sulfoxide has a much smaller degree of autoprotolysis than does water [Eqs. (1-9) and (1-10)] (27–29), and it has been pointed out by Bordwell (30) and others that DMSO as a reference state allows direct measurements to be made on nitrogen and carbon acids whose degree of ionization is far too small to be directly measured in even the most alkaline aqueous solutions. Likewise, DMSO solutions containing strong acids are highly acidic and can protonate weak organic bases.

$$2\,H_2O \rightleftharpoons H_3O^+ + HO^- \tag{1-9}$$

$$K_{auto} = [H_3O^+][HO^-] = 10^{-14}$$

$$2\,CH_3SCH_3 \rightleftharpoons H_3C\overset{\overset{+OH}{\|}}{-}S{-}CH_3 + H_3C\overset{\overset{O^-}{|}}{-}S{=}CH_2 \tag{1-10}$$

$$K_{auto} = \left[CH_3\overset{\overset{+OH}{\|}}{S}CH_3\right]\left[H_3C\overset{\overset{O^-}{|}}{-}S{=}CH_2\right] = 10^{-33}$$

Despite this advantage the DMSO standard state and its attendant pK scale are unlikely to supplant the long-established aqueous scale, and the most that might be expected would be an indefinite period of confusion if attempts were made to supplant aqueous pK values with those based on DMSO medium, even though in the case of very weak acids the latter would probably be the more firmly established of the two. [Particular care, however, is required in making measurements in DMSO because of the great effect that small amounts of water and other likely impurities have on the acidity or basicity of the medium (31).]

A distinctive feature of pK_{HA} measurements in aqueous solution is the absence of ion-pairing effects, except at very high solute concentrations. Although solvation is a powerful influence, it is at least comparatively uniform from compound to compound. The situation is much more difficult in most other solvents, where ion pairing can be a serious problem; even in DMSO, ion pairing is significant in all but the most dilute solutions.

If it could be shown that pK values in water and in DMSO are related in some direct way, one could use DMSO to determine the acidity constants of extremely weak acids and then simply convert them to the familiar scale based on water. It turns out that not all acids respond in the same way to solvent changes, and so such an approach will not be applicable in many cases. Nonetheless, measurements in DMSO or other media can provide us with a wealth of useful information, particularly if pK values determined in this medium are always clearly labeled as such. The procedure followed in this work will be to indicate the standard state for a pK value, provided that state is *not* water. A pK referred to the standard state of water will be simply given as pK_{HA} for a neutral acid and pK_{BH^+} for the conjugate acid of a neutral base, whereas, for example, a pK_{HA} referred to the standard state of DMSO will be given as pK_{HA}^{DMSO}.

A compound whose acidity constant is listed as pK_{HA}^{DMSO} will almost certainly have had this quantity determined by direct measurement in DMSO, whereas the analogous situation with respect to pK_{HA} and measurements in water frequently does not apply. Indeed, a pK_{HA} that is given as significantly less than zero or significantly greater than 14 could not have been determined directly in water because of the leveling effect imposed by the autoprotolysis of the solvent. In these cases the actual measurements of ionization would have been made in a mixed aqueous or even nonaqueous medium and the acid strength referred to aqueous conditions by one of the means to be discussed later.

It is important to recognize that, regardless of the medium actually used to measure the degree of ionization, the pK_{HA} is that of the acid in water. Thus, if the pK_{HA} of aniline is indeed 28, as has been estimated, its degree of dissociation to give the anilide ion, $C_6H_5NH^-$, in water at pH 7 is 1 molecule in 10^{21}.

How can the problem of the limited range of water for direct measurement be overcome? That is, how can the pK_{HA} of an extremely weak acid or the pK_{BH^+} of an extremely weak base be referred to water? There have been a number of attempts to overcome this problem, with the acidity function, a concept introduced by Hammett in 1932 (*32, 33*), being the most widely applied and probably the most successful. As with so many of the powerful generalizations that have periodically transformed physical–organic chemistry, the acidity function approach has been found on close examination to have serious flaws, though it remains an extremely effective means of studying acid–base chemistry in regions outside the aqueous state, the state to which the reference pK values are firmly anchored. The use of this approach to determine the acidity constants of very strong and very weak organic acids (pK_{HA}) is considered in Chapter 2 and of very weak organic bases (pK_{BH^+}) in Chapter 3.

References

1. J. N. Brønsted, *Rec. Trav. Chim.* **42**, 718 (1923).
2. J. N. Brønsted, *Chem. Ber.* **61**, 2049 (1928).
3. See also T. M. Lowry, *Chem. Ind.* **42**, 43 (1923) and ref. *4.*
4. R. P. Bell, "The Proton in Chemistry," 2nd ed., p. 4. Cornell Univ. Press, Ithaca, New York, 1973.
5. H. L. Finston and A. C. Rychtman, "A New View of Current Acid–Base Theories." Wiley (Interscience), New York, 1982.
6. R. J. Gillespie, *in* "Friedel–Crafts and Related Reactions" (G. Olah, ed.), Vol. 1, p. 169. Wiley (Interscience), New York, 1963.
7. D. P. N. Satchell and R. S. Satchell, *J. Chem. Soc. Q. Rev.* **25**, 171 (1971).
8. E. M. Arnett and G. Scorrano, *in* "Advances in Physical Organic Chemistry" (V. Gold and D. Bethell, eds.), Vol. 13, p. 85. Academic Press, London, 1976.
9. F. Hibbert and K. P. P. Hunte, *J. Chem. Soc., Perkin Trans. 2* p. 1895 (1983).
10. See Chapter 2, Section IV,C,5, for the source of the K value given here.
11. See Chapter 2, Section II,A.
12. See pp. 207, 223, and 224, herein.
13. M. Fujio, R. T. McIver, and R. W. Taft, *J. Am. Chem. Soc.* **103**, 4017 (1981) and refs. therein.
14. J. A. Bartmess, J. A. Scott, and R. T. McIver, *J. Am. Chem. Soc.* **101**, 6046 (1979) and refs. therein.
15. G. I. McKay, M. H. Lien, A. C. Hopkinson, and D. K. Boehme, *Can. J. Chem.* **56**, 131 (1978) and refs. therein.
16. S. Ikuta, P. Kebarle, G. M. Bancroft, T. Chan, and R. C. Puddephatt, *J. Am. Chem. Soc.* **104**, 5899 (1982) and refs. therein.
17. R. W. Taft, *Prog. Phys. Org. Chem.* **14**, 247 (1983) and refs. therein.
17a. R. N. McDonald, A. K. Chowdhury, and W. D. McGhee, *J. Am. Chem. Soc.* **106**, 4112 (1984).
18. C. H. DePuy, V. M. Bierbaum, and R. Damrauer, *J. Am. Chem. Soc.* **106**, 4051 (1984).
19. L. P. Hammett and R. P. Chapman, *J. Am. Chem. Soc.* **56**, 1282 (1934).
20. S.-J. Ye and H. H. Jaffé, *J. Am. Chem. Soc.* **81**, 3274 (1959).
21. D. Dolman and R. Stewart, *Can. J. Chem.* **45**, 903 (1967).
22. G. Janata and G. Jansen, *J. Chem. Soc. Faraday Trans. 1* p. 1656 (1972).
23. A. J. Kresge and H. J. Chen, *J. Am. Chem. Soc.* **94**, 8192 (1972).
24. C. Capobianca, F. Magno, and G. Scorrano, *J. Org. Chem.* **44**, 1654 (1979).
25. U. Quintily, G. Scorrano, and F. Magno, *Gazz. Chim Ital.* **111**, 401 (1981).
26. C. H. Rochester, "Acidity Functions." Academic Press, London, 1970.
27. E. C. Steiner and J. M. Gilbert, *J. Am. Chem. Soc.* **87**, 382 (1965).
28. E. C. Steiner and J. D. Starkey, *J. Am. Chem. Soc.* **89**, 2751 (1967).
29. R. Stewart and J. R. Jones, *J. Am. Chem. Soc.* **89**, 5069 (1967).
30. F. G. Bordwell, *Pure Appl. Chem.* **49**, 963 (1977).
31. C. D. Ritchie, *in* "Solute–Solvent Interactions" (J. F. Coetzee and C. D. Ritchie, eds.), p. 231. Dekker, New York, 1969.
32. L. P. Hammett, *Chem. Rev.* **16**, 67 (1935).
33. L. P. Hammett, "Physical Organic Chemistry," 1st ed. McGraw-Hill, New York, 1940.

2

Strengths of Neutral Organic Acids

I. Introduction

The strongest organic acid is trifluoromethanesulfonic acid, CF_3SO_3H, commonly known as triflic acid (*1*). It is one of the strongest monoprotic acids of any sort and, indeed, there are those who regard it as *the* strongest monoprotic acid (*2, 3*). Its chief rival for this distinction is the inorganic compound fluorosulfuric acid, FSO_3H, which does, indeed, appear to be somewhat the weaker of the two when these acids are compared in anhydrous acetic acid. Engelbrecht and Rode (*4*) used conductivity measurements to show that trifluoromethanesulfonic acid protonates this solvent [Eq. (2-1)]

$$CF_3SO_3H + CH_3CO_2H \qquad CF_3SO_3^- + CH_3CO_2H_2^+ \qquad (2\text{-}1)$$

to a greater degree than do any of the other common acids tested. The order of protonating capacity was found to be $CF_3SO_3H > HClO_4 > HBr > HI > FSO_3H > H_2SO_4 > HCl$ (*3–5*).

On the other hand, Russell and Senior have shown clearly that in the sulfuric acid solvent system fluorosulfuric acid is slightly stronger than trifluoromethanesulfonic acid (*6*). Since ion pairing is much more extensive in acetic acid than in sulfuric acid the latter solvent system is probably the better one to use when estimating orders of acid strength in water, which would deprive trifluoromethanesulfonic acid of the honor of primacy among monoprotic acids in water, the solvent that we are using as our reference medium. [The strongest acid in sulfuric acid medium, as indicated by the extent of protonation of nitroarenes, is $HB(HSO_4)_4$ (*7*).] The precise order may not seem of great consequence since all the acids just referred to are completely ionized in water for all practical purposes. It can, however, influence the rate at which the acid anions are displaced in nucleophilic processes.

Trifluoromethanesulfonic acid is a thermally stable, readily distilled, colorless liquid. Its relatively low boiling point, 162°C, is in accord with the usual effect of fluorine substitution on volatility. (Methanesulfonic acid boils at this temperature only when the pressure is reduced to 8 torr). It absorbs water readily to form the crystalline hydrate $CF_3SO_3^- \cdot H_3O^+$ (*8*), which melts at 34°C.

The organic sulfonic acids are strong acids in water, with methanesulfonic acid, one of the weakest of the class, having an ionization constant that has been estimated from Raman measurements to be 83 (see Section II,A), corresponding to a degree of ionization in 0.1 M aqueous solution of $>99.8\%$. The only other organic acids of comparable strength to the sulfonic acids are the cyanocarbons, some of which have extremely high acidities (Section IV,B).

For purposes of subsequent discussion organic acids have been divided into three groups: (*a*) oxygen acids, (*b*) sulfur acids, and (*c*) nitrogen and carbon acids. The reason for this, apart from organizational convenience, is the fact that members of these classes to a great extent occupy different regions of the acid–base spectrum and to some degree different techniques have been used to determine or estimate their acid strengths. The oxygen acids (ionizable proton attached to oxygen in the acid form) range in strength from the weakest, the alcohols, whose pK_{HA} values are comparable to that of water, to the sulfonic acids, some of which have acidities comparable to that of sulfuric acid. The sulfur acids (ionizable proton attached to sulfur) occupy a narrower range of acidity, with none being as weak as alcohols and none as strong as the sulfonic acids. Nitrogen and carbon acids have their ionizable protons attached to elements of lower electronegativity, and many of these compounds are extremely weak acids whose degree of ionization in water is undetectable. There are exceptions, however; hexanitrodiphenylamine (**1**) has a pK_{HA} of 2.63 (*9*), meaning that it is stronger than chloroacetic acid, and pentacyanocyclopentadiene (**2**), which is isolable only in the form of its salt, appears to have an acidity comparable to that of the stronger mineral acids (*10, 11*).

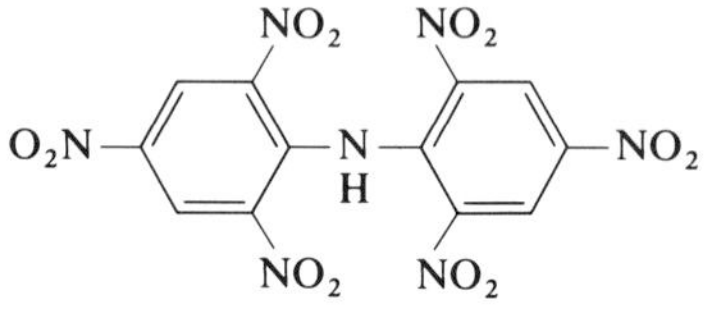

1 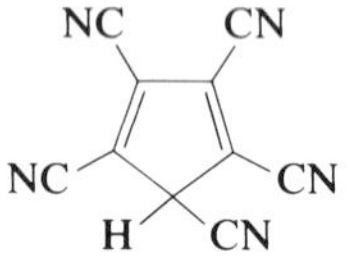   2

II. Oxygen Acids

Of the commonly encountered organic compounds that have protons attached to oxygen, alcohols are the weakest and sulfonic acids the strongest, and both present some difficulties with regard to determining their acid strengths. Carboxylic acids and phenols are of intermediate strength; their ionization constants in water can usually be obtained with little difficulty using the glass electrode and standard techniques. The trihaloacetic acids are exceptional in being highly ionized in water, and the ionization behavior of these and other strong or moderately strong acids shows signs of differing qualitatively as well as quantitatively from that of weaker acids. In addition to the trihaloacetic acids this group includes sulfonic acids, alkylsulfuric acids, phosphonic acids, and alkylphosphoric acids. (The lower-valence analogues, the sulfinic and phosphinic acids, are weak acids in water; the relation between oxidation level and acid and base strength is discussed in Chapter 3, Section VIII.)

A. Strong and Moderately Strong Oxygen Acids

Evidence has accumulated in recent years that strong and moderately strong acids have a somewhat different ionization and dissociation pattern than weak acids. Indeed, the terms *ionization* and *dissociation*, which are used interchangeably to describe the behavior in water of such compounds as carboxylic acids, appear to be anything but synonymous as far as sulfonic and other such acids are concerned.

$$CH_3CO_2H + H_2O \underset{}{\overset{K_i}{\rightleftharpoons}} CH_3CO_2^- \cdot H_3O^+ \overset{K_d}{\rightleftharpoons} CH_3CO_2^- + H_3O^+ \quad (2\text{-}2)$$

$$CH_3SO_3H + H_2O \overset{K_i}{\rightleftharpoons} CH_3SO_3^- \cdot H_3O^+ \overset{K_d}{\rightleftharpoons} CH_3SO_3^- + H_3O^+ \quad (2\text{-}3)$$

The first step in Eqs. (2-2) and (2-3) is ionization, transfer of a proton to a water molecule to form an ion-pair complex containing highly polarizable protons (*12*); the second step is dissociation of the complex to form the separated ions. Some time ago Eigen (*13*) described acid–base processes in these terms, but only more recently has it been recognized that the undissociated forms of moderately strong acids, such as methanesulfonic acid, are present in water to a great extent in the form of ion pairs; that is, K_i in Eq. (2-3) is fairly large. For carboxylic and other weak acids, on the other hand, undissociated ion pairs are no more than transient intermediates in the overall proton transfer process; that is, K_i in Eq. (2-2) is very small, and Eq. (2-2) reduces to the familiar form shown in Eq. (2-4).

$$CH_3CO_2H + H_2O \rightleftharpoons CH_3CO_2^- + H_3O^+ \quad (2\text{-}4)$$

For moderately strong acids discrepancies often appear between the hydrogen ion concentration determined by potentiometric or conductimetric techniques and the concentration of the acid anion determined by spectroscopic and other techniques (*14–19*). This phenomenon escaped detection for many years because it occurs only with those acids whose degree of ionization in water is high and whose ionization constants are notoriously difficult to measure accurately. Why should it be more difficult to determine the ionization constant of a highly ionized acid than of a slightly ionized acid? Because with very few exceptions it is very much easier to measure accurately the concentration of ions, particularly the hydrogen ion, than that of neutral molecules. The glass electrode provides an extremely accurate means of measuring small concentrations of hydrogen ion in water, and it has long been assumed that the concentration of the other components in the equilibrium could be obtained with a high degree of precision from this quantity. Various measurements of the concentration of the anion by such techniques as Raman spectroscopy support this long-held view in the case of weak acids; that is, K_d in Eq. (2-2) is very much greater than K_i. With moderately strong acids, however, the concentrations of anion and proton appear to diverge, although the degree of divergence in some cases is difficult to evaluate quantitatively because with such acids the relative amount of un-ionized acid is rather low and the anion and proton concentrations are often only marginally less than the stoichiometric concentrations. Furthermore, with strong acids higher concentrations of substrate are needed, and this can require activity corrections.

If accurate means existed of detecting small concentrations of un-ionized neutral acid in the presence of larger amounts of its ionized or dissociated products, values of the ionization constants K_i and dissociation constants K_d in Eqs. (2-2) and (2-3) could be obtained. Apart from the glass electrode the laser Raman spectrometer appears to be the most effective instrument for measuring the concentrations of components of acid–base equilibria. Here again, however, the neutral acid is often less amenable to quantitative measurement than is its anion. This point is illustrated in the extensive studies that have been made of methanesulfonic acid.

Clarke and Woodward (20) and Covington and Thompson (21) measured the degree of ionization of this acid by following the appearance of the intense sharp band at 1051 cm^{-1}, characteristic of the methanesulfonate ion, that accompanies dilution of methanesulfonic acid with water. At the same time the band at 1358 cm^{-1}, characteristic of the un-ionized molecule, disappears. Unfortunately, this band is broader and much less intense than other bands (many of which are common to both the ion and the neutral molecule), and it is undetectable in a 1 M solution and only barely detectable in a 7 M solution. By measuring the intensity of the 1051 cm^{-1} band at various concentrations of aqueous methanesulfonic acid and extrapolating to infinite dilution in water, an estimate of the ionization constant in water was obtained. The extrapolations were done using both molarity and molality concentration units, both giving curved lines converging on the value of 83 for the ionization constant of methanesulfonic acid, corresponding to a pK_{HA} of -1.92 in water. Unfortunately, the precise extrapolated value depends to a great extent on the degree of ionization measured in the most dilute solutions used, ~ 0.5 M, and these are just the ones for which the stoichiometric concentration of acid and the experimentally measured concentration of anion are very close. That is, a very small error in the measured concentration of ion will substantially change the value of the degree of ionization and the ionization constant. Nonetheless, it is clear from the trends in the more accurately measurable degrees of ionization in more concentrated solutions that the pK_{HA} of methanesulfonic acid in water should not be more than 0.06 pK unit away from the value of -1.92 given by Covington and Thompson, and this is the value that Guthrie has taken as the starting point for his estimate of the strengths of stronger acids (22).

Taking -1.92 as the pK_{HA} of methanesulfonic acid, the question arises, Does this correspond to ionization [K_i in Eq. (2-3)] or dissociation [K_d in Eq. (2-3)] or both? Clearly, it must include both, since Raman measurements detect the presence of the anion, whether it is free or complexed. A problem arises, however, in reconciling the dimensions of these quantities; K_i is dimensionless, whereas K_d has units of moles per liter. The expression for the apparent equilibrium constant determined by Raman spectroscopy has the form of Eq. (2-4a), whereas that determined by techniques that measure the concentration or activity of free hydrogen ion (e.g., the glass electrode) has the form of Eq. (2-4b) (14, 16).

$$K = \frac{([A^-] + [A^- \cdot H_3O^+])^2}{[HA]} \tag{2-4a}$$

$$K = \frac{[H_3O^+][A^-]}{[HA] + [A^- \cdot H_3O^+]} \tag{2-4b}$$

In practical terms, does incomplete dissociation of the complex reduce the effective acidity of a dilute aqueous solution of methanesulfonic acid? That is, will the pH of, say, a 10^{-2} M solution of methanesulfonic acid be higher than expected for an acid of such strength? At this and lower concentrations the degree of ionization and the degree of dissociation of the complex are both very high and so the difference between the free hydrogen ion concentration and the stoichiometric concentration of acid will be too small to measure.

We shall use pK_{HA} as the symbol for the ionization constant of both strong and weak acids in water. For weak acids the concentration of the undissociated ion-pair complex is extremely low at all concentrations, and it can be effectively ignored as far as acid–base equilibria are concerned; that is, ionization and dissociation are synonymous. For strong acids the proportion of complex is very low in dilute solution but becomes significant at higher concentrations. This effect compounds, to some extent, the problems associated with activity effects that also arise as one departs from the standard state of dilute aqueous solution.

Other moderately strong organic acids that have been observed to form ion-pair complexes at higher concentrations include trifluoroacetic acid, trichloroacetic acid, butynedioic acid, squaric acid (1,2-dihydroxycyclo-butenedione), dichloroacetic acid, and difluoroacetic acid (14–19). In addition, it is known that in concentrated solutions of sulfuric acid extensive ion pairing occurs. Thus, the Raman spectrum of 74 wt % aqueous sulfuric acid, which has the composition $H_2SO_4 \cdot 2H_2O$, shows it to consist largely of the ion-pair species $HSO_4^- \cdot H_2O \cdot H_3O^+$ (23). It has also been suggested that the unexpectedly low acid strength of hydrogen fluoride in water is due to its being ionized but not dissociated (23a).

B. Estimating the Strength in Water of Strong Acids

Of the common mineral acids that are generally designated "strong," perchloric acid is the only one that retains that rank when acetonitrile is used as solvent (24–26). Hydrochloric, hydrobromic, sulfuric, and nitric acids are all weakly ionized in this medium, hydrochloric acid, for example, having a pK of 8.9; that is, the equilibrium shown in Eq. (2-5) is well to the left.

$$HCl + CH_3C{\equiv}N \rightleftharpoons CH_3C{\equiv}NH^+ + Cl^- \qquad (2\text{-}5)$$

The strongest organic acids, the sulfonic acids, are also weak in acetonitrile, except for trifluoromethanesulfonic acid. Although its strength appears not to have been measured in this medium, it is stronger than perchloric acid when sulfuric acid is the medium, and so it can be expected to be highly ionized in acetonitrile.

The same order of acid strengths of the common mineral acids ($HClO_4 >$ $HBr > H_2SO_4 > HCl > HNO_3$) is generally found in other solvents, but discrepancies in the order of acid strength are frequently found when a wider selection of acids is examined. Table 2-1 shows the order of acid strengths observed in a number of solvents. It can be seen that there are several acids whose relative acid strengths depend on the solvent in which they are dissolved and one in particular, fluorosulfuric acid, whose position in the order of precedence seems to be particularly capricious. The inversion of strengths of fluorosulfuric and trifluoromethanesulfonic acids in acetic and trifluoroacetic acids has already been referred to. In addition, fluorosulfuric acid and perchloric acid have different orders of strength in acetic acid or sulfolane (perchloric acid stronger) and sulfuric acid (fluorosulfuric acid stronger). Comparing fluorosulfuric acid and disulfuric acid, we see that fluorosulfuric acid is the stronger of the two in sulfolane but the weaker of the two in both sulfuric acid and propylene carbonate.

Though it is clear that there is no order of acid strength that will apply to all solvent systems, it is possible to produce a probable ordering in inert high-dielectric media by using as the criterion for resolving discrepancies the dielectric constant of the medium (Scheme 2-1). This would correspond to the

Table 2-1

ORDER OF ACID STRENGTHS IN VARIOUS SOLVENTS

Solvent	Dielectric constant	pK_{auto}	Order	Ref.
Acetonitrile	36	28.5	$HClO_4 > HBr > H_2SO_4 > HCl > HNO_3$	24–26
Acetic acid	6.1	14.5	$CF_3SO_3H > HClO_4 > HBr > HI > FSO_3H > H_2SO_4 > HCl$	4
Trifluoroacetic acid	7.8		$HClO_4 > H_2SO_4 > CH_3SO_3H \sim 4\text{-}CH_3C_6H_4SO_3H$	27
Sulfuric acid	101	3.6	$H_2S_2O_7 > FSO_3H > CF_3SO_3H \sim ClSO_3H > HClO_4$	6, 28, 29
Sulfolane	43.3		$HSbCl_6 > HClO_4 > FSO_3H > H_2S_2O_7$	30
Dimethyl sulfoxide	46.7	33	$CF_3SO_3H > H_2SO_4 > CH_3SO_3H > HCl > CF_3CO_2H$	31
Propylene carbonate	64.7	29	$HPF_6 > H_2S_2O_7 > FSO_3H$	32, 32a

HPF_6, $HSbCl_6$ > $H_2S_2O_7$ > FSO_3H > CF_3SO_3H, $ClSO_3H$ > $HClO_4$ > HBr > HI >

H_2SO_4 > HCl > CH_3SO_3H > HNO_3 > CF_3CO_2H

Scheme 2-1   General order of acid strength in high-dielectric media.

ordering in water, the medium taken as the standard state herein, except that several of these compounds are unstable in aqueous solution.

Of the acids listed in Scheme 2-1 only trifluoroacetic acid, nitric acid, and methanesulfonic acid show any indications of being less than completely ionized in aqueous solution. If the pK_{HA} of methanesulfonic acid is taken as -1.9, a 0.1 M solution contains less than 0.2% of the solute in the un-ionized form. The bulk of the material is present either as dissociated ions or as the ion-paired species $CH_3SO_3^- \cdot H_3O^+$.

Early attempts by Schwarzenbach (33), Kossiakoff and Harker (34), and Pauling (35) to estimate the strengths of strong inorganic acids were based on consideration of formal charges and the number of oxygen atoms that could bear the negative charge in the anion. All these approaches give a pK_1 for sulfuric acid, for example, of about -3. Following a somewhat different approach and using methanesulfonic acid ($pK_{HA} \equiv -1.9$) as an anchor, Guthrie estimated the strengths in water of a number of strong acids, both organic and inorganic (22). Two approaches to the problem were tried and gave results in fair agreement with one another. The first approach, which is applicable principally to sulfonic and sulfuric acids, is based on substituent effects. A simplified version of this approach is as follows.

It had been shown by Kresge and Tang (36) that pK_2 of phosphonic acids, which corresponds to an ionization process occurring near pH 7, is governed by the Hammett–Taft equation [Eq. (2-6)], where σ^* is the substituent constant for the Y group in Eq. (2-7) (37, 38).

$$pK_2 = 8.10 - 1.26\sigma^* \qquad (2\text{-}6)$$

$$YPO_3H^- \rightleftharpoons H^+ + YPO_3^{2-} \qquad (2\text{-}7)$$

The first ionization constants of these acids are not known to the same degree of precision, but the available evidence (39–41) indicates that the slope of the line in Eq. (2-6) (the ρ^* value in Hammett–Taft nomenclature) also correlates these ionization constants fairly well. That is, a substituent Y in YPO_3H_2 has roughly the same effect on pK_1 and on pK_2. Assuming that the same substituent effect applies to the structurally related sulfonic acids enables one to calculate pK_{HA} values for those sulfonic acids whose degree of ionization is too great to be measured directly. Since CH_3SO_3H is 9.9 pK units more acidic than $CH_3PO_3H^-$, FSO_3H should be 9.9 units more acidic than FPO_3H^- (pK 5.1) (42), CH_3OSO_3H 9.9 units more acidic than $CH_3OPO_3H^-$ (pK 6.6, 6.3) (43, 49), and $C_6H_5SO_3H$ 9.9 units more acidic than $C_6H_5PO_3H^-$

(pK 7.1) (39). [It is not, in fact, necessary to have experimental data for each compound of formula YPO_3H^- in order to estimate the strength of the acid YSO_3H, since one can use with some confidence Eq. (2-6) and the σ^* value of the Y group in question, which for most groups is a well-established quantity.] This procedure gives a pK_{HA} of -3.6 for methylsulfuric acid and a pK_{HA} of -2.8 for benzenesulfonic acid.

The second method used by Guthrie (22), and by others (42–47), to estimate the strengths of strong acids makes use of the differentiating properties of acidic solvents such as acetic acid, trifluoroacetic acid, and sulfuric acids. As we have seen, even fluorosulfuric acid and perchloric acid are incompletely ionized in these solvents (Table 2-1), and this allows ΔpK values to be obtained for such acids in these and other solvents. As also noted previously, inversions in order of acid strength are observed in some cases in going from one solvent to another; thus, it is not surprising that ΔpK for a pair of acids may vary somewhat in magnitude and, indeed, occasionally change sign. Nonetheless, averaging out the differences leads to a fairly consistent set of acid strength values that for most acids agrees quite well with the pK values determined by the method of substituent effects outlined earlier. In both approaches methanesulfonic acid has been the anchor compound and its pK has been taken as -1.9.

Table 2-2 lists the pK_{HA} values of a number of strong acids determined by both methods. It is significant that the greatest discrepancies occur with

Table 2-2

ESTIMATED pK_{HA} VALUES IN WATER OF ACIDS OF STRUCTURE ZSO_3H

Acid (ZSO_3H)	Method 1 (pK_{HA} of ZPO_3H^- $-$ 9.9)[a]	Method 2 (solvent differentiation)[b]	Mean
$C_2H_5SO_3H$	-1.7	-1.7	-1.7
CH_3SO_3H	-1.9	-1.9	-1.9
$4\text{-}CH_3C_6H_4SO_3H$	-2.7	—	-2.7
$C_6H_5SO_3H$[c]	-2.8	-2.8	-2.8
$HOSO_3H$	-3.1	-2.8	-3.0
CH_3OSO_3H	-3.6	-3.4	-3.5
$4\text{-}O_2NC_6H_4SO_3H$	-3.7	-4.0	-3.8
CF_3SO_3H	-5.1	-5.9	-5.5
FSO_3H	-4.8	-6.4	-5.6

[a] References for pK_{HA} values of acids of structure ZPO_3H^-: $Z = CH_3$ (36), aryl (39), HO and CH_3O (49), CF_3 ($36, 37$), F (42).

[b] J. P. Guthrie (22).

[c] An acidity function measurement of the acidity of this compound in sulfuric acid gave a pK_{HA} of -6.6 (51); this value appears to be too negative by several pK units.

fluorosulfuric acid and trifluoromethanesulfonic acid, two acids whose strengths can often be inverted by changes in solvent. The final column in Table 2-2 gives the mean value for each acid listed, and these pK's will be used in later discussions.

It should be pointed out that a number of attempts have been made to measure the strength of sulfonic acids by means of the acidity function technique, using sulfuric acid as solvent. In some cases the extent of dissociation was measured by nmr or ultraviolet spectroscopy and in other cases by means of solubility or kinetic measurements (51, 52). All these determinations give pK values that are much more negative than those given in Table 2-2, usually in the range of -6 to -7 for arylsulfonic acids. [Benzenesulfonate ion is half-protonated in 83% sulfuric acid (53, 54).]

It is very unlikely that arylsulfonic acids are as strong as these acidity function results would suggest. The evidence presented earlier shows that benzenesulfonic acid is slightly weaker than sulfuric acid. Thus, if benzenesulfonic acid were to have a pK between -6 and -7, sulfuric acid would have to have a pK in the neighborhood of -7. Since the pK of the second dissociation constant of sulfuric acid is 2.0, the difference between its first and second dissociation constants would be some 9 pK units, a much larger difference than is found for any other acid of analogous structure. Perrin's tabulation of the strengths of inorganic acids lists 14 compounds of formula $(HO)_m YO_n$ whose first and second dissociation constants are known with some degree of confidence (55). [One of these is carbonic acid, whose pK_1 requires correcting from 6.35 to 3.85 to account for the undissociated compound being largely present in aqueous solution as carbon dioxide (62).] The average value of ΔpK for these 14 acids is 4.7 units, the two largest being for periodic acid ($\Delta pK = 6.7$) and carbonic acid ($\Delta pK = 6.5$). It is thus extremely unlikely that sulfuric acid has a pK much more negative than -4, and the same conclusion can be reached with regard to the slightly weaker acid, benzenesulfonic acid.

It would appear, therefore, that the equilibrium governing the protonation of sulfonate anions is not governed by any of the acidity functions that have been used for the purpose.

C. Phosphonic Acids

Phosphonic acids (3) are analogous in structure to sulfonic acids (4). Their

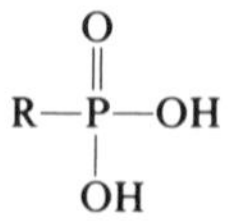

3 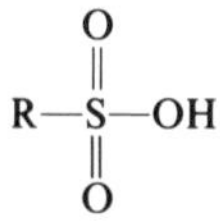   4

nomenclature differs slightly, however; CH_3SO_3H is called meth*ane*sulfonic acid, whereas $CH_3PO_3H_2$ usually goes by the name of meth*yl*phosphonic acid rather than methanephosphonic acid, although the latter name appears extensively in the older literature. The latter compound can be considered a derivative of phosphoric acid (**5**) with the hydroxyl group replaced by methyl or, alternatively, as a derivative of phosphonic acid (**6**) with the hydrogen bound to phosphorus replaced by methyl (*48*). [Phosphonic acid (**6**) is the more stable tautomer of phosphorous acid, $P(OH)_3$; *Chemical Abstracts* uses the term *phosphorous acid* only when all three hydroxyl groups are engaged, as in anhydride or ester formation.] In any case, alkylphosphonic acids, such as **7** or **8**, would be expected to be weaker than either phosphoric or phosphonic acid and, as can be seen by comparing the p*K* values of **5** through **8**, that is what in fact is found (*49, 50, 56*).

$$
\begin{array}{ccccc}
\underset{\displaystyle \overset{\textstyle O}{\parallel}}{\text{H—O—P—OH}} & \underset{\displaystyle \overset{\textstyle O}{\parallel}}{\text{H—P—OH}} & \underset{\displaystyle \overset{\textstyle O}{\parallel}}{\text{H}_3\text{C—P—OH}} & \underset{\displaystyle \overset{\textstyle O}{\parallel}}{\text{(CH}_3)_3\text{C—P—OH}} & \underset{\displaystyle \overset{\textstyle O}{\parallel}}{\text{H}_5\text{C}_6\text{—P—OH}} \\
\text{OH} & \text{OH} & \text{OH} & \text{OH} & \text{OH} \\
\mathbf{5} & \mathbf{6} & \mathbf{7} & \mathbf{8} & \mathbf{9}
\end{array}
$$

pK_{HA}	1.97	1.85	2.38	2.79	1.83

Curiously, the opposite is found to be the case when a *hydroxylic* hydrogen atom in phosphoric or phosphonic acid is replaced by an alkyl group, that is, when one of the acid groups is esterified (Table 2-3) (*57–59*). This is a general phenomenon and, as can be seen in the table, applies to both pK_1 and pK_2. Note, too, that in the case of the two aryloxymethylphosphonic acids shown in the table the ethyl ester is the stronger acid, despite the absence of a chloro group in the aromatic ring. The same result is found with a variety of other polyhydroxylic acids, including sulfuric acid (Table 2-2), carbonic acid, arsenious acid, and silicic acid (*22, 55*); esterification invariably increases the strength of the acid.

Alkyl is generally considered to be electron donating relative to hydrogen, and this effect is most marked when the alkyl group and hydrogen atom to be compared are attached to an sp^2 carbon. When the attached carbon is sp^3, on the other hand, there is some doubt regarding their relative effects (*63*). When the attached atom is oxygen, as in the acids and esters referred to earlier, alkyl is acid strengthening, suggesting that it is electron withdrawing relative to hydrogen, although solvation effects could well be important here (*63a*). It is interesting that σ and σ^* values of methoxyl and hydroxyl show the same trend, the values for methoxyl being generally more positive than those for hydroxyl (*60, 64*). (The *base* strengths of carboxylic acids and esters and also alcohols and ethers show analogous effects; see Chapter 3, Sections V,A and B.)

Table 2-3

ACID STRENGTH OF SOME PHOSPHORIC AND PHOSPHONIC
ACIDS AND THEIR METHYL OR ETHYL ESTERS

Acid	pK_1	pK_2	Ref.
$HO-\overset{\overset{O}{\|\|}}{\underset{\underset{OH}{\|}}{P}}-OH$	1.97	6.82	*49*
$H_3CO-\overset{\overset{O}{\|\|}}{\underset{\underset{OH}{\|}}{P}}-OH$	1.54	6.31	*49*
$H_3CO-\overset{\overset{O}{\|\|}}{\underset{\underset{OCH_3}{\|}}{P}}-OH$	1.29	—	*49*
$H_5C_2O-\overset{\overset{O}{\|\|}}{\underset{\underset{OH}{\|}}{P}}-OH$	1.60	6.62	*49*
$H_5C_2O-\overset{\overset{O}{\|\|}}{\underset{\underset{OC_2H_5}{\|}}{P}}-OH$	1.39	—	*49*
Cl—(ring, Cl)—O—CH$_2$—$\overset{\overset{O}{\|\|}}{\underset{\underset{OH}{\|}}{P}}$—OH	1.25	—	*61*
Cl—(ring)—O—CH$_2$—$\overset{\overset{O}{\|\|}}{\underset{\underset{OC_2H_5}{\|}}{P}}$—OH	0.91	—	*61*

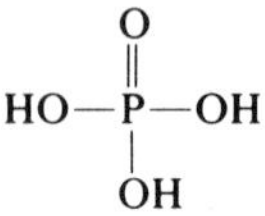

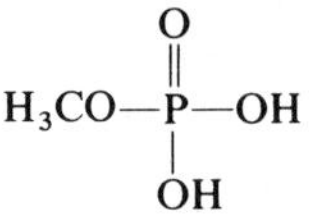

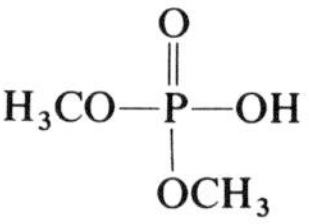

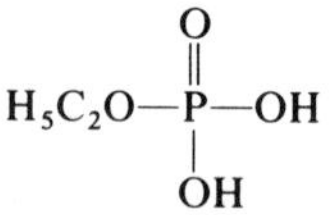

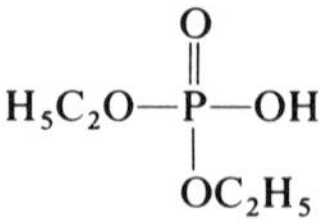

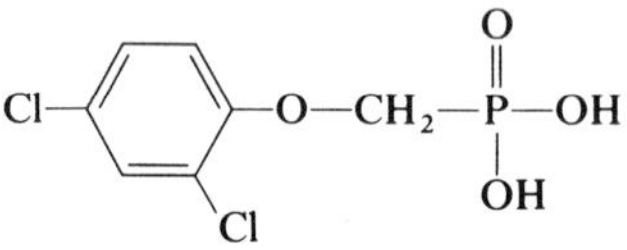

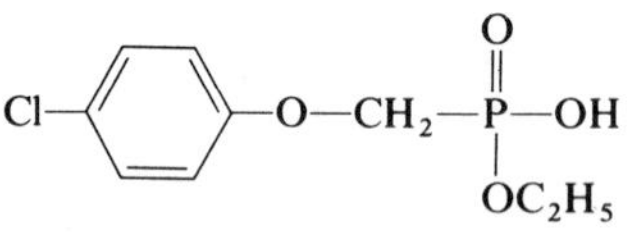

Phosphoric and phosphonic acids are considerably weaker than the analogous sulfuric and sulfonic acids, methylphosphonic acid being some four pK units less acidic than methanesulfonic acid, and methylphosphoric acid, $CH_3OPO_3H_2$, being about five pK units less acidic than methylsulfuric acid, CH_3OSO_3H (Tables 2-2 and 2-3). Nonetheless, chlorophenyl derivatives of phosphorus acids are sufficiently acidic to require the presence of a strong mineral acid, such as hydrochloric acid, to repress their ionization, and this presents some difficulties in determining their acidity constants. Phillips found that plotting the apparent pK of these acids against the square root of the hydrochloric acid molarity gave straight lines (61). Extrapolation to zero concentration of hydrochloric acid gave the following results (see compounds **10** and **11** and also Table 2-3).

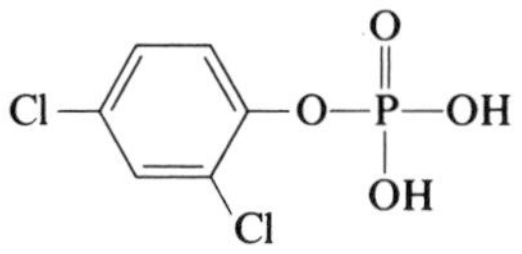

10 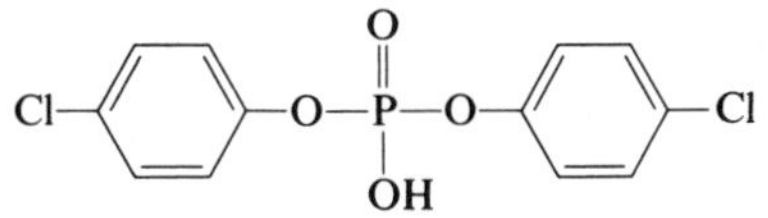   **11**

pK_{HA} 0.57   −0.12

Phenylphosphonic acid (benzenephosphonic acid), **9**, has a pK_1 of 1.83 and a pK_2 of 7.07, making it somewhat stronger than methylphosphonic acid (**7**), whose pK_1 is 2.38 and pK_2 is 7.74 (50). The strongest monosubstituted arylphosphonic acid is the p-nitro compound (pK_1 = 1.24, pK_2 = 6.23), and the weakest are the p-alkoxy compounds, the pK_1 and pK_2 for p-ethoxy, for example, being 2.06 and 7.28. A p-amino group, which is normally a stronger electron-donating group than a p-alkoxy group, does not in fact raise the pK_1 of an arylphosphonic acid, because it is protonated at the acidities at which such pK values are measured. The second dissociation constant of 4-amino-phenylphosphonic acid can be measured, however; it is 7.53 (39).

The effect of a —PO_3H_2 ring substituent on an acidic or basic site elsewhere in the molecule is one of electron withdrawal, as expected. Even the —PO_3H^- group is electron withdrawing, with the —$PO_3{}^{2-}$ group being electron donating, though only slightly so.

D. Sulfinic and Phosphinic Acids

Sulfinic and phosphinic acids (**12** and **14**) are the lower oxidation state analogues of sulfonic and phosphonic acids (**13** and **15**). The two series differ,

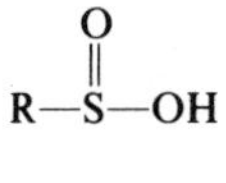

12 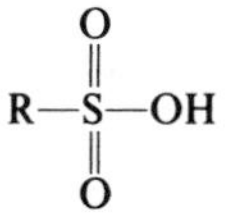   **13** 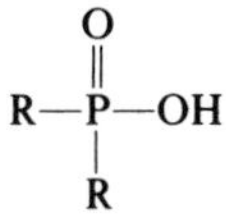   **14** 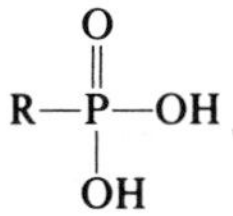   **15**

however, in an important respect. Sulfinic acids have one nonhydroxylic oxygen atom and sulfonic acids have two, whereas both phosphinic and phosphonic acids have one such atom. It has long been known that for acids containing elements of variable oxidation level the acid strength is determined to a great extent by the number of nonhydroxylic oxygen atoms that are present (65–69). Consequently, phosphinic and phosphonic acids (and also

Table 2-4

COMPARISON OF ACID STRENGTHS OF SULFINIC AND SULFONIC ACIDS

Sulfinic acid	pK_{HA}	Sulfonic acid	pK_{HA}	ΔpK	Ref.
CH_3SO_2H	2.28	CH_3SO_3H	-1.9	4.2	70[a]
$C_6H_5SO_2H$	1.21	$C_6H_5SO_3H$	-2.8	4.0	71[a,b]
p-$O_2NC_6H_4SO_2H$	0.64	p-$O_2NC_6H_4SO_3H$	-3.8	4.4	71[a]
p-$CH_3C_6H_4SO_2H$	1.24	p-$CH_3C_6H_4SO_3H$	-2.7	3.9	71[a]
$C_6H_5CH_2SO_2H$	1.45	—	—	—	72
$C_6H_5(CH_2)_4SO_2H$	2.23	—	—	—	72

[a] See Table 2-2.
[b] Reference 72 gives 1.29 for the pK of benzenesulfinic acid and ref. 73 gives 1.3, whereas ref. 74 gives a much higher value and reports substituent effects that are bizarre.

phosphoric acid) have roughly the same acid strengths, whereas sulfinic acids are some four pK units weaker than their sulfonic acid analogues (Table 2-4). Sulfenic acids, which have the formula RSOH and possess no nonhydroxylic oxygen atoms (75), are still weaker; because of their inherent instability (75a) their pK values are not known with any certainty, though they are believed to be somewhat more acidic compounds than phenols (76).

Methanesulfinic acid (12, R = CH$_3$) has very limited stability; it begins to decompose after a day or so at room temperature, even under a nitrogen atmosphere (70). Arenesulfinic acids are much more stable, although they are usually stored and supplied commercially in the form of their sodium salts.

Organic phosphinic compounds are derivatives of phosphinic acid (16) in which organic groups replace either or both of the hydrogen atoms attached to phosphorus. Phosphinic acid is sometimes called hypophosphorous acid, but *Chemical Abstracts* uses the latter name for the compound $[(HO)_2P]_2$ and its derivatives.

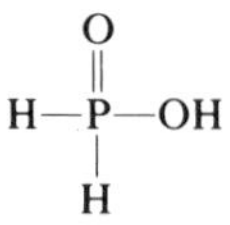

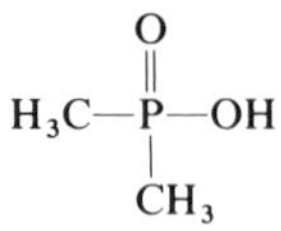

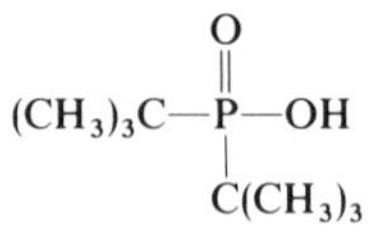

$$
\begin{array}{ccc}
\overset{\displaystyle O}{\underset{\displaystyle H}{\overset{\|}{H-P-OH}}} & \overset{\displaystyle O}{\underset{\displaystyle CH_3}{\overset{\|}{H_3C-P-OH}}} & \overset{\displaystyle O}{\underset{\displaystyle C(CH_3)_3}{\overset{\|}{(CH_3)_3C-P-OH}}} \\
\mathbf{16} & \mathbf{17} & \mathbf{18} \\
pK_{HA} \quad <1.5 & 3.1 & 4.2
\end{array}
$$

$$
\begin{array}{cc}
\overset{\displaystyle O}{\underset{\displaystyle H}{\overset{\|}{H_5C_6-P-OH}}} & \overset{\displaystyle O}{\underset{\displaystyle CH_3}{\overset{\|}{H_5C_6-P-OH}}} \\
\mathbf{19} & \mathbf{20} \\
pK_{HA} \quad \sim 1.7 & 2.9
\end{array}
$$

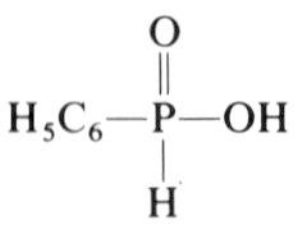

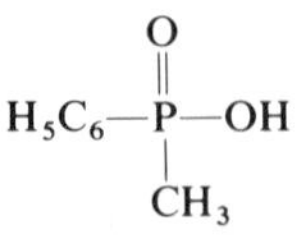

the order of 0.5 to 1.0 pK unit, are due to the intrinsic difficulty of making measurements at high ionic concentration and to what extent they are due to there being a distinction between ionization and dissociation.

Perrin *et al.* have used the Hammett and Taft relations to predict the acidities of a large number of organic acids (*60*). Using their approach a pK_{HA} of 0.4 can be calculated for both trifluoroacetic and trichloroacetic acids, the σ^* values of the CF_3 and CCl_3 groups being very close. However, fluorine is more electronegative than chlorine, and CF_3 might have been expected to have a greater acid-strengthening effect than CCl_3. Indeed, the σ and σ^* values of some groups containing these elements are consistent with such a difference (*60*), and it is possible that the calculated pK for CF_3CO_2H should be somewhat closer to zero. There nonetheless remains a discrepancy of at least 0.6 pK unit between this value and that inferred from Raman and related measurements. It is tempting to attribute this difference to the ionization and dissociation processes not being identical; one might expect, however, that Hammett–Taft calculations would yield ionization rather than dissociation constants in those cases (strong or moderately strong acids) where such a distinction appears to be significant.

The strongest monosubstituted acetic acid is nitroacetic acid (pK_{HA} = 1.48). The instability of dinitroacetic acid with respect to decarboxylation prevents its pK from being determined but, if stable, it and the trinitro compound would undoubtedly be strong acids. Some idea of the strength to be expected of trinitroacetic acid can be obtained by using the published σ^* value for the $(NO_2)_3C$ group (4.54) (*84*) and the method of Perrin *et al.* (*60*). This gives a pK_{HA} of -2.7 for $(NO_2)_3C{-}CO_2H$.

The only groups that are acid weakening when substituted for one of the hydrogen atoms in an aliphatic carboxylic acid are alkyl, alkylsilyl, and negatively charged groups such as $-CO_2^-$ and $-SO_3^-$. In the case of the $-SO_3^-$ group the acid-weakening effect becomes apparent only when the substituent and the carboxylic acid groups are separated by two or more methylene groups. That is, $^-O_3SCH_2CO_2H$ is stronger than acetic acid, but acids of formula $^-O_3S(CH_2)_nCO_2H$, where $n > 2$, are weaker than the corresponding compounds $CH_3(CH_2)_{n-1}CO_2H$. Bell has examined this phenomenon in detail and has attributed the acid-strengthening effect when $n = 1$ to the proximity to the carboxyl group of the partially positively charged sulfur atom, which outweighs the effect of the more distant full negative charge (*89*).

Charged substituents produce anomalous effects in aromatic carboxylic acids as well. Wepster *et al.* (*90*, *91*) assembled voluminous data showing that the Hammett equation fails in the case of meta and para groups that carry a charge, whether it be positive or negative. That is, σ values for groups such as $(CH_3)_3N^+$ or SO_3^- that are obtained from the dissociation constants of the appropriately substituted benzoic acids in water fail quite conspicuously to correlate the reactivities of other compounds containing these groups.

Listed in Table 2-6 are a large number of monosubstituted benzoic acids together with their pK_{HA} values. All uncharged ortho substituents increase the strength of benzoic acid except, possibly, several of the alkoxyl groups, which may have a marginal weakening effect.

Table 2-6

pK_{HA} VALUES OF MONOSUBSTITUTED
BENZOIC ACIDS IN WATER AT 25°C[a]

Benzoic acid	Ortho	Meta	Para
Unsubstituted	4.20	4.20	4.20
CH_3	3.91	4.27	4.37
C_2H_5	3.79	—	4.35
$(CH_3)_2CH$	3.64	—	4.35
$(CH_3)_3C$	3.54	4.28	4.40
$(CH_3)_3Si$	—	4.09	4.20
C_6H_5	3.46	—	—
CO_2H[b,c]	2.95	3.62	3.54
$CO_2{}^-$[b,d]	5.41	4.60	4.46
F	3.27	3.86	4.14
Cl	2.91	3.83	3.97
Br	2.85	3.81	4.00
I	2.86	3.85	4.00
CN	3.14	3.60	3.55
HO	2.98	4.07	4.58
NO_2	2.21	3.49	3.44
CH_3O	4.09	4.09	4.47
C_2H_5O	4.21[e]	4.17[e]	4.80[e]
$n\text{-}C_3H_7O$	4.24[e]	4.20[e]	4.78[e]
$i\text{-}C_3H_7O$	4.24[e]	4.15[e]	4.68[e]
C_6H_5O	3.53	3.95	4.52
CH_3SO_2	—	3.52	3.64
H_2NSO_2	—	3.54	3.47
CH_3CO_2	3.38	—	3.92
CH_3CONH	3.61	4.03	—
$(CH_3)_3N^+$	—	3.18[f]	3.23[f]

[a] See footnote a, Table 2-5.
[b] Uncorrected for symmetry (Section II,E,3).
[c] pK_1.
[d] pK_2.
[e] 20°C. These values need to be confirmed; see N. H. P. Smith, *Nature (London)* **211**, 186 (1966).
[f] A. J. Hoefnagel, M. A. Hoefnagel, and B. M. Wepster, *J. Org. Chem.* **43**, 4720 (1978).

Two explanations have been given for the increase in acid strength that virtually any ortho substituent, regardless of its electronic effect, confers on benzoic acid. The generally accepted explanation invokes steric inhibition of resonance between carboxyl group and ring in the neutral acid molecule (92, 93). The second explanation, which has received little attention, attributes the effect to hindrance to solvation of the neutral molecule, the basis of this view being the virtual disappearance of acidity differences between ortho- and para-substituted acids in media of extremely high ionic strength (94).

Unlike o-alkoxyl groups an o-hydroxyl group has a very large acid-strengthening effect on benzoic acid, undoubtedly because of the stabilization of the carboxylate anion by hydrogen bonding. Thus, 2-hydroxybenzoic acid (salicylic acid), whose pK_{HA} is 2.98, is ~1.2 units stronger than benzoic acid. The effect is even more striking in the case of 2,6-dihydroxybenzoic acid [Eq. (2-7a)], whose pK_{HA} is 1.2, making it ~3.0 units stronger than benzoic acid.

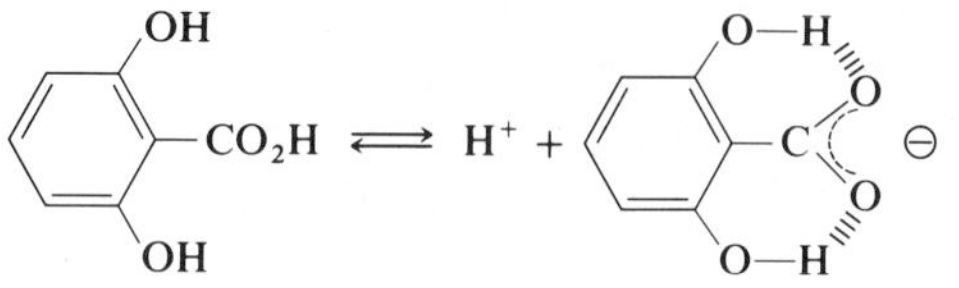

$$\tag{2-7a}$$

The strongest aromatic carboxylic acid whose pK has been determined is 2,4,6-trinitrobenzoic acid (95). It has a pK_{HA} of ~0.65, making it considerably stronger than pentafluorobenzoic acid ($pK_{HA} = 1.75$) or 4-trifluoromethyl-2,3,5,6-tetrafluorobenzoic acid ($pK_{HA} = 1.44$) (96). If the effects of substituents on acidity are additive (60), then 4-nitro-2,6-dihydroxybenzoic acid should have a pK_{HA} of 0.4, making it stronger than 2,4,6-trinitrobenzoic acid. Indeed, the effect may be more than additive since the nitro group will increase the hydrogen-bonding capacity of the hydroxyl groups.

In view of trimethylsilylacetic acid being markedly weaker than acetic acid (Table 2-5) one would expect to find 3- and 4-trimethylsilylbenzoic acids to be weaker than benzoic acid. Such is not the case, as can be seen from the results shown in Table 2-6. This phenomenon has been rationalized in a number of ways; $d-\pi$ interactions between silicon and ring carbon have been invoked (97), as have solvation effects in which the bulk of the aromatic substituent plays a role (98). In the latter connection it is interesting that 3-trimethylsilyl-and 3-tert-butylbenzoic acids have anomalous, but similar values of ionization enthalpy and entropy.

1. Medium Effects on Carboxylic Acid Strength

It is well known that the extent of dissociation of a neutral acid in a particular solvent depends on both the basicity and the dielectric constant of the solvent. It turns out that the dielectric constant is critical as far as carboxylic acids are concerned. Even the basicity of a solvent such as pyridine cannot compensate for the medium's lower dielectric constant ($D = 12.3$, compared with 78.3 for water); the degree of ionization of benzoic acid, for example, is $\sim 10^7$ times lower in pyridine than in water [Eqs. (2-8) and (2-9) (99)].

$$C_6H_5CO_2H + C_5H_5N \rightleftharpoons C_6H_5CO_2^- + C_5H_5NH^+ \qquad K = 10^{-11} \qquad (2\text{-}8)$$
$$\text{(solvent)}$$

$$C_6H_5CO_2H + H_2O \rightleftharpoons C_6H_5CO_2^- + H_3O^+ \qquad K = 10^{-4} \qquad (2\text{-}9)$$

Addition of methanol or ethanol to water decreases the ionization constant of carboxylic acids as expected. Benzoic acid, for example, is some two pK units weaker in a 2:1 ethanol–water mixture than in water and some six pK units weaker in ethanol than in water. In this connection it is not uncommon to find that addition of a small amount of water to an organic solvent has a greater effect on solute acidity than does addition of the corresponding amount of the organic solvent to water.

In addition to having a general acid-weakening effect on carboxylic acids, most organic solvents have the effect of spreading out the acidity constants. That is, the pK_{HA} values in a series of carboxylic acids (e.g., substituted benzoic acids) will span a wider range in organic or mixed solvents than in water. This is reflected in the general increase in the Hammett reaction constant ρ that accompanies addition of solvents such as ethanol and dioxane to water (Table 2-7). The combination of these two factors, the acid-weakening and spreading-out effects, means that the weaker the acid the greater will be the discrepancy between the ionization constants in water and in the mixed or pure organic medium. For moderately strong carboxylic acids, on the other

Table 2-7

IONIZATION CONSTANT OF BENZOIC ACID AND HAMMETT REACTION CONSTANT ρ IN AQUEOUS ALCOHOL MIXTURES[a]

	Water	50 wt % methanol	50 vol % ethanol	Ethanol
pK_{HA}	4.20	5.43[b]	5.50[c]	10.25[c]
ρ	1.00	1.28[b]	1.46[d]	1.63[e]

[a] See also A. J. Hoefnagel, M. A. Hoefnagel, and B. M. Wepster, *J. Org. Chem.* **43**, 4720 (1978).

[b] Reference *130*.

[c] E. Grunwald and B. J. Berkowitz, *J. Am. Chem. Soc.* **73**, 4939 (1951).

[d] J. D. Roberts, E. A. McElhill, and R. Armstrong, *J. Am. Chem. Soc.* **71**, 2923 (1949).

[e] M. Kilpatrick and W. H. Mears, *J. Am. Chem. Soc.* **62**, 3047, 3051 (1940).

hand, the acid-weakening effect will be less. Thus, 3,5-dinitrobenzoic acid is one pK unit weaker in 65% aqueous ethanol than in water, whereas benzoic acid is two pK units weaker. It is, of course, reasonable that delocalization of charge in the acid anion should make the ionization equilibrium constant less vulnerable to dielectric effects.

The dielectric effect on acid strengths that is associated with a change in solvent can be compensated for to a considerable extent by considering relative acid strengths in various media (*100*). Thus, Eq. (2-10) can be replaced by Eq. (2-11), in which there are the same number of ions on both sides of the equation.

$$RCO_2H + S \rightleftharpoons RCO_2^- + SH^+ \qquad (2\text{-}10)$$

$$RCO_2H + R'CO_2^- \rightleftharpoons RCO_2^- + R'CO_2H \qquad (2\text{-}11)$$

The use of Eq. (2-11) is helpful but it does not guarantee a consistent order of acid strengths, even among closely related compounds. Thus, though 2-iodobenzoic acid is 0.6 pK unit stronger than 3-nitrobenzoic acid in water, it is 0.1 unit weaker in ethanol (*101*). Such inversions of strengths of substituted benzoic acids almost always involve an ortho substituent. The order of acid strength of meta- and para-substituted benzoic acids is virtually independent of solvent, as would be expected from the known success of the Hammett equation in correlating the vast body of rate and equilibrium data that has been collected over the years for such compounds.

Measurements of the degree of ionization of carboxylic acids in the gas phase show a much wider range of acid strengths than is the case in solution. Table 2-8 lists the free energy of dissociation in the gas phase of a number of carboxylic acids as determined by Cumming and Kebarle (*102*); the free-energy difference for the dissociation of acetic and trifluoroacetic acids is ~ 20 kcal mol^{-1} greater than in aqueous solution. The order of acid strengths for carboxylic acids is somewhat different in the gas and condensed phases because of the much greater role that molecular polarizability plays in gas phase processes. A charged atom in an isolated ion has only its attached group to interact with, whereas in the condensed phase the polarizability of the surrounding molecules of the medium is a factor in stabilizing the ion. As a result the order of acid strengths in the gas phase is *n*-butyric > propionic > acetic, and bromoacetic > chloroacetic > fluoroacetic, in contrast to the orders found in aqueous solution (*103*).

2. Temperature Effects on Carboxylic Acid Strength

The carboxylic acid ionization equilibrium is unusual in having an extremely small temperature dependence. Thus, careful measurements show that the pK_{HA} values of acetic acid in water at 0, 25, and 60°C are within 0.05 pK

Table 2-8

FREE-ENERGY CHANGES FOR GAS
PHASE DISSOCIATION OF
CARBOXYLIC ACIDS[a]

Acid	ΔG^0 (300 K), kcal mol^{-1}
CH_3CO_2H	341.5
$C_2H_5CO_2H$	340.3
$n\text{-}C_3H_7CO_2H$	339.5
$ClCH_2CH_2CH_2CO_2H$	338.4
HCO_2H	338.2
$CH_3CHClCH_2CO_2H$	334.9
$C_6H_5CO_2H$	331.7
FCH_2CO_2H	331.0
$CH_3CH_2CHClCO_2H$	330.9
$ClCH_2CO_2H$	328.8
$BrCH_2CO_2H$	327.4
F_2CHCO_2H	323.5
Cl_2CHCO_2H	321.9
CF_3CO_2H	316.3

[a] Data from ref. *102*.

unit of one another (Table 2-9). In fact, the dissociation constant of this acid reaches a maximum near 25°C, declining slightly as the temperature is raised or lowered. Other aliphatic monocarboxylic acids are similar, though some show a somewhat larger temperature dependence than does acetic acid (*104–106*).

It follows from the van't Hoff equation [Eq. (2-12)] and its integrated form [Eq. (2-13)] that a reaction whose equilibrium constant is temperature

$$\frac{d \ln K}{dT} = \frac{\Delta H}{RT^2} \tag{2-12}$$

$$R \ln K = \Delta S - \frac{\Delta H}{T} \tag{2-13}$$

independent must have a reaction enthalpy of zero; the position of equilibrium is then determined only by the entropy of reaction. Dissociation constants of carboxylic acids can be conveniently determined with rather high precision using the glass electrode. This makes it possible to detect slight curvature in plots of pK_{HA} against $1/T$ for many carboxylic acids. Such curvature would not be noticed in many other chemical reactions where the precision of measurement is not as great. Furthermore, the change in sign of

Table 2-9

TEMPERATURE DEPENDENCE OF DISSOCIATION CONSTANTS OF CARBOXYLIC
ACIDS IN WATER

Acid	0°C	15°C	25°C	40°C	60°C	Ref.[a]
Formic	3.79	3.76	3.75	3.77	3.81	1
Acetic	4.78	4.76	4.76	4.77	4.81	2
Butyric	4.81	4.81	4.82	4.86	4.92	3
Benzoic	—	4.22	4.20	4.22	—	4
2-Methylbenzoic	—	3.87	3.89	3.93	—	4
4-Methylbenzoic	—	4.39	4.36	4.32	—	4
2,6-Dimethylbenzoic	—	3.23	3.36	3.47	—	4
4-Cyano	—	3.56	3.55	3.56	—	4

[a] References: 1, H. S. Harned and N. D. Embree, *J. Am. Chem. Soc.* **56**, 1042 (1934); 2, H. S. Harned and R. W. Ehlers, *J. Am. Chem. Soc.* **55**, 652 (1933); 3, H. S. Harned and R. O. Sutherland, *J. Am. Chem. Soc.* **56**, 2039 (1934); 4, J. M. Wilson, N. E. Gore, J. E. Sawbridge, and F. Cardenas-Cruz, *J. Am. Chem. Soc. B* p. 852 (1967).

the slope that often occurs near room temperature (as it does with acetic acid) makes the slight curvature more obvious than it would otherwise be.

For those reactions that do not give linear plots of log K against $1/T$ one can calculate ΔH values at each temperature from the slope of the curve. This is effectively what was done by Wilson *et al.* (*107*) using the method of Clarke and Glew (*108*). One might expect that the slight dependence of pK_{HA} on temperature for benzoic acid would give rise to inconsequential changes in ΔH and ΔS. That, however, is not the case; $\Delta H^{15°} = 1.0$ kcal mol^{-1} and $\Delta S^{15°} = -15.9$ cal deg^{-1} mol^{-1}, whereas $\Delta H^{40°} = -1.1$ kcal mol^{-1} and $\Delta S^{40°} = -22.8$ cal deg^{-1} mol^{-1}.

Despite an apparent variation of more than 2.1 kcal in the enthalpy and 6.9 cal deg^{-1} in the entropy of dissociation of benzoic acid over a temperature range of only 25°C, it turns out that making the usual assumption of ΔH and ΔS being temperature independent introduces only small errors in the calculation of pK_{HA} values at various temperatures. Thus, taking ΔH equal to 150 cal deg^{-1} mol^{-1} and ΔS equal to -18.7 cal deg^{-1} mol^{-1} (the values obtained at 25°C) we obtain the following pK_{HA} values (experimental values in parentheses): 15°C, 4.20 (4.22); 25°C, 4.20 (4.20); 40°C, 4.19 (4.22).

In the case of 2,6-dimethylbenzoic acid the ΔH value taken from the slope of the van't Hoff curve increases from -5.8 kcal mol^{-1} at 15°C to -1.9 kcal mol^{-1} at 40°C. Even in this case, where the curvature of the van't Hoff plot is quite pronounced, the assumption of a temperature-independent ΔH introduces errors of only 0.02 pK unit at 15°C and 0.04 pK unit at 40°C.

Curiously, 2-methylbenzoic acid gives a linear van't Hoff plot of high precision.

Apart from the importance of ΔH and ΔS (whether constant or variable) in linking experimentally determined pK_{HA} and temperature, what chemical insight into the ionization or dissociation process is provided by these quantities? Blandamer, Scott, and Robertson have analyzed the data and have concluded that it supports a two-stage reversible process, the dissociation proceeding from ion-pair intermediates (109, 110). It will be recalled that in the case of strong and moderately strong acids there is evidence to suggest that ion pairs are the predominant forms of the undissociated acids, though that is not the case with weak acids.

It has been frequently noted that ionization of carboxylic acids is controlled primarily by the entropy rather than the enthalpy of reaction. For most carboxylic acids of pK 4 to 5, ΔH is near zero and ΔS is in the general neighborhood of -20 cal deg^{-1} mol^{-1}, the latter quantity reflecting the high degree of solvation of the ions that takes place. With the introduction of electron-withdrawing groups pK_{HA} drops, as expected, but this is primarily the result of the entropy of ionization becoming less negative. In the cases of the trihaloacetic acids the entropy accompanying dissociation is near zero (Table 2-10). Such acids are presumably highly solvated in their undissociated forms. It will be recalled that these are precisely the acids whose undissociated forms are believed to exist largely as ion pairs (Section II,A), and it is perhaps not surprising that their undissociated forms are highly solvated.

The increase in the strength of benzoic acid that is brought about by o-alkyl substituents clearly has a different basis than that brought about by electron-withdrawing groups. This difference is reflected in the thermodynamics of ionization. All of the benzoic acids listed in Table 2.10 that contain an o-methyl group have ΔH values between -1 and -4 kcal mol^{-1} and entropies that are quite negative (-21 to -30 cal deg^{-1} mol^{-1}). Virtually all other benzoic acids have small positive values of ΔH and values of ΔS that are markedly less negative (greater than -19 cal deg^{-1} mol^{-1}).

Electronic effects on chemical equilibria might be expected to be manifested chiefly or wholly in the enthalpy of reaction term. Except for the reactivities of many meta and para compounds it turns out that this expectation is seldom realized, ionization of carboxylic acids being particularly conspicuous in this regard. The reasons for this situation are clear. Enthalpy changes at absolute zero, which presumably would reflect static effects such as induction and resonance, can be obtained from the effect of temperature on the equilibrium constant only by means of a lengthy extrapolation, even for those reactions that give apparently linear van't Hoff plots [Eq. (2-12)]. Since the ionization of carboxylic acids is subject to large solvation effects and since these acids

Table 2-10

THERMODYNAMIC DATA FOR DISSOCIATION OF CARBOXYLIC ACIDS IN WATER AT 25°C

Acid	pK	ΔH^0 (kcal mol^{-1})	ΔS^0 (cal deg^{-1} mol^{-1})
Aliphatic carboxylic acids[a]			
Formic	3.75	0	−17.2
Acetic	4.76	−0.10	−22.1
Hexanoic	4.89	−0.60	−24.2
Chloroacetic	2.86	−1.14	−16.9
Dichloroacetic	1.35	−0.10	−6
Trichloroacetic	(0.52)[b]	0.28	−1.4
Trifluoroacetic	(0.50)[b]	0.40	−1.0
Meta- and para-substituted benzoic acids[c]			
Unsubstituted	4.20	0.15	−18.7
3-Methyl	4.27	1.20	−15.3
4-Methyl	4.36	1.10	−16.3
3-Chloro	3.83	0.24	−16.7
4-Chloro	3.99	0.24	−16.7
4-Nitro	3.44	0.12	−15.3
Ortho-substituted benzoic acids[c]			
2-Methyl	3.89	−1.02	−21.2
2,3-Dimethyl	3.77	−2.01	−23.7
2,4-Dimethyl	4.22	−2.32	−27.1
2,5-Dimethyl	3.99	−2.72	−27.4
2,6-Dimethyl	3.36	−4.28	−29.7
2,4,6-Trimethyl	3.45	−4.37	−30.4
2,3,5,6-Tetramethyl	3.42	−3.50	−27.3

[a] From compilations of J. W. Larson and L. G. Hepler, in "Solvent–Solute Interactions" (J. F. Coetzee and C. D. Ritchie, eds.), Dekker, New York, 1969, and from J. L. Kurz and J. M. Farrar, J. Am. Chem. Soc. **91,** 6057 (1969).

[b] Dissociation constants obtained potentiometrically; see Table 2-5 for a different indication of the strengths of these acids.

[c] From J. M. Wilson, N. E. Gore, J. E. Sawbridge, and F. Cardenas-Cruz, J. Am. Chem. Soc. B p. 852 (1967).

usually give nonlinear van't Hoff plots, it is clear that enthalpies measured near room temperature will be poor indicators of simple electronic effects, the latter being overlaid by large amounts of thermal energy.

Bell has frequently made the point that dissociation constants at a single temperature (i.e., ΔG) are better indicators of electronic and similar effects in acids and bases than is ΔH (111). There is a general tendency for changes in

ΔH and ΔS to oppose one another with the result that abrupt, and often mysterious, changes in these terms are smoothed out when pK values are compared.

3. Dicarboxylic Acids: Symmetry Effects

Simple symmetry arguments show that ΔpK for the first and second dissociation constants of a dicarboxylic acid should be 0.6 (i.e., $K_1/K_2 = 4$), provided that the two carboxyl units are sufficiently far apart that one has no effect on the other. This follows from the fact that K_1 will be twice as large as it would otherwise be [because of the presence of two carboxyl groups in the reactant and one carboxylate group in the product of Eq. (2-14)] and K_2 is half as large as it would otherwise be [because of the presence of one carboxyl group in the reactant and two carboxylate groups in the product of Eq. (2-15)].

$$HO_2C(CH_2)_nCO_2H \rightleftharpoons H^+ + HO_2C(CH_2)_nCO_2^- \qquad (2\text{-}14)$$

$$HO_2C(CH_2)_nCO_2^- \rightleftharpoons H^+ + {}^-O_2C(CH_2)_nCO_2^- \qquad (2\text{-}15)$$

There are two reasons for K_1/K_2 being always greater than 4. First is the simple electrostatic effect on the molecule's second carboxyl group that is due to the appearance of one unit of negative charge in the molecule, an effect that depends on the distance between the two carboxyl groups. The second factor is hydrogen bonding that may occur between carboxyl and carboxylate groups in the monoanion; this stabilizes the monoanion, making K_1 larger and K_2 smaller than would otherwise be the case (112–114). The departure of $K_1/4K_2$ from unity (Table 2-11) is a measure of these combined effects.

A number of attempts have been made to account quantitatively for these two factors, which can act singly or together to raise K_1/K_2 from the statistical value of 4 to a value as high as 10^8 in the case of $(\pm)$-2,3-di-*tert*-butylsuccinic acid. The treatment of the electrostatic effect, first by Bjerrum (115) and later by Eucken (117) and Kirkwood and Westheimer (118), was fairly successful in relating K_1/K_2 to dielectric constant and distance between carboxyl groups for molecules in which hydrogen bonding in the monoanion is absent (115–120).

Westheimer and Benfey in 1956 suggested that the hydrogen-bonding effect could be evaluated by a comparison of the first dissociation constant of a dicarboxylic acid with the dissociation constant of the monomethyl ester (121). In the absence of all hydrogen bonding, K_1 should be twice as great (the statistical factor) as K_{HA} for the ester, assuming that the electronic effects of carboxyl and carbomethoxy are the same. Thus, departure of the quantity $K_1/2K_{ester}$ from unity can be used as a measure of the hydrogen-bonding effect, most of which will reside in the monoanion. An examination of this

Table 2-11

DISSOCIATION DATA FOR DICARBOXYLIC ACIDS IN WATER[a]

Acid	pK_1	pK_2	$K_1/4K_2$	$K_1/2K_{ester}$
Malonic	2.83	5.69	181	1.8
Methylmalonic	3.05	5.76	128	1.1
Dimethylmalonic	3.17	6.06	194	1.2
Succinic	4.19	5.48	4.9	1.1
($\pm$)-2,3-Dimethylsuccinic	3.94	6.20	46	1.4
meso-2,3-Dimethylsuccinic	3.77	5.94	37	1.6
Tetramethylsuccinic	3.56	7.41	1780	13.2
($\pm$)-2,3-Diethylsuccinic	3.51	6.60	309	8.9[b]
meso-2,3-Diethylsuccinic	3.63	6.46	170	2.3[b]
($\pm$)-2,3-Diisopropylsuccinic	—	—	—	470[b]
meso-2,3-Diisopropylsuccinic	—	—	—	1.3[b]
($\pm$)-2,3-Di-tert-butylsuccinic	2.20	10.3	3.1×10^7	1170 (1800)[b]
meso-2,3-Di-tert-butylsuccinic	—	—	—	1.3[b]
Fumaric	3.02	4.38	5.8	1.0
Maleic	1.92	6.34	6600	5.3
Phthalic	2.95	5.41	72	1.1

[a] Dissociation constants from refs. 92, 124, 126–129.
[b] 1:1 water–ethanol by weight.

quantity for the compounds in Table 2-11 shows that it is indeed near unity for many dicarboxylic acids, including malonic, succinic, and phthalic acids.

Substitution with bulky alkyl groups can, in some cases, cause $K_1/2K_{ester}$ to become very large, indicating that hydrogen bonding in the monoanion has become important [Eq. (2-16)]. This is a special case of the Thorpe–Ingold

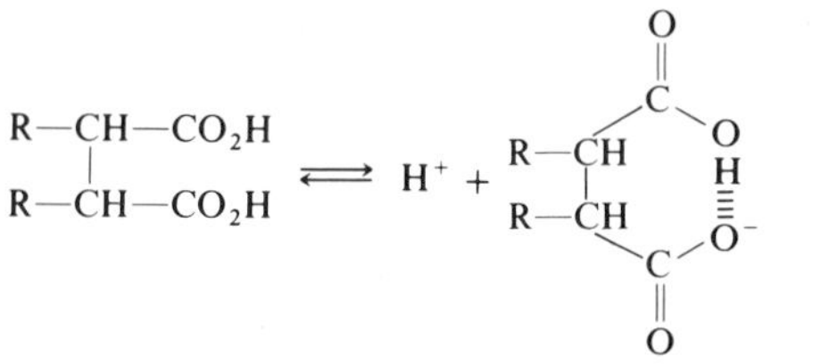

$$
\begin{array}{l}
\text{R—CH—CO}_2\text{H} \\
\quad | \\
\text{R—CH—CO}_2\text{H}
\end{array}
\rightleftharpoons \text{H}^+ +
\qquad\qquad (2\text{-}16)
$$

effect, which recognizes the relief of steric strain that accompanies cyclization in highly substituted bifunctional compounds (122, 123).

Eberson has shown that, although the meso forms of 2,3-dialkylsuccinic acids have values of $K_1/2K_{ester}$ that are close to unity, the values for the enantiomeric isomers [designated ($\pm$) in Table 2-11] increase from 1.4 for the dimethyl compound to greater than 10^3 for the di-tert-butyl compound (124).

The reasons for the drastically different behavior of the diastereomeric forms of these dicarboxylic acids can be seen by an examination of the most stable conformers of the two groups of molecules. When the alkyl groups are antiperiplanar to one another in the meso compound, as in **22**, the two carboxyl groups will also be antiperiplanar, preventing interaction between them. On the other hand, when the alkyl groups are antiperiplanar in the $(+)$ or $(-)$ forms, as in **23**, the carboxyl groups will be gauche and direct

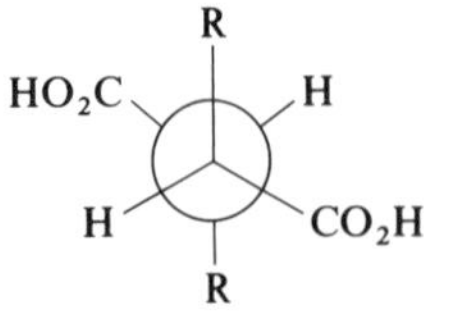

22 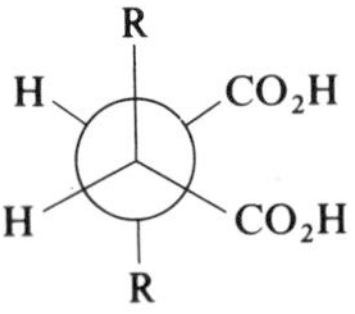   23

interaction can occur. The larger the alkyl groups the greater will be the requirement that they be antiperiplanar and the greater the tendency for the monoanion of the $(+)$ or $(-)$ isomer to form strong internal hydrogen bonds, resulting in elevated values of $K_1/2K_{ester}$ and $K_1/4K_2$ (*92, 124, 125*).

In the case of the isophthalic acids **23a**, it has been shown that a substituent Z has virtually the same effect on the second dissociation as on the first (*130*).

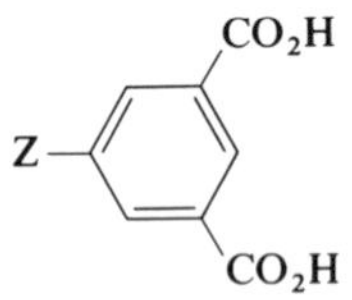

23a

F. Phenols and Enols

Phenols span more than 12 pK units in acid strength; the weakest are the 2,6-di-*tert*-butylphenols possessing alkyl or alkoxyl groups in the 4 position. Their water solubilities are extremely low, but indirect determinations put their pK_{HA} values near 12.2 (*132–134*). The strongest phenol that has been prepared appears to be dichloropicric acid (**24**); it is a strong acid in water, with a pK_{HA} that appears to be in the range of 0 to -0.5 (*135*).

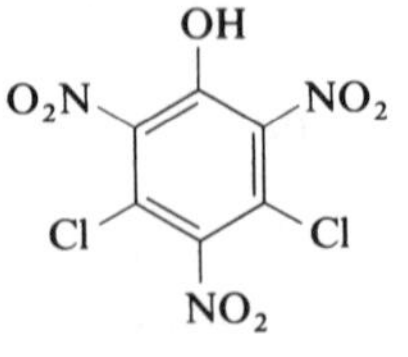

24

Table 2-12

ACIDITY CONSTANTS IN WATER
OF o-ALKYLPHENOLS AT 25°C[a]

Phenol	pK_{HA}
Unsubstituted	9.99
2-Methyl	10.26
2-tert-Butyl	11.34
2,6-Dimethyl	10.59
2,6-Di-tert-butyl	11.7
4-Methyl-2,6-di-tert-butyl	12.2
2,4,6-Tri-tert-butyl	12.2

[a] From refs. 132, 133, 136–138.

Unlike benzoic acids, in which o-alkyl substituents invariably increase acid strength, phenols have their acidities *lowered* by o-alkyl groups (Table 2-12). Only part of this result can be ascribed to the inductive effect; a large part of it must be due to steric hindrance to solvation of the anion, particularly in the case of bulky alkyl groups.

As with benzoic acids, changing the medium from water to alcohol or a similar solvent has a twofold effect on the degree of dissociation of a series of substituted phenols. Acid strengths are both decreased and spread out. That is, the series will display a wider span of pK values in ethanol or higher alcohols than in water, with the most acidic phenols being the ones least affected by the change in solvent (*139, 140*). [This effect is much less marked with methanol (*141*).] In the case of strongly acidic phenols such as picric acid and its derivatives, which have highly delocalized negative charges in their anionic forms, dispersion forces between anion and solvent molecule provide a considerable measure of stability of the anion in the case of alcoholic solvents (*142*). This compensates to a considerable extent for the dielectric effect that invariably causes the strengths of acids to be lower in media of lower polarity.

The order of strength of meta- and para-substituted phenols is the same in the gas phase as in aqueous solution, but their acidities span a much wider range in the gas phase (*143, 144*). The Hammett ρ value for phenolic acidities in the gas phase is ∼ 14, compared with 2.23 in water (*136*). Interestingly enough, the first dissociation of 4-hydroxybenzoic acid in the gas phase involves the phenolic group rather than the carboxyl group (*144*).

Many phenols have entropies of ionization that are in the range of -20 to -25 cal deg^{-1} mol^{-1}, and in this respect they resemble carboxylic acids. The difference in the strengths of phenols and carboxylic acids stems mainly from the enthalpy term. Most carboxylic acids have heats of ionization that are

Table 2-13

THERMODYNAMIC DATA FOR IONIZATION OF PHENOLS IN
WATER AT 25°C[a,b]

Phenol	pK	ΔH^0 (kcal mol^{-1})	ΔS^0 (cal deg^{-1} mol^{-1})
Unsubstituted	9.99	5.65	−26.7
2-Methyl	10.33[c]	5.73	−28.1
3-Methyl	10.10	5.52	−27.7
4-Methyl	10.28	5.50	−28.6
2-Chloro	8.48	4.63	−23.5
4-Chloro	9.38	5.80	−23.5
2-Nitro	7.22	4.66	−17.4
3-Nitro	8.35	4.71	−22.5
4-Nitro	7.14	4.70	−16.9

[a] Data of Fernandez and Hepler (*145, 146*) and Chen and Laidler (*147*).
[b] Thermodynamic parameters and pK_{HA} values for a large number of
   phenols have been tabulated by Rochester (*148*).
[c] Reference *132* gives 10.26 for this compound.

close to zero (Section II,E,2), whereas this term is positive in the case of
phenols (Table 2-13).

As with carboxylic acids the increase in acid strength that accompanies the
introduction of electron-withdrawing groups is largely concentrated in the
entropy term. In the case of carboxylic acids the enthalpy of ionization does
not stray far from zero; in the case of phenols the enthalpy remains near 5 kcal
mol^{-1}. This means that phenols, unlike most carboxylic acids, have an
appreciable temperature coefficient of ionization. (At ambient temperature an
enthalpy of 5 kcal corresponds to an increase of $\sim 30\%$ in the ionization
constant for a 10°C rise in temperature.)

Phenols are often regarded as enols of exceptional stability whose existence
is due to the special stability of the aromatic ring. It turns out that enols in
general are more stable than has hitherto been believed (*149*) and that their
acidities span much the same pK_{HA} range as do those of phenols. Table 2-14
lists a number of enols and their pK_{HA} values. The most acidic enols (e.g.,
squaric acid) (*150*) are comparable to picric acid in strength, whereas the
weakest (the enols of aliphatic aldehydes or ketones) are comparable to alkyl-
substituted phenols.

Delocalization of negative charge in the enolate anion is the principal factor
determining the acid strength of the enol. In the case of ascorbic acid (**31**) the
most acidic hydroxyl group is that attached to C-3, because in this case the
charge can be most effectively delocalized [Eq. (2-17)].

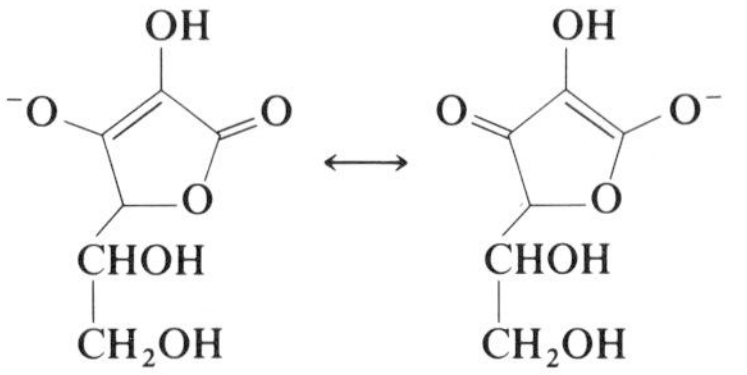

$$(2\text{-}17)$$

In the case of compounds in Table 2-14 that contain more than one carbonyl group (**30, 32, 34, 35,** and **36**), only one of these at a time can take part formally in charge delocalization in the anion. This point is illustrated in the case of croconic acid (**34**) by the resonance shown in Eq. (2-18).

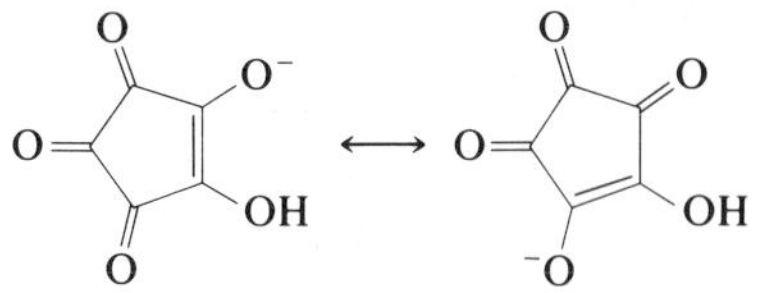

$$(2\text{-}18)$$

In the case of **36**, the second most acidic enol in Table 2-14, formal resonance in the anion involves additionally one of the nitro groups [Eq. (2-19)].

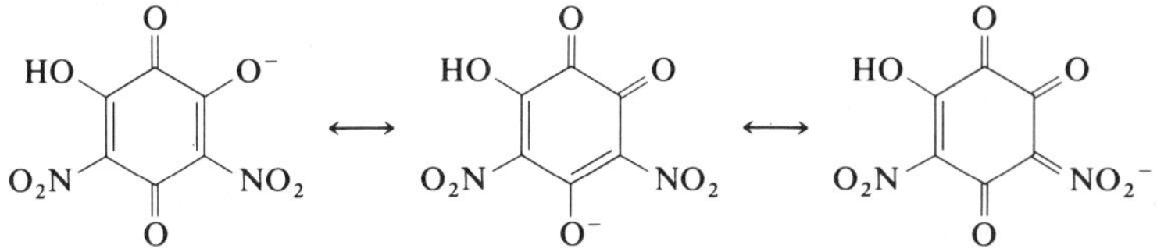

$$(2\text{-}19)$$

Kresge et al developed two techniques for generating aliphatic enols and measuring their acid strengths in water (*151, 151a*). In one a small quantity of lithium enolate dissolved in tetrahydrofuran is added to a large excess of water. The enol so generated isomerizes to ketone via an ionization step, followed by rate-controlling protonation at carbon [Eq. (2-19a)]. The reac-

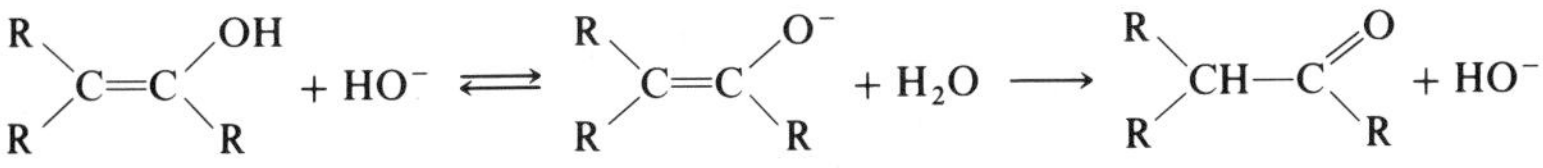

$$(2\text{-}19a)$$

tion is thus base-catalyzed since the concentration of the enolate ion depends on the basicity of the medium. It is found that the rate levels off above pH 12, presumably as a result of the substrate being almost entirely in the anionic form. It is a simple matter to determine the pK of the enol since this will correspond to the pH at which the rate is half that at the plateau. This turns

Table 2-14

ACID STRENGTHS OF ENOLS IN WATER

Name (structure number)	Formula	pK_{HA}	Ref.
Isobutyraldehyde enol (**25**)	$(CH_3)_2C{=}CHOH$	11.6	*151*
Acetone enol (**26**)	$CH_3COH{=}CH_2$	10.8	*152*
Acetophenone enol (**27**)	$H_5C_6{-}COH{=}CH_2$	10.3	*153*
Maltol (**28**)	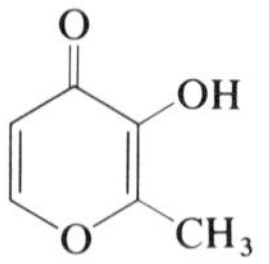	8.4, 8.7	*155, 155a*
Tropolone (**29**)	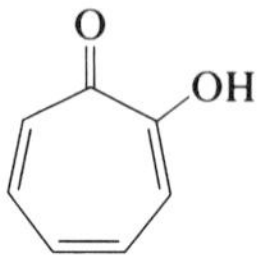	6.7	*155, 156*
Rhodizonic acid dihydrate (**30**)	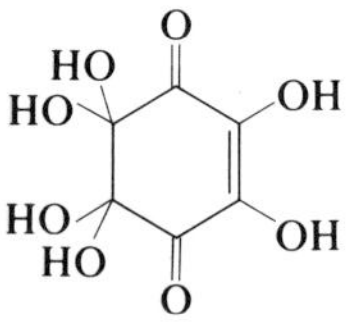	4.4	*157*
Ascorbic acid (**31**)	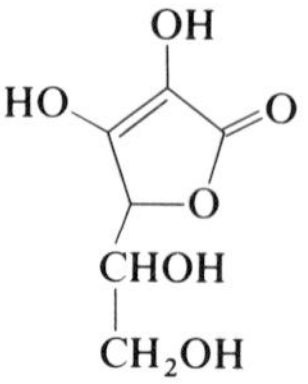	4.1	*161, 162*
2,5-Dihydroxy-1,4-benzoquinone (**32**)	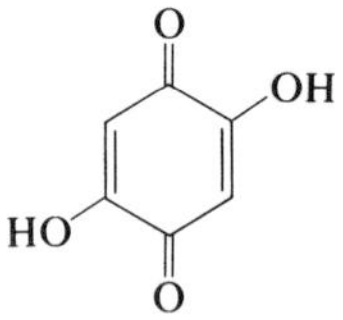	2.7	*163*
Deltic acid (**33**)	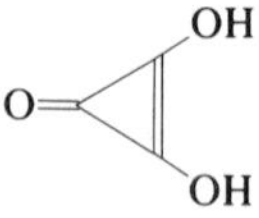	2.6	*164*
Croconic acid (**34**)	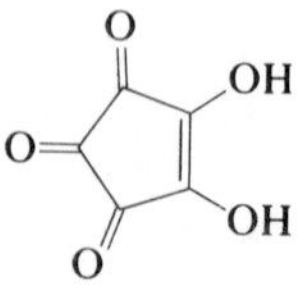	0.8	*150*

Table 2-14 (cont.)

Name (structure number)	Formula	pK_{HA}	Ref.
Squaric acid (**35**)	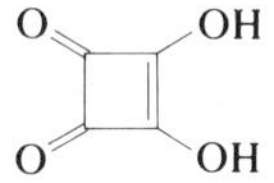	0.5	*165, 166*
2,6-Dihydroxy-3,5-dinitro-1,4-benzoquinone (**36**)	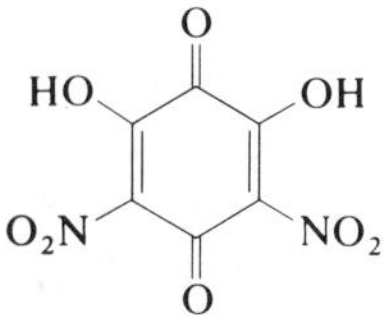	<0	*163, 167*
Tricyanovinyl alcohol	$(CN)_2C{=}C{-}OH$ $\quad\quad\quad\ \ \overset{|}{CN}$	≪0	*168*

out to be 11.6 for the enol of isobutyraldehyde, a value close to that estimated earlier by Guthrie and Cullimore (*169, 169a*).

Kresge *et al.* also generated enols by means of flash photolysis and again measured their rates of decay as a function of pH. They found a value of 10.94 for the pK_{HA} of the enol of acetone (*151a*), a value very close to that of Taphui and Jencks (*152*), who used the rate of halogenation of acetone to estimate this quantity. Wirz *et al.* had earlier used flash photolysis to generate the enol of acetophenone, whose pK_{HA} they determined to be 10.34 (*153*).

There appears to be much less regularity in the thermodynamic parameters that govern the ionization of enols than is the case with phenols or carboxylic acids. Both ΔH and ΔS are negative for squaric acid (**35**, $\Delta H = -1.5$ kcal mol^{-1}, $\Delta S = -7.5$ cal deg^{-1} mol^{-1}) (*165*), whereas they are both positive for the closely related compound croconic acid (**34**, $\Delta H = 3.5$–3.9 kcal mol^{-1}, $\Delta S = 8.9$–9.8 cal deg^{-1} mol^{-1}) (*150, 170*). It is sometimes difficult to obtain good agreement between values of ΔH determined, on the one hand, calorimetrically and, on the other, by use of the van't Hoff equation [Eq. (2-12)] (*171*). Since ΔS values are usually calculated using one or the other of these values and K, errors in ΔH will usually be duplicated in errors in ΔS. Thus, values of ΔH and ΔS reported in the literature (and herein) are sometimes given to quite unrealistic degrees of precision. The differences noted for squaric acid and croconic acid, however, seem far too great to be attributed entirely to experimental error.

Oximes, which are enols of nitroso compounds, resemble the enols of ordinary aldehydes and ketones in having pK_{HA} values in the region of 10 to 12 (*171a*).

G. Alcohols

The acid strengths of alcohols span ~ 14 pK units. The weakest alcohol whose acid strength in water has been evaluated is *tert*-butyl alcohol, the pK_{HA} of which is estimated to be 19.2 (*172*). The strongest alcohol that has been prepared is its nitrooctafluoro derivative, $(CF_3)_2C(OH)CF_2NO_2$ (*184*), whose pK_{HA} of 3.9 makes it only slightly weaker than formic acid (*185*).

The acid strengths of alcohols whose pK_{HA} values are in the range of 5 to 10 can be determined with reasonable precision by means of standard titration techniques using the glass electrode. Such methods are unsuitable for very weak acids, however, and recourse must be made to indicator (*178*), conductimetric (*175–177*), thermochemical (*180*), or other techniques. Unfortunately, ionization of alcohols is seldom accompanied by a sufficiently pronounced spectral change to allow ultraviolet spectroscopy to be used effectively for this purpose, except for some aromatic alcohols (*174*).

The most effective approach to determining the acidities of weakly acidic alcohols has been that of Ballinger and Long, who measured with great precision the conductivities of aqueous sodium hydroxide solutions containing variable amounts of alcohol (*175–177*). By noting the change in conductance with concentration they were able to determine the acidities of a number of the alcohols listed in Table 2-15.

In 1952 Hine and Hine measured the strengths of a large number of alcohols in isopropyl alcohol solvent by an indicator method (*178*). Most of the results for aqueous solution have been gathered since that time, but the order of acid strengths is similar for the two solvent systems. Murto combined these results with some nucleophilic displacement data to estimate the acid strengths of the two weakest acids listed in Table 2-15, *tert*-butyl alcohol and *s*-butyl alcohol (2-butanol) (*172*).

Reeve *et al.* more recently used the competitive reactivity of alkoxides in mixed hydroxylic solvents to measure the acid strengths of aliphatic alcohols (*185a*). Their values are close to those shown in Table 2-15, except for *tert*-butyl alcohol, which they found to be only slightly less acidic than isopropyl alcohol.

The acidities of cyanohydrins and hemiacetals can be estimated using the Taft equation and the ρ^* of 1.42 evaluated by Ballinger and Long for the dissociation of compounds of structure ZCH_2OH (*175*). Using 3.30 for σ^* of CN (*60*) and 15.9 for the pK_{HA} of ethanol ($Z = CH_3$, $\sigma^* = 0$) gives a value of 11.2 for $CNCH_2OH$, the cyanohydrin of formaldehyde [Eq. (2-20)].

$$\log K_{CNCH_2OH} = \log K_{CH_3CH_2OH} - \rho\sigma^*$$

$$= 15.9 - 4.7$$

$$= 11.2 \qquad\qquad (2\text{-}20)$$

Table 2-15

ACID STRENGTHS OF ALCOHOLS AND RELATED COMPOUNDS IN WATER

Alcohol	pK_{HA}	Ref.	Alcohol	pK_{HA}	Ref.
$(CH_3)_3COH$	19.2	172	CF_3CH_2OH	12.4	176
$CH_3CH_2CHOHCH_3$	17.6	172	Fructose	12.3	180
$(CH_3)_2CHOH$	17.1	172	CCl_3CH_2OH	12.2	175
$CH_3CH_2CH_2CH_2OH$	16.1	172	Ribose	12.2	180
$CH_3CH_2CH_2OH$	16.1	172	$4\text{-}CH_3OC_6H_4CHOHCF_3$	12.2	181
CH_3CH_2OH	15.9	172	$C_6H_5CHOHCF_3$	11.9	181
H_2O	15.74		$NCCH_2OH$	11.3	—[a]
$H_2C{=}CHCH_2OH$	15.5	175	$3\text{-}O_2NC_6H_4CHOHCF_3$	11.2	181
CH_3OH	15.5	175	$CF_3CH(OH)_2$	10.0	173
$C_6H_5CH_2OH$	15.4	172	$CCl_3CH(OH)_2$	10.0	179
$HOCH_2CH_2OH$	15.4	175	$C_6H_5C(OH)_2CF_3$	10.0	181
$CH_3OCH_2CH_2OH$	14.8	175	$CH_3(CF_3)_2COH$	9.6	173
Glycerol	14.4	175	$(CF_3)_2CHOH$	9.3	185
$ClCH_2CH_2OH$	14.3	175	$3\text{-}O_2NC_6H_4C(OH)_2CF_3$	9.2	181
$C(CH_2OH)_4$	14.1	175	$CHF_2C(OH)_2CHF_2$	8.9	183
$HC{\equiv}CCH_2OH$	13.6	175	$CF_3C(OH)_2CHF_2$	7.7	183
$CH_3CH(OH)_2$	13.6	179	$CF_3C(OH)_2CF_3$	6.5	182, 183, 185
$HOCH_2OH$	13.3	179	$(CF_3)_3COH$	5.1	173
Cl_2CHCH_2OH	12.9	175	$(CF_3)_2C(OH)CF_2NO_2$	3.9	185
Glucose	12.5	180			

[a] Calculated from data in refs. 60 and 175; see text.

On the basis that phenyl groups only marginally increase the acid strengths of alcohols (see Table 2-15) we can predict that the cyanohydrin of benzaldehyde, $C_6H_5CHOHCN$, will have a pK_{HA} near 11.

A dissociation constant of 7.8×10^{-7} has been reported for 2,2,2-trinitroethanol, which would make it the strongest primary alcohol by a wide margin. The dissociation involves more than simple proton loss, however, as shown in Eq. (2-21) (186).

$$(NO_2)_3CCH_2OH \rightleftharpoons H^+ + CH_2O + (NO_2)_3C^- \qquad (2\text{-}21)$$

The pK_{HA} of CH_3OCH_2OH can be calculated from the σ^* value of the methoxyl group (1.81) to be 13.3. Other hemiacetals, particularly the reducing sugars, may be somewhat more acidic because of the presence of electron-withdrawing oxygen atoms elsewhere in the molecule. Glucose ($pK_{HA} = 12.5$) and fructose ($pK_{HA} = 12.3$) are illustrative. In all these cases the hemiacetal hydroxyl group appears to be the most acidic site in the molecule (180, 187, 188). It should be noted, however, that many carbohydrates that have their hemiacetal hydroxyl groups tied up, either as acetals or as nucleosides, have

Table 2-16

EFFECT OF ADDITIONAL OXYGEN ATOM ON ALCOHOL ACID STRENGTH[a]

Compound	pK_{HA}	Analogous alcohol	pK_{HA}	ΔpK_{HA}
Hydroperoxy compounds				
$(CH_3)_3COOH$	12.8	$(CH_3)_3COH$	19.2	6.4
$(CH_3)_2CHOOH$	12.1	$(CH_3)_2CHOH$	17.1	5.0
$HOOH$	11.6	HOH	15.7	4.1
C_2H_5OOH	11.8	C_2H_5OH	15.9	4.1
CH_3OOH	11.5	CH_3OH	15.5	4.0
CF_3OOH	6.4	(CF_3OH)[b]	—	—
gem-Diols				
$CF_3CH(OH)_2$	10.0	CF_3CH_2OH	12.4	2.4
$C_6H_5C(OH)_2CF_3$	10.0	$C_6H_5CHOHCF_3$	11.9	1.9
$HOCH_2OH$	13.3	CH_3OH	15.5	2.2
$(CF_3)_2C(OH)_2$[c]	6.5	$(CF_3)_2CHOH$	9.3	2.8

[a] pK_{HA} values taken from Table 2-15 and refs. *193* and *195*.
[b] Unstable.
[c] The dilithium salt of this *gem*-diol is a stable compound (*196*).

acidities comparable to those with free hydroxyl groups at these positions. Thus, the pK_{HA} of adenosine (**37**) is 12.4 (*180, 189–191*) and that for guanosine (**38**) is ~ 12.3 (*189, 190, 192*). Presumably, the presence of numerous oxygen and nitrogen atoms in the molecules (and three ionizable hydroxyl groups) is sufficient to make these compounds several pK units stronger than simple aliphatic alcohols.

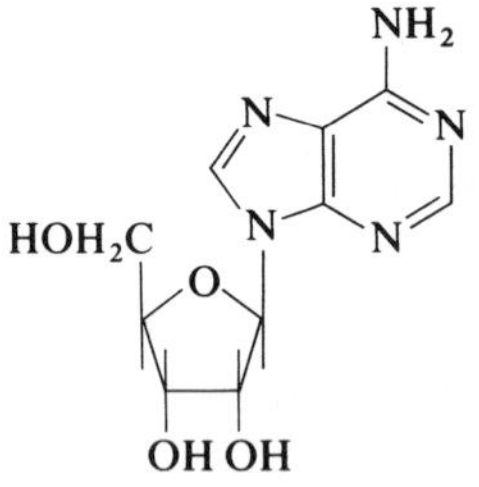

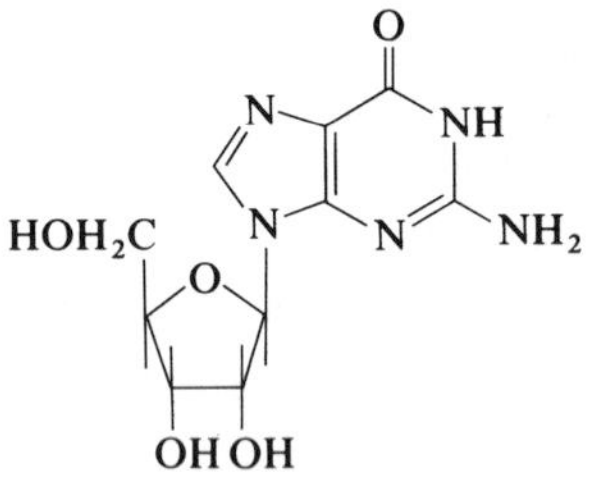

The acid-strengthening inductive effect of nearby oxygen can be seen by comparing the strengths of alcohols with those of hydroperoxy compounds and *gem*-diols (carbonyl hydrates). Table 2-16 shows that most hydroperoxy compounds are four to five pK units stronger than the corresponding alcohols. *tert*-Butyl hydroperoxide is markedly weaker than other hydroperoxides, but the differences are not as great as is the case with alcohols, where *tert*-butyl alcohol is almost four pK units weaker than water or methanol. Presumably, steric hindrance to solvation of the *tert*-butoxide ion is present, which will be much less pronounced in the case of the hydroperoxy anion.

It is interesting that peroxy acids are *weaker* than their carboxylic analogues; for example, the pK_{HA} of CH_3CO_3H is 8.2 (*193*) and that of CH_3CO_2H is 4.8. This can be accounted for by the opportunity for internal hydrogen bonding in the neutral molecule and by the absence of formal resonance structures in the peroxy acid anions [Eq. (2-22)].

$$R-C \overset{\displaystyle O''''H}{\underset{\displaystyle O-O}{\big<}} \rightleftharpoons H^+ + R-C \overset{\displaystyle O}{\underset{\displaystyle O-O^-}{\big<}} \tag{2-22}$$

It is less easy to account for the lack of effect on acidity of the additional oxygen atom that carbonic acid possesses relative to formic acid. Formic acid has a pK_{HA} of 3.75, and carbonic acid a pK_{HA} of 3.85 (*62*). The latter value takes account of the equilibrium with carbon dioxide but is not corrected for the presence of two hydroxyl groups in the molecule; if even a partial symmetry correction were applied (*194*), it would put the relative acidities of formic and carbonic acids farther out of joint.

III. Sulfur Acids

There are three general classes of organic compound containing the S—H bond: alkanethiols, arenethiols, and thiocarboxylic acids. Each occupies a distinct region of the pK spectrum, and each compound is considerably stronger than the analogous compound containing an O—H bond.

Alkanethiols have acidities comparable to those of phenols and are some five to seven pK units stronger than the analogous alcohols. The order of acid strength $CH_3OH > C_2H_5OH > (CH_3)_2CHOH > (CH_3)_3COH$ is duplicated in the thiol series, that is, $CH_3SH > C_2H_5SH > (CH_3)_2CHSH > (CH_3)_3CSH$ (Table 2-17), but the decline is much less marked. Thus, while *tert*-butyl alcohol is almost four pK units weaker than methanol, the difference in the thiols is less than one unit. Presumably, thiolate ions are affected by factors such as steric hindrance to solvation to a smaller extent than are alkoxide ions.

The pK_{HA} of thiols can be readily determined spectrophotometrically, unlike the situation with alcohols. The thiolate ion absorbs strongly in the ultraviolet region of the spectrum near 250 nm, where absorption of the thiols is low (*197*).

Arenethiols are roughly four pK units stronger than their phenolic analogues. The Hammett ρ value for thiophenols is 1.81 (*198*), compared with

Table 2-17

ACID STRENGTHS OF SULFUR ACIDS IN WATER
AT 25°C

Acid	pK_{HA}	Ref.
Thiols		
$(CH_3)_3CSH$	11.22	*197*
$(CH_3)_2CHSH$	10.86	*197*
$CH_3CH_2CH_2SH$	10.65	*205, 206*
C_2H_5SH	10.61	*197*
CH_3SH	10.33	*202*
$H_2C{=}CHCH_2SH$	9.96	*205*
$HOCH_2CH_2SH$	9.72	*197*
$C_6H_5CH_2SH$	9.43	*205*
$HSCH_2CH_2SH$	9.0^a	*203, 204*
Thiophenols		
$2,6\text{-}(CH_3)_2C_6H_3SH$	7.38	*207*
$2\text{-}CH_3C_6H_4SH$	7.00	*207*
(H_2S)	6.99	*55, 208, 209*
$4\text{-}CH_3C_6H_4SH$	6.82	*198*
C_6H_5SH	6.61	*198*
$4\text{-}ClC_6H_4SH$	6.14	*210*
$3\text{-}ClC_6H_4SH$	5.78	*198*
$2\text{-}ClC_6H_4SH$	5.68	*207*
$2\text{-}O_2NC_6H_4SH$	5.45	*207*
$3\text{-}O_2NC_6H_4SH$	5.24	*198*
$4\text{-}O_2NC_6H_4SH$	4.71	*210*
Thiocarboxylic acids		
CH_3COSH	3.43^b	*202*
C_6H_5COSH	2.48	*212, 213, 215*
$HCOSH$	2.06	*200*
Dithiocarboxylic acids		
$C_2H_5OC{\displaystyle{\overset{S}{\underset{SH}{\diagup\diagdown}}}}$	2.74^c	*201*
$CH_3C{\displaystyle{\overset{S}{\underset{SH}{\diagup\diagdown}}}}$	2.57	*214*
$C_6H_5C{\displaystyle{\overset{S}{\underset{SH}{\diagup\diagdown}}}}$	1.92	*211*
$H{-}C{\displaystyle{\overset{S}{\underset{SH}{\diagup\diagdown}}}}$	0.85	*200*

[a] Second dissociation constant, 10.6.
[b] Ref. *197* gives 3.62.
[c] The values for this compound listed in ref. *213* are
approximately one pK unit lower.

2.23 for phenols (*136*) meaning that the acid strengths of thiophenols are affected by substituents to a somewhat smaller extent than those of phenols.

The enthalpy of ionization of benzenethiol [Eq. (2-23)] is 2.65 kcal mol^{-1} and the entropy of ionization -21.4 cal deg^{-1} mol^{-1} (*197*). The corresponding figures for phenol are 5.65 kcal mol^{-1} and -26.7 cal deg^{-1} mol^{-1} (Table 2-13), and so both thermodynamic parameters contribute to making thiophenol the stronger acid, although the effect resides principally in the enthalpy term.

$$C_6H_5SH \rightleftharpoons H^+ + C_6H_5S^- \qquad (2\text{-}23)$$

Monothiocarboxylic acids exist predominantly in the thiol form (**39**) rather than in the thion form (**40**). Despite thiocarboxylic acids being stronger than

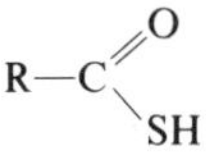

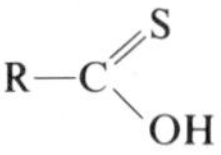

carboxylic acids the hydrogen-bonding capacity of the thio proton is low, as shown by the fact that there is no association of thioacetic acid in the gas phase (*199*). Thioacetic acid (ethanethioic acid) is just over one pK unit stronger than acetic acid, most of this difference being due to a less negative entropy of ionization [$\Delta S = -14.7$ cal deg^{-1} mol^{-1} for CH_3COSH (*197*) and -22.1 cal deg^{-1} mol^{-1} for CH_3CO_2H (Table 2-10)].

Dithiocarboxylic acids are the strongest of the sulfur acids. The effect of successive replacement of oxygen by sulfur in formic acid can be seen by comparing the acid strengths of **41** through **43**.

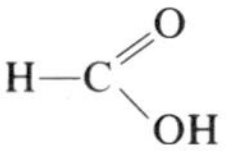

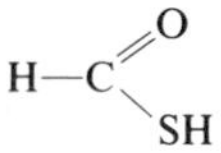

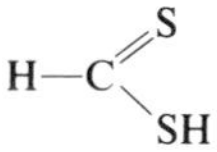

The salts of dithiocarboxylic acids containing an alkoxyl group at the carbonyl carbon atom, known as xanthates, are stable but the free acids decompose at room temperature into carbon dioxide and alcohol. Gavrish has estimated the pK_{HA} values of ethylxanthic acid (**44**) to be 2.7 and that of the methyl analogue 2.3 (*201*) although lower values appeared in the earlier literature (*211, 213*).

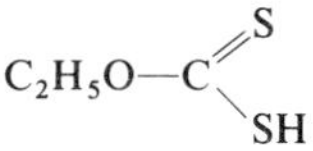

The same general effects are present in a series of sulfur acids ZSH as in the corresponding oxygen acid series, regardless of whether Z is alkyl, aryl, or acyl. Figure 2-1 shows a plot of pK_{HA}^{ZSH} for all the monothio compounds of Table 2-17 against pK_{HA}^{ZOH} (ordinate), covering some 16 pK units for ZOH and 9 pK units for ZSH. The slope of the line, 1.62, shows that sulfur acids in general are less sensitive to substituent effects than are oxygen acids. The most deviant point in Fig. 2-1 is that for the unsubstituted compounds H_2O and H_2S. The anomaly appears to be due to H_2S being stronger, rather than H_2O being weaker, than expected. In the gas phase the following order of acidities is found: $(CH_3)_3COH > (CH_3)_2CHOH > C_2H_5OH > CH_3OH > H_2O$ (*216–219*). This accords with the notion that the size of the attached group, and hence its polarizability (*220–222*), is the dominant factor in determining the order of acid and base strengths of compounds of closely similar chemical type. In the analogous sulfur series H_2S is misplaced, the order of acidities in the gas phase being $H_2S > (CH_3)_3CSH > (CH_3)_2CHSH > C_2H_5SH > CH_3SH$ (*216*). The greater than expected acidity of H_2S appears to have its origins in a rather high electron affinity of the HS· radical. Thus, H_2S is stronger than the alkanethiols in both water and the gas phase.

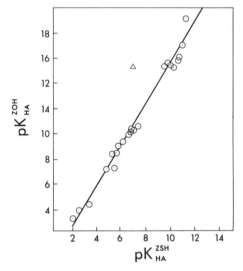

Fig. 2-1   Comparison of acidities of oxygen acids ZOH with the corresponding sulfur acids ZSH. Data are from Tables 2-10, 2-13, 2-15, and 2-17, with appropriate statistical corrections being made in the case of ethylene glycol and the carboxylic acids. The triangle represents H_2O/H_2S (uncorrected).

IV. Carbon and Nitrogen Acids

Carbon acids cover an extraordinarily wide range of acid strengths. The strength of the weakest, the alkanes, can only be roughly estimated but their pK_{HA} values in water appear to be in the region of 55 to 70. On the other hand, the strongest carbon acids that have been prepared, the cyanocarbons, can be very strong, indeed; many of them are completely ionized in water and are comparable in strength to the strong mineral acids.

Neutral nitrogen acids do not cover as wide a range of acid strengths as carbon acids, but they nonetheless span ~ 30 pK units, a range considerably greater than that found for organic oxygen acids (Section II). The weakest of the nitrogen acids are the aliphatic amines, whose strengths must be not far from that of ammonia, which has a pK_{HA} of ~ 35. Two of the strongest nitrogen acids that have been prepared in pure form are hexanitro-diphenylamine (1, $pK_{HA} = 2.63$) and bis(tricyanovinyl)amine (45), which is isolated as the hydrate containing 3.5 molecules of water (223, 224). This dark maroon crystalline substance presumably exists in an ionic form (e.g., 45b) rather than in the covalent form 45a.

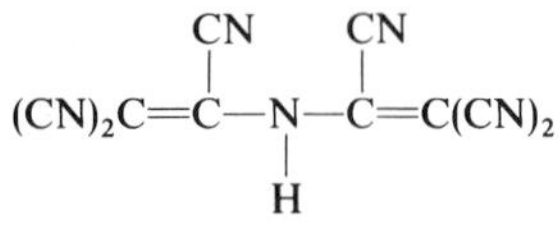

45a 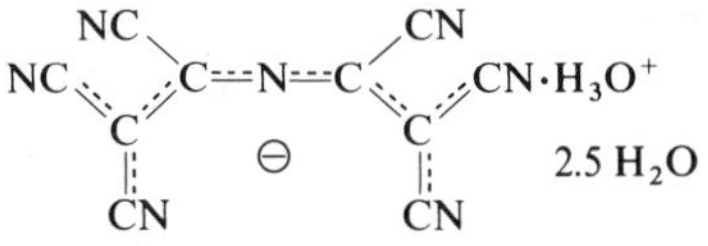   45b

Despite such examples of moderately strong nitrogen acids it is clear that the overwhelming majority of nitrogen compounds are very weak acids, which produce undetectable concentrations of anion in aqueous solution. Carbon acids are similar, and many of the techniques used to estimate the pK_{HA} values of carbon and nitrogen acids are the same. Furthermore, for those relatively few carbon and nitrogen acids that are detectably ionized in water or whose pK_{HA} values in water can be reliably estimated, we shall see that there is a fairly small difference in acid strength between compounds of analogous structure, whereas the analogous oxygen acids are usually considerably more acidic.

A. Comparison of the Strengths of Carbon, Nitrogen, and Oxygen Acids

Listed in Table 2-18 are pK_{HA} values of the relatively few examples available of compounds that differ only in the identity of the atom that loses the proton (carbon, nitrogen, or oxygen) and whose acid strengths in water have been directly measured. It can be seen that the nitrogen acids are much closer in

Table 2-18

COMPARISON OF THE ACID STRENGTHS IN WATER OF ANALOGOUS CARBON, NITROGEN, AND OXYGEN ACIDS[a]

Carbon acid, pK_{HA}	ΔpK_{HA}	Nitrogen acid, pK_{HA}	ΔpK_{HA}	Oxygen acid, pK_{HA}
CH_3NO_2 10.2	← 3.7 →	NH_2NO_2 6.5	← 7.8 →	$HONO_2$ −1.3
(2-methyl-1,3,5-trinitrobenzene structure) 14.4	← 2.2 →	(2-amino-1,3,5-trinitrobenzene structure) 12.2	← 11.9 →	(2-hydroxy-1,3,5-trinitrobenzene structure) 0.3
(2-methyl-nitrobenzene structure) 17.1	← 2.1 →	(2-amino-nitrobenzene structure) 15.0	← 10.9 →	(2-hydroxy-nitrobenzene structure) 4.1
$H_3C-\overset{\overset{O}{\|\|}}{C}-CH_3$ 19.0	← 3.9 →	$H_3C-\overset{\overset{O}{\|\|}}{C}-NH_2$ ∼15.1	← 10.3 →	$H_3C-\overset{\overset{O}{\|\|}}{C}-OH$ 4.8
(4-nitrotoluene structure) —		(4-nitroaniline structure) 18.4	← 11.2 →	(4-nitrophenol structure) 7.2
$C_6H_5SO_2CH_3$ —	← —[b] →	$C_6H_5SO_2NH_2$ 10.0	← 12.8 →	$C_6H_5SO_2OH$ −2.8

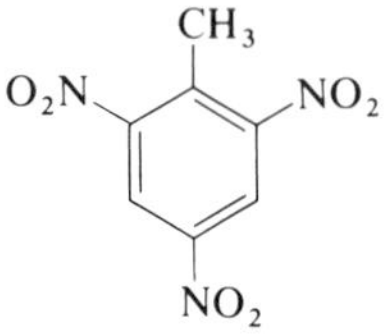

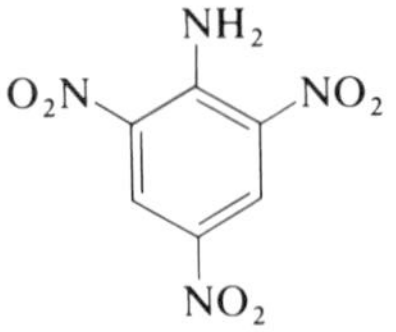

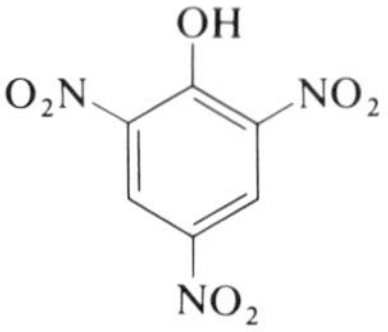

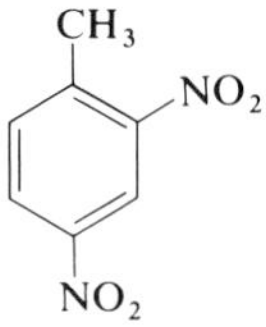

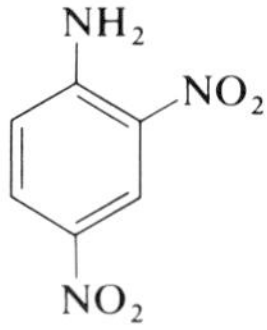

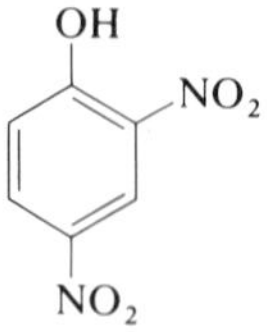

[a] Data for the nitroarenes come from R. Schaal, C. R. Hebd. Seances Acad. Sci. **238**, 2156 (1954); other compounds are listed elsewhere in this chapter.
[b] Probably large; see text.

strength to their carbon analogues than to their oxygen analogues. A probable exception is the family of compounds $RSO_2OH/RSO_2NH_2/RSO_2CH_3$; although pK_{HA} values of the sulfones have not been accurately determined in water, indications are that they are much weaker than the analogous amides [e.g., see the work of Bordwell (225, 226) and Jones (227)].

It is unlikely that this general pattern is maintained in the very weakly acidic region. Thus, toluene is undoubtedly a much weaker acid than aniline. The latter has a pK_{HA} of ~ 28, and estimates of pK of toluene in water have ranged from 32 to > 50. In the gas phase aniline is more acidic than toluene by 12.5 kcal mol^{-1} and less acidic than phenol by 14.7 kcal mol^{-1} (216). Bordwell et al. compared the strengths of toluene and aniline in dimethyl sulfoxide (DMSO) solution; toluene is too weak an acid to be measurably ionized in this medium, whereas aniline can be comfortably converted to its anion by sodium dimsyl in DMSO (228). By making use of substituent effects the pK_{HA}^{DMSO} of toluene has been estimated to be nearly 42, which would make it ~ 11 pK units weaker than aniline and ~ 25 units weaker than phenol in this medium.

There are indications that many weak carbon acids tend to be slightly more acidic in water than in DMSO (229) and a value for toluene in the range 40–41 would be consistent with this notion.

Interestingly, 4-nitrotoluene is slightly *stronger* than 4-nitroaniline in DMSO (228), an indication of the need for caution in making generalizations about the strengths of carbon, nitrogen, and, indeed, oxygen acids, particularly in nonaqueous systems. Hallé et al. (230) have shown that an inversion of acid strengths occurs between the two ionizable protons in **46** as the medium changes from water to DMSO.

46

In aqueous solution the carboxyl group of **46** is some six pK units stronger than the amino group. [The amino group in the carboxylate anion of **46** is half-ionized in water at pH 10.1, whereas the carboxyl acidity is close to that of benzoic acid, 4.2 (231).] In DMSO solution containing very little water, however, initial proton loss from **46** takes place at the amino group.

Other cases of carbon acids being more acidic than their nitrogen analogues in DMSO include the ketone **47**, which is more than one pK unit stronger than the amide **48** (226), and cyclopentadiene (**49**), which is about five pK units stronger than pyrrole (**50**) (232). Cyclopentadiene is also more acidic than pyrrole in the gas phase (216).

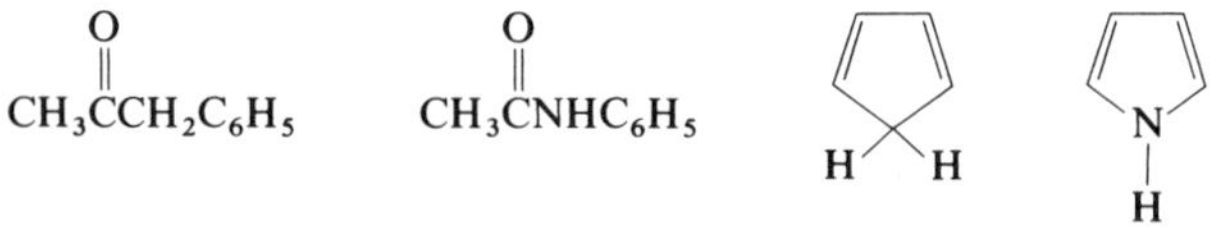

$$CH_3\overset{\overset{O}{\|}}{C}CH_2C_6H_5 \qquad CH_3\overset{\overset{O}{\|}}{C}NHC_6H_5$$

47   48   49   50

Bordwell (226) attributed the relative weakness of **48** as an acid to strong resonance stabilization of the carboxamide unit in the neutral form of the amide, such resonance being completely absent, of course, in the ketone. Whether this unusual order of acidity also prevails in aqueous solution is not yet known, but it seems doubtful that this will turn out to be the case since acetamide is almost four pK units stronger than acetone in water (Table 2-18), whereas it is only one pK unit stronger in DMSO. Homer and Johnson commented in their 1970 review on the paucity of aqueous ionization data for amides and expressed the belief that the situation would be remedied shortly (233). This has not occurred to any significant extent, although a considerable number of heterocyclic compounds that are formally amides have been studied. Some, such as **50a**, have considerable acid strength as a result of extensive charge delocalization in the anion (see also Table 2-21).

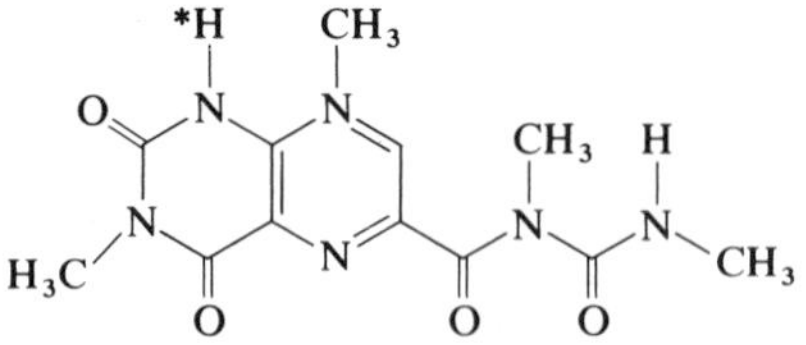

50a

pK_{HA} = 2.79 (acidic proton indicated by asterisk) (233a)

The small difference in pK_{HA} between analogous carbon and nitrogen acids that is apparent in Table 2-18 can be traced to the difference in stabilities of the neutral compounds and not, as is so often the case in ionic processes, to stabilization of the anions. Taking the ketone, amide, and carboxylic acid series, for example, resonance stabilization in the anion follows the order

$$RCO_2^- > RCONH^- > RCOCH_2^-$$

whereas resonance stabilization of the neutral compounds follows the order

$$RCONH_2 > RCO_2H > RCOCH_3$$

Thus, amide acidities are displaced toward those of ketones and away from those of carboxylic acids. The same situation applies to the other series shown in Table 2-18. This effect will diminish as resonance becomes less important, and it might be expected in the case of saturated systems, where there is no resonance at all, that the acidities of alkanes, alkylamines, and alcohols will reflect electronegativity differences between carbon, nitrogen, and oxygen and that there will be comparable differences between the strengths of these compounds as acids.

In the series $CH_3OH/CH_3NH_2/CH_3CH_3$ there are ~ 20 pK units difference between methanol and methylamine. [The value of 15.5 for methanol is firm; Bell (234) has estimated the pK_{HA} of ammonia in water to be 35, and methylamine is likely to be only slightly weaker.] If we assume that ethane would in turn be ~ 20 pK units weaker than methylamine, we have a pK_{HA} near 55 for ethane, neglecting solvation effects. The latter are not likely to affect the argument here for the following reason.

Water will have a large hydrogen-bonding interaction with all three anions and, unlike the cases discussed earlier, there is no resonance to disperse the charge in the anions. Thus, solvation differences are likely to be the greatest for the neutral compounds, with the alcohol being solvated most and the alkane least. Assuming a uniform decrease in the degree of solvation in the series methanol–methylamine–ethane, there should not be a major effect on the pK differences in the series.

A number of estimates of alkane acidities have been previously made, ranging from values below 40 to others of well over 70 (235–239).

It was pointed out earlier that pyrrole is a weaker acid than cyclopentadiene in DMSO, their pK_{HA}^{DMSO} values being 23.0 and 18.0, respectively (232). As with the amides considered earlier the decreased strength of the nitrogen acid can be attributed to resonance in the neutral molecule, though in this case it is the aromatic sextet that is involved. Such resonance is completely lacking in the carbon analogue, cyclopentadiene. The anions of both compounds, of course, are stabilized to a large degree by resonance (232a).

B. Strong Carbon Acids

The three substituent groups that have been frequently used to produce strong carbon acids, that is, compounds that are highly ionized in aqueous solution, are nitro, cyano, and trifluoromethyl. Of these the nitro group has the strongest electron-withdrawing character; however, its use is limited by two factors. First, an accumulation of nitro groups in an organic compound makes for instability. [This property has been turned to advantage in the determination of the pK_{HA} of trinitromethane, which can be generated by the radiolytic decomposition of tetranitromethane in water (240, 241).] Second, nitro has a strong saturation effect in almost all systems that have been studied; that is, the effect drops off as additional nitro groups are added. (Nitrophenols are exceptional in this respect; Table 2-18.)

Table 2-19

pK_{HA} VALUES OF METHANE DERIVATIVES[a]

CH_4	CH_3NO_2	$CH_2(NO_2)_2$	$CH(NO_2)_3$
~ 58	10.2	3.6	0.2
CH_4	$H_3C-\overset{\overset{\displaystyle O}{\|}}{C}-CH_3$	$CH_2(\overset{\overset{\displaystyle O}{\|}}{C}CH_3)_2$	$CH(\overset{\overset{\displaystyle O}{\|}}{C}CH_3)_3$
~ 58	19.0	8.8	5.9
CH_4	CH_3CN	$CH_2(CN)_2$	$CH(CN)_3$
~ 58	24	11.2	-5

[a] See refs. *240* and *242* and Table 2-22.

Carbonyl groups are similar to nitro groups in having a saturation effect, whereas the cyano group, largely because of its linear geometry, exerts an effect that is closer to being additive. Even so, the first cyano group almost certainly has a substantially greater effect than subsequent groups since a single cyano unit will change the geometry of the anion from pyramidal to trigonal, a geometry that will be retained upon addition of further cyano groups (Table 2-19).

Of the family of compounds CH_4, CH_3CN, $CH_2(CN)_2$, and $CH(CN)_3$, the pK_{HA} of only $CH_2(CN)_2$ is accurately known. Alkane mononitriles are very weak, as indicated by the powerfully basic conditions required to make such compounds react as active methylene units in the Thorpe–Ziegler reaction (*243*). An early estimate of the acidity of acetonitrile, CH_3CN, was that made by Bonhoeffer *et al.* (*244*), who measured the rate of exchange in KOD/D_2O and concluded that the pK_{HA} is ~ 25, which would make it much weaker than acetone, for example, whose pK_{HA} is 19.0. Judging from its pK in DMSO and from the effect of aryl substituents on acetonitrile and aniline, the pK_{HA} of acetonitrile must, indeed, be close to Bonhoeffer's estimate of 25. [The pK_{HA}^{DMSO} of acetonitrile is 31.3, some five units higher than that of acetone (*245*); assuming the same ΔpK in water gives a pK_{HA} of 24 for acetonitrile. 3-Chloro- and 4-nitrodiphenylamine are between two and three pK units weaker than 3-chloro and 4-nitrophenylacetonitrile; assuming a comparable ΔpK for the parent compounds, aniline (p$K_{HA} = 27.7$) and acetonitrile, would give a value near 25 for the pK_{HA} of the latter compound.]

Cyanoform, $CH(CN)_3$, is a strong acid, being half-ionized in 60% aqueous sulfuric acid and unstable in pure form. Indeed, when isolated it exists predominantly as the isomeric dicyanoketimine, $(CN)_2C=C=NH$ (*246–247*), although spectral evidence points to $CH(CN)_3$ being the form that is in equilibrium with the anion in aqueous sulfuric acid (*168*).

The cyanocarbons as a group are remarkably strong acids. Boyd used an acidity function technique to determine the strengths in aqueous sulfuric and perchloric acids of a number of these compounds (248). He did not simply measure the degree of ionization of such compounds and use an acidity function such as H_0 to calculate pK, but rather overlapped a series of cyanocarbons starting in dilute aqueous solution. Although the overlap was imperfect, it is clear that hexacyanoheptatriene and cyanoform, which are half-ionized in 50 and 60% sulfuric acid, respectively, will be essentially completely ionized in water and that their pK_{HA} values cannot be far from the average values of -3.7 and -5.1 given by Boyd.

It may seem surprising that the gap between the acidities of malonitrile and cyanoform is as large as it is, greater than the gap between malonitrile and acetonitrile. It is likely that relief of steric strain in going from the tetragonal neutral compound to the trigonal anion is responsible for this effect. [The analogous nitro and carbonyl compounds do not behave in this way, since they would be required to adopt a planar arrangement to achieve maximum charge dispersion in the anion, thus creating steric strain in the trisubstituted anions (249).]

Pentacyanopropene (**51**) and hexacyanoisobutene (**52**) are completely in the anionic form in 12 M sulfuric acid. The latter substance can be isolated only in the form of its sodium or other salt, whereas pentacyanopropene can be obtained as a rather unstable yellow-brown crystalline ionic dihydrate,, which decomposes to a black residue on being dried completely (168, 223, 250). The high acid strength of this compound and its ionic nature suggest that it has the structure **53** and contains a molecule of water of crystallization.

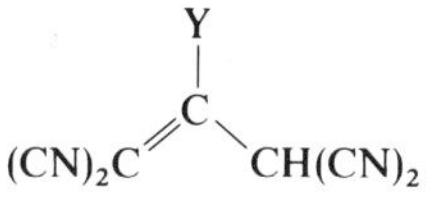

51 Y = CN
52 Y = CH(CN)₂   **53**

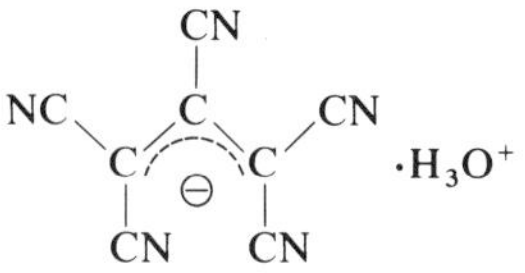

Possibly the strongest of the cyanocarbons are the polycyanocyclopentadienes, which are prepared by the stepwise cyanation of cyclopentadiene by cyanogen chloride (250, 251). Whereas the dicyano compound **54** is a stable crystalline compound with a pK_{HA} of 2.4 [making it slightly stronger than the mononitro compound, pK_{HA} = 3.25 (251a)] the higher members of the series are all strong acids. The anomalously low solubility of the polycyanocyclopentadiene salts in sulfuric acid has precluded direct measurement of their pK_{HA} values by the Hammett overlap procedure used by Boyd for the acyclic cyanocarbons; however, on the basis of measurements in acetonitrile solution Webster concluded that the pK_{HA} of **55** is less than -11.

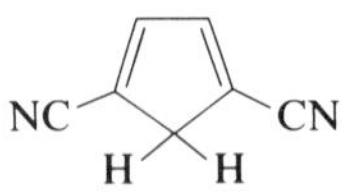

The strongest cyclopentadiene containing only nonconjugating substituents is pentakis(trifluoromethyl)cyclopentadiene (**56**) (*252*). It is a volatile liquid that is corrosive to glass; it is freely soluble in water but, like the polycyclopentadienes, it has a low solubility in sulfuric acid.

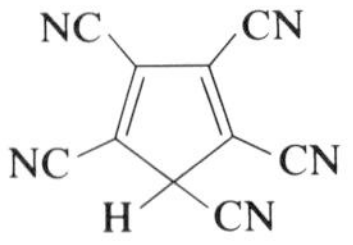

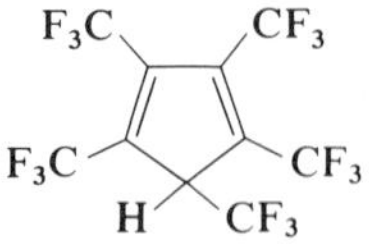

Tee and Iyengar pointed out that the keto forms of phenols must be highly acidic (*252a*). They estimated the pK_{HA} of 2,5-cyclohexadienone to be near -1 and noted that even in 1 *M* hydrochloric acid the pathway from the keto form to phenol via the anion is of lower energy than that via the cation.

The distinction between ionization and dissociation that applies to moderately strong acids (Section II,A) is unlikely to be important in the case of cyanocarbon acids. The charge on the anion is so delocalized that hydrogen bonding in solution between the hydronium ion and the anion is likely to be negligible.

C. Weak Carbon and Nitrogen Acids

The vast majority of carbon and nitrogen acids are far too weak to be detectably ionized in aqueous solution, and so special techniques have to be devised to measure or estimate their pK_{HA} values.

1. Nonaqueous Systems

During the early 1930s Conant and Wheland and then McEwen took advantage of the spectral changes that accompany the ionization of many weak carbon acids to develop a scale of acidities using diethyl ether as solvent (*253, 254*). They measured the equilibrium concentration of a carbon acid and its sodium or potassium salt in the presence of a second carbon acid and its salt and thus obtained the acidity in ether of one with respect to the other, that is, ΔpK_{HA}^{ether}. (The usual criterion for such indicator measurements is that ΔpK be less than 2.0, i.e., the percentage of ionization be between 10 and 90%.)

McEwen assigned a pK of 16 to methanol and then built up, in a stepwise manner, a scale of acidities for about 30 compounds, each weaker than its predecessor. The series ended with isopropylbenzene, for which a pK_{HA}^{ether} of 37 was assigned. The equilibrium between adjacent pairs of carbon acids is attained only slowly at this point. These acidity constants are relative thermodynamic quantities, with ether being the standard state. Several other solvent systems have subsequently been used in a similar manner, including cyclohexylamine by Streitwieser (239), isopropyl alcohol by Hine (178), DMSO and N-methylpyrrolidin-2-one by Bordwell (255), and dimethoxy-ethane by Shatenshtein (256). All produce relative thermodynamic pK values for the particular solvent system used, since the medium does not change significantly as the system becomes more basic. Only the identity of the bases changes in successive measurements, and their concentrations are low. (The term *relative thermodynamic* pK is used here to represent a quantity that differs from the thermodynamic pK in the particular solvent by an unknown but constant amount for all acids in the series.)

There are two problems to be faced in relating such pK values to those for aqueous solution. First, is the matter of assigning a pK to an anchor compound, an acid whose degree of ionization is to be compared with that of its neighbors in the series and whose pK_{HA} is known. Of course, if the absolute dissociation constant of such an anchor compound is known in the solvent in question, then absolute thermodynamic pK values in that solvent are obtained for the whole series as well.

Second, it must be shown that ΔpK values are the same in water and in the solvent in question; that is, a plot of $pK_{HA}^{solvent}$ against any pK_{HA} values that are available must be linear and of unit slope (257). If this condition is met, the other pK_{HA} values become available. In practice, this is not often found to be the case, although with nitrogen and carbon acids that produce charge-delocalized anions the condition appears to be very nearly met.

McEwen used 16.0 for the pK of methanol, a value that is close to the best value now available for aqueous solution (15.5). Streitwieser's system uses cyclohexylamine as solvent with lithium or cesium cyclohexylamide as the base, a system that is capable of ionizing extremely weak acids. It suffers somewhat from having a low dielectric constant, and thus ion pairing is extensive in this medium. Any specific solute interactions, such as ion pairing, can make for irregularities in the strengths of a series of acids and may make it difficult to use such data to deduce acidity constants in other media, water, of course, being the medium in which we are particularly interested. It turns out that many weak acids that produce highly delocalized anions give ion-pair pK values that are comparable to those found in water. Streitwieser adopted 18.5 as the pK of 9-phenylfluorene (a value that had been determined by the acidity function technique and refers to water as the standard state) and took this as the standard to which other weak acids could be compared, using the cyclohexylamine–cesium cyclohexylamide system and the method of indicators (258–261).

Bordwell's approach has been somewhat different, in that his pK values refer to the standard state of DMSO, the medium also used for the equilibrium measurements (*229*). This approach could be taken because absolute values of dissociation constants of a number of acids (e.g., acetic acid) have been determined by potentiometric techniques. Extremely basic DMSO systems can be obtained by using as the solute base the sodium salt of the solvent $CH_3SOCH_2^-Na^+$, known as sodium dimsyl. Although ion pairing does occur in this medium, it is not significant at the low solute concentrations employed in indicator studies. Shatenshtein *et al.* have also used DMSO (*262*) [and 1,2-dimethoxyethane (*262a*)] to measure the equilibrium acid strength of numerous organic compounds.

The polar aprotic solvent hexamethylphosphoramide also produces powerfully basic systems when alkaline hydroxides or alkoxides are used as solutes (*263*).

One of the most powerfully basic media is liquid ammonia containing potassium amide; it is capable of fully ionizing diphenylmethane and similar hydrocarbons (*264*). This medium has been used by a number of investigators for equilibrium (*265*) and kinetic measurements (Section IV,C,3) but suffers from the low temperature required and from a rather low dielectric constant, 22.4, and a consequently high degree of ion pairing.

Lithium dialkylamides dissolved in ether and other solvents are widely used in synthetic organic chemistry when extremely basic systems are needed. Lithium tetramethylpiperidide (**57**), which is ～1.6 units more basic in

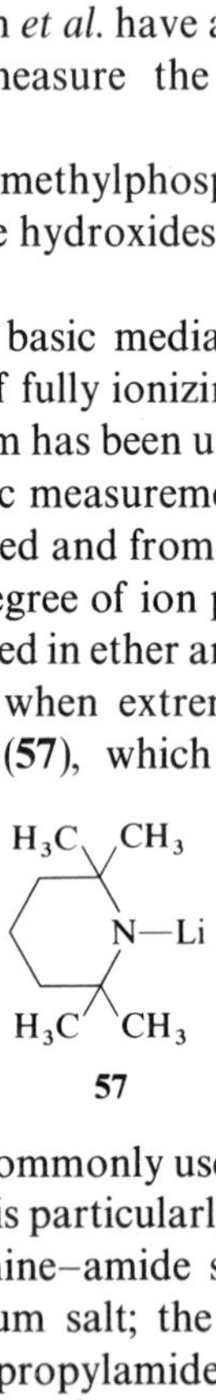

57

tetrahydrofuran than the more commonly used lithium diisopropylamide and which has a low nucleophilicity, is particularly useful in this respect (*266, 267*).

Another powerfully basic amine–amide system consists of 1,3-diamino-propane containing its potassium salt; the latter is known as KAPA, an acronym for potassium 3-aminopropylamide (*268, 269*). Arnett has estimated that such systems are some 10^5-10^6 times as basic as potassium dimsyl in DMSO (*270*).

Ethers or ether–alkane mixtures have also been used as solvent systems, particularly for studies of the equilibrium between various metal-complexed carbanions (*271–273, 274a*). Although these results might appear not to be particularly pertinent to aqueous protolytic equilibria, they provided a useful indication of relative carbanion stability and were combined by Cram with the work of McEwen and Streitwieser to produce a frequently cited scale of acidities (235, 273a).

The monographs by Jones (227), Reutov et al. (274), and Cram (235) provide more detailed descriptions of the use of nonaqueous solvents to assess the strengths of carbon acids.

2. Electrochemical and Thermochemical Methods

Three important electrochemical methods have been used over the years to study chemical equilibria; they are potentiometry, conductivity, and polarography.

Potentiometry in its modern application to acid–base equilibria makes use of the glass electrode, and this is the technique invariably used for spectrophotometric pK determinations in aqueous solution. It has been found that in certain other solvents the glass electrode also responds reversibly, if sometimes sluggishly, to acidity changes. Kolthoff and Reddy showed that the glass electrode could be used in DMSO, and they determined pK_{HA}^{DMSO} for a number of weak acids, up to values of ~ 12 (275). Ritchie and Uschold used a similar apparatus that allowed them to titrate some extremely weak acids with sodium dimsyl in DMSO using a glass electrode and a silver reference electrode (276). They found that the Nernst equation was obeyed, indicating that absolute values of pH^{DMSO} and pK_{HA}^{DMSO} can be obtained, which serve in turn as reference points for Bordwell's indicator determinations (Section IV,C,1). As Ritchie pointed out, however, large errors can result from the presence of small amounts of water and other impurities in the DMSO used in such measurements (277).

Electrochemical reduction, including polarography, has been widely used to estimate carbanion stabilities (274, 278–282). Reutov's method consists of measuring the polarographic half-wave reduction potentials for a series of symmetrical organomercury compounds in such solvents as dimethylformamide or dimethoxyethane [Eq. (2-24)], thereby making it possible to obtain the affinity of carbanions for the mercuric ion.

$$2 R^- + Hg^{2+} \rightleftharpoons R_2Hg \qquad (2\text{-}24)$$

$$R^- + H^+ \rightleftharpoons RH \qquad (2\text{-}25)$$

There is evidence that the equilibrium constants for this reaction are related, though not linearly, to those for the corresponding reaction between carbanions and the proton [Eq. (2-25)] and thus are a measure of the acidity constant of the compound RH.

Breslow's method makes use of the basic thermodynamic principle that the energetics of an equilibrium process are independent of the path by which equilibrium is attained (283–285). This principle is illustrated by the cycle shown in Eq. (2-26).

$$
\begin{array}{ccc}
RH & \rightleftharpoons & H^+ + R^- \\
+\frac{1}{4}O_2 \Big\Updownarrow & & \Big\Updownarrow + e^- \\
ROH & \underset{+H^+, \, -H_2O}{\rightleftharpoons} R^+ & \underset{+e^-}{\rightleftharpoons} R\cdot
\end{array}
\qquad (2\text{-}26)
$$

If the equilibrium constants can be evaluated for the four reactions in Eq. (2-26) that constitute the alternative path from RH to R^-, then the equilibrium constant for simple dissociation is obtained. Although the first of these steps, the oxidation of alkane to alcohol, does not take place reversibly, the associated free-energy changes can be estimated from known thermodynamic quantities. In any case there are reasons for believing that the energetics of this step are not strongly dependent on the identity of R. The second step, dissociation of the alcohol to carbonium ion, is, of course, strongly dependent on the identity of R. Fortunately, pK_{R^+} values for numerous alcohols are available, they having been determined in conjunction with the development of the H_R (J_0) acidity scale [Eq. (2-27)] (286).

$$
ROH + H^+ \rightleftharpoons R^+ + H_2O \qquad (2\text{-}27)
$$

The electrochemical potentials of the third and fourth steps of the alternative route can be obtained by polarography or cyclic voltammetry. The sum of the other changes shown in Eq. (2-26) is $\frac{1}{2}O_2 + 2H^+ + 2e^- \rightleftharpoons H_2O$, a known quantity, and thus the cycle is complete.

For weaker carbon acids, such as alkanes, Breslow used the thermodynamic cycle shown in Eq. (2-28). The first step is simple bond dissociation, and it is

$$
\begin{array}{ccc}
RH & \longrightarrow & H^+ + R^- \\
\Big\downarrow & & \Big\uparrow -e^- \\
R\cdot + H\cdot & \xrightarrow{+e^-} & H\cdot + R^-
\end{array}
\qquad (2\text{-}28)
$$

assumed that gas phase data for this reaction can be applied to solution; the other data are acquired electrochemically. The method is particularly useful for obtaining relative strengths of extremely weak acids, such as alkanes (287–290).

Arnett measured the heat of neutralization of a large number of weak oxygen, nitrogen, and carbon acids in DMSO containing sodium dimsyl as base (291, 291a). There is a good correlation between the reaction enthalpies and the acidity constants determined by other means. Indeed, the correlation can be extended to cover protonation of many of these same compounds, mostly anilines and diphenylamines, by powerfully acidic media, thus covering a very wide range of pK values. This correlation, which is remarkable for being a linear enthalpy relationship, provides support for the acidity function method used to determine the pK_{HA} (and pK_{BH^+}) values of the compounds in question and will be discussed in Section IV,C,4.

3. Kinetic Methods

In terms of equilibrium acid strength we have seen that nitrogen acids are often closer to their carbon than to their oxygen analogues. By contrast, the *rates* of ionization of nitrogen acids tend to resemble those of oxygen acids, being instantaneous for most practical purposes. Carbon acids, particularly those that are very weak or undergo significant geometrical changes on ionizing (*292–294*), are usually very slow to reach equilibrium. The term *pseudoacid* (*295*) or *secondary acid* (*296*) has sometimes been used to describe such compounds. Both nitrogen and oxygen acids have the advantage, of course, of having their ionizable protons hydrogen-bonded to water. Furthermore, they usually require a smaller degree of geometrical reorganization in going from neutral compound to anion.

The rate of ionization of carbon acids has been much used in the past as an indication of equilibrium acid strength (*242, 297, 298*), though it has to a large extent been superseded by direct or indirect equilibrium measurements of the type considered in other sections of this chapter. The basis for the kinetic method is the linear relationship that often exists in a series of closely related compounds between the logarithm of the rate of proton loss and the logarithm of the equilibrium constant for the same reaction, which is the well-known Brønsted relation. By extending such plots to cover compounds whose rates of ionization in water can be measured but whose equilibrium constants cannot, one can obtain an estimate of the strength of such weak acids, although such plots may not always be linear in the region of interest (*299*). In the case of acetone and other ketones this method has usually given pK_{HA} values that are close to those obtained by other methods (*152, 300*).

The main problem associated with the kinetic method is the limited structural variation that can be tolerated if rate and equilibrium data are to be compared. In Table 2-20 such data for three carbon acids are compared. Although their equilibrium acid strengths are fairly close, their rates of ionization vary enormously, with the strongest acid of the three, nitromethane, having the lowest rate of ionization. Nitroalkanes are notorious in this respect; their rates of ionization are much less than that which would be expected of acids of their strength.

Ionization rates of a large number of very weak carbon acids have been measured in a number of solvents, including liquid ammonia containing potassium amide by Shatenshtein and his group (*301*) and cyclohexylamine containing its lithium or cesium salt by Streitwieser and his group (*302, 303*).

Table 2-20

RATE AND EQUILIBRIUM DATA FOR THREE
CARBON ACIDS IN WATER AT 25°C[a]

Compound	pK_{HA}	k_{ioniz} (sec^{-1})
CH_3NO_2	10.2	4.3×10^{-8}
$\underset{\displaystyle CH_3}{CH_3\overset{O}{\overset{\|}{C}}-CH-\overset{O}{\overset{\|}{C}}CH_3}$	11.0	8.3×10^{-5}
$NCCH_2CN$	11.2	1.5×10^{-2}

[a] Data of Pearson and Dillon (242).

The kinetic isotope effects associated with the ionization of carbon acids have
been critically analyzed by Kresge and Powell (304).

A novel application of rate measurements in the determination of equilibrium constants is that of Jones et al. (305). They studied the detritiation
of a weak carbon acid (e.g., a 9-alkylfluorene containing tritium at the acidic 9
position) using a powerfully basic medium (e.g., DMSO containing base). This
reaction is monitored in the presence and in the absence of a second acid
whose degree of ionization is of interest. If this second acid, often a nitrogen
acid that ionizes instantaneously, is partially ionized under the experimental
conditions employed, it will have lowered the basicity of the medium to the
extent that it has consumed some of the base and hence reduced the
detritiation rate of the carbon acid indicator. By this means the extent of
ionization of the second acid can be determined. The difference between a
kinetic indicator as used here and a conventional equilibrium indicator is that
rates can be measured over an enormous range of values, whereas equilibrium
constants for conventional indicators can usually be determined accurately
only in the range of 10 to 90% ionization.

Kinetic studies of enolization in acidic and basic media have been made by
many workers over the years, and these results combined with the somewhat sparser data available for the rates of ketonization lead to keto–enol
equilibrium constants for a number of ketones and other carbonyl compounds. In order to determine the pK_{HA} of such compounds all that is needed
is the acidity constant of the enol, since these quantities are related, as shown
in Eq. (2-29).

$$pK_{HA} = pK_{HA}^{enol} + \log \frac{[ketone]}{[enol]} \qquad (2-29)$$

Using Kresge's data for the pK_{HA}^{enol} of isobutyraldehyde, that is, the pK_{HA} of $(CH_3)_2C{=}CHOH$, 11.63, and the keto–enol equilibrium ratio, 7.8×10^3, leads to a pK_{HA} of 15.53 for isobutyraldehyde (306).

Guthrie (169, 169a, 307, 307a) used both kinetic and thermochemical methods to determine the acidities of carbonyl compounds and, where comparisons can be made his values are in good agreement with those estimated in other ways. In 1982 Toullec reviewed the studies that had been made over the years on the carbonyl enolization process and on the pK_{HA} values derived for such carbon acids (308). More recently, Chiang et al. have obtained a value of 19.16 for the pK_{HA} of acetone in water (151a), a value that is very close to that given in Table 2-22.

The base-catalyzed racemization of several α-amino acids has been studied and the rates used to estimate the strengths of these compounds as carbon acids. Values of pK_{HA} in the range 16–17 for dissociation of the proton at the α-carbon atom were obtained (307b).

4. Acidity Functions for Strongly Basic Media

At about the same time that Conant and Wheland and McEwen were examining the strengths of very weak acids in diethyl ether and other solvents, Hammett was approaching the problem of determining the strengths of weak acids and bases in a different way. He developed the concept of the acidity function, in which the medium is gradually changed from water to one that is highly acidic or, in the present context, highly basic. If certain conditions are met, the pK values so determined refer to the standard state of water, even though the actual measurements are conducted in mixed media (286, 309–311). Acidity functions were first used to determine the strengths of very weak organic bases using aqueous sulfuric acid or similar systems, and this approach is considered in detail in Chapter 3.

Acidity functions for both acidic and basic media have been the subject of numerous reviews and monographs (286, 311–315). For basic systems, in which the ionization of interest is $HA \rightleftharpoons H^+ + A^-$, the function takes the form of Eq. (2-30a,b), the negative subscript in H_- designating the charge on the base; activity, activity coefficient, and concentration are designated, respectively, by the symbols a, f, and brackets:

$$H_- = pK_{HA} + \log \frac{[A^-]}{[HA]} \qquad (2\text{-}30a)$$

$$H_- = -\log a_{H^+} \frac{f_{A^-}}{f_{HA}} \qquad (2\text{-}30b)$$

Beginning in water, the standard state, where it can be seen from both Eqs. (2-30a) and (2-30b) that $H_- = pH$, the pK_{HA} of an anchor compound such as 2,4-dinitrodiphenylamine is measured. This compound, which has a pK_{HA} of 13.84, is just strong enough for its acidity to be measured under conditions in which the activity coefficients of all the species involved in the reaction are essentially unity. More basic solutions are then prepared by the addition of larger amounts of bases such as sodium hydroxide, or by the addition to water of soluble liquids such as hydrazine that are basic in their own right, or by the addition of polar aprotic solvents such as DMSO to water containing a small amount of hydroxide ion, whose concentration is kept constant.

Any of these three devices to increase basicity causes activities and concentrations to diverge. The essence of the acidity function method is the assumption that the activity coefficient ratio for members of a series of weak acids (or bases) will be similarly affected by the change in medium; that is, Eq. (2-31) holds in any of the mixtures that are used to produce conditions

$$f_{HA_1}/f_{A_1^-} = f_{HA_2}/f_{A_2^-} = f_{HA_x}/f_{A_x^-} \tag{2-31}$$

basic enough to ionize the weaker acids of the series. This requirement arises because the term $a_{H^+}f_{A^-}/f_{HA}$ in Eq. (2-30b) must be independent of the identity of the acid HA, although it will, of course, vary with the composition of the medium. If this condition is met, the pK_{HA} values (standard state water) of successively weaker acids can be measured. The practical test of this requirement is to see if plots of $\log([A^-]/[HA])$, usually measured spectroscopically, against some measure of basicity are parallel for two or more acids in the regions in which their ionizations overlap.

The first attempt to apply the acidity function technique to basic media was that of Schwarzenbach and Sulzberger, who used high concentrations of alkali metal hydroxides in water for the purpose (*316*). There are severe salting-out problems with such solutions, and even spectroscopic concentrations of weak organic acids of interest can seldom be achieved when the concentration of hydroxide is high (*315*). Edward and Wang (*317*) used thioacetamide as an indicator and found that 4.5 M aqueous sodium hydroxide has an H_- value of 14.95, compared with a value of 14.66 that derives from the hydroxide concentration; that is, the solution is about two times (0.29 logarithmic unit) as basic as would be expected by considering the effect of concentration alone. The discrepancy is much greater still with saturated aqueous solutions of potassium hydroxide (*318*). The term *excess basicity* has been used by Cox and Stewart (*319*) to describe the difference between the observed basicity and that which the system would have if it were ideal (*314, 320*).

There is a dearth of liquids that are bases that can be mixed with water to produce systems analogous to the mixtures of water and sulfuric acid that are often used to measure the strengths of very weak bases. Probably the best systems in this regard are aqueous ethylenediamine mixtures used by Schaal (*321*) and aqueous hydrazine mixtures first used by Deno (*322, 323*). These systems are only two to three logarithmic units more basic than 0.1 M aqueous sodium hydroxide, and in order to determine the pK_{HA} values of very weak carbon and nitrogen acids it has proved necessary to follow the third approach, which makes use of the powerful effect of polar aprotic solvents on the equilibrium shown in Eq. (2-32).

$$Z—H + OH^- \rightleftarrows Z^- + H_2O \qquad (2\text{-}32)$$

Addition of a polar aprotic solvent such as pyridine, DMSO, or sulfolane (tetramethylene sulfone) has the effect of pushing the equilibrium of Eq. (2-32) to the right by increasing the activity of hydroxide ion. Since Z—H will usually be an amine or an alkane derivative, the anion Z^- will almost always have a delocalized charge and hence will not be similarly affected. Indeed, polarization effects may actually stabilize the anion in the mixed medium compared with water.

In 1960 Langford and Burwell showed that a 95 mol % sulfolane–5 mol % water solution containing 0.01 M phenyltrimethylammonium hydroxide was $\sim 10^8$ times as basic toward carbon and nitrogen acids as the purely aqueous system. That is, the pH of 12 was raised to an H_- value near 20 (*324*).

Dimethyl sulfoxide is even more effective than sulfolane, and Stewart and O'Donnell and Dolman and Stewart constructed a scale using aromatic amine indicators and aqueous DMSO containing 0.011 M tetramethylammonium hydroxide that extended to an H_- value of 26.2 for 99.6 mol % DMSO, which is sufficiently basic to convert 3-chloroaniline predominantly to its anion (*325–327*). The only water in this system is that introduced with the base, which was added as crystalline tetramethylammonium hydroxide pentahydrate. Although excellent overlap was obtained for the two dozen or so amine indicators used to span this enormous range of basicities, two fundamental questions regarding the results must be addressed. First, despite the good indicator overlap obtained between adjacent pairs of indicators, might not small errors accumulate to produce an appreciable error in the pK_{HA} of the final indicator, 3-chloroaniline? Second, even if the pK_{HA} values of all the indicators have in fact been accurately determined, how general is the H_- scale that they generate? That is, does it govern the ionization of other weak acids, particularly weak carbon acids?

Perhaps the best test of the first question is to see if the differences in acid strength of the aromatic amine indicators measured by the acidity function technique in water–DMSO mixtures remain the same when measurements are conducted in pure DMSO. The pK_{HA} values determined in the latter medium (Section IV,C,1) refer to the standard state of DMSO, whereas the acidity function measurements, which are conducted in media that become ever richer in DMSO as the indicator acid strength declines, refer to the standard state of water. If the fundamental acidity function assumption holds, that is, Eq. (2-31) is valid, then changing the medium will not cause a change in the pK *difference* for any pair of indicators measured in those media. This means that a plot of pK_{HA} against p$K_{HA}^{solvent}$, where the latter refers to the second component used in developing the H_- scale, must be linear and of unit slope (257). Figure 2-2 shows a plot of pK_{HA} against pK_{HA}^{DMSO} for all aromatic amines for which data are available, excluding polynitro compounds. It can be seen that a satisfactory linear relationship exists, the slope of the line being 1.11. In applying this test to the acidity function approach it should be remembered that the greatest change in the properties of the solvent is likely to occur as one approaches pure DMSO and the last traces of water are removed from the system. The acidity function measurements that give rise to the pK_{HA} values shown in Fig. 2-2 are made before this point is reached, and it therefore appears likely that such pK_{HA} values are not greatly in error, though there might be a slight drift that is undetectable even when strict inspections for

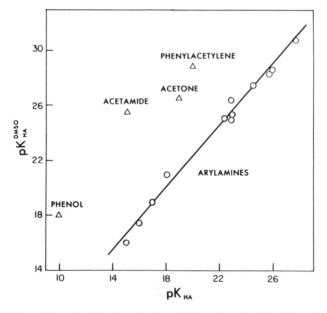

Fig. 2-2 Comparison of pK values in water (pK_{HA}, determined by the acidity function method) and in DMSO (pK_{HA}^{DMSO}). Arylamines in order of increasing pK_{HA} are 2,4-dinitroaniline, 2,5-dichloro-4-nitroaniline, 4-chloro-2-nitroaniline, 4-nitroaniline, diphenylamine, 2,6-dichloroaniline, 2,4-dichloroaniline, 4-cyanoaniline, 3-cyanoaniline, 3-trifluoromethylaniline, 3-chloroaniline, and aniline. Data from refs. *226, 229, 328*, and Tables 2-21 and 2-22.

good overlap between adjacent indicators are made. Any such drift would cause the measured pK_{HA} values for the weakest acids to be slightly low. [That Eq. (2-31) is obeyed in the accessible region of overlap but is flouted when the degree of ionization is very low is always a possibility, though there is no reason to suspect that this is a serious source of error (*319*).]

Schaal and Hallé *et al.* measured the degree of ionization of various compounds, including aromatic amines, using buffered aqueous DMSO solutions and obtained thereby pK values that refer to a standard state of infinite dilution in that solvent mixture (*329–331*). More recently, Edward *et al.* subjected this data to factor analysis and raised doubts regarding the strict validity of the acidity function approach (and related extrapolative techniques discussed later) when major solvent changes are involved (*332*).

It is clear from the presence of wayward points in Fig. 2-2 that the pK_{HA} value of weak acids that are structurally unrelated to arylamines cannot simply be taken from the H_- value of the medium in which they are half-ionized. The answer to the second question posed earlier regarding the generality of the scale is thus partially answered. Of course, with such compounds it will be seen that the fundamental Hammett postulate as embodied in Eq. (2-31) is not obeyed. In practice this means that a plot of $\log([A^-]/[HA])$ against H_- will not yield a straight line of unit slope. Even in these cases, however, acidity functions can often be used to obtain reasonable pK_{HA} values. First, if there are enough deviant compounds that are structurally related, a separate acidity function can be constructed, even though such a scale may be valid only for the compounds with which it was constructed. A number of such distinctive scales exist (*333–338*). Second, use can be made of extrapolative techniques that have been developed in recent years and that are described in more detail in Chapter 3. Their application to strongly basic systems is described later.

Bunnett and Olsen appear to have been the first investigators to grapple successfully with the problem of compounds whose log ionization plots are not parallel to those of the indicators used to construct the acidity function scale (*339*). They derived an equation that has been widely used to estimate the strengths of weak bases in aqueous sulfuric and other acids and that is discussed in Chapter 3. This general approach has been more recently applied to strongly basic media, where deviant ionization behavior is at least as common as in acidic solution.

The acidity function approach requires that the activity coefficient ratio for a weak acid HA be equal in any basic medium to that of any of the scale indicators HIn that are partially ionized in that particular medium [Eq. (2-33)]. This leads to plots of $\log([A^-]/[HA])$ against $\log([In^-]/[HIn])$

$$f_{A^-}/f_{HA} = f_{In^-}/f_{HIn} \qquad (2\text{-}33)$$

being linear and of unit slope. In those cases where the plots are not of unit slope they are, nonetheless, very often found to be highly linear. As a consequence, a plot of $\log([A^-]/[HA])$ against the acidity function H_- will also often be linear, as a result of the relationship shown in Eq. (2-34) holding.

$$\log\frac{f_{A^-}}{f_{HA}} = m\log\frac{f_{In^-}}{f_{HIn}} \qquad (2\text{-}34)$$

This condition, that the logarithms of the activity quotients be linearly related, is less stringent than that of the original Hammett assumption, which requires that these terms cancel. Cox and Yates referred to the latter condition as the zeroth-order approximation and the former, less stringent requirement as the first-order approximation (*314*).

The empirical observation that plots of $\log([A^-]/[HA])$ against H_- are usually linear, even when the slope is not unity, allows one to estimate the pK_{HA} in the standard state of water by simply extrapolating such a plot to the origin, since if the relationship is truly linear the fractional degree of ionization in the standard state is given (Fig. 2-3). In practice the extrapolated pK_{HA} can be obtained by using Eq. (2-35), where m is the slope of the ionization plot,

$$pK_{HA} = pH_{aq} + m(H_-^{1/2} - pH_{aq}) \qquad (2\text{-}35)$$

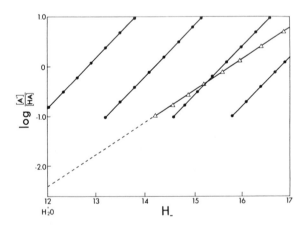

Fig. 2-3 Schematic ionization plots for four well-behaved indicators (circles) used to construct an H_- scale for mixtures of water and a polar aprotic solvent containing 10^{-2} M hydroxide ion. The triangles represent the ionization of a weak acid that does not follow H_- but the ionization plot of which is apparently linear. Extrapolation to the origin gives the degree of ionization in water at pH 12, from which pK_{HA} can be calculated.

$H_{-}^{1/2}$ is the H_{-} where the compound is half-ionized, and pH_{aq} is the pH of water containing the fixed amount of hydroxide ion that has been used throughout (usually 12.0, the pH of 10^{-2} M hydroxide ion).

It may be preferable to use Eq. (2-36), which is the analogue of the Bunnett–Olsen equation used in acidic media and which has been derived in the same way as the Bunnett–Olsen equation (319):

$$H_{-} - \log\frac{[A^{-}]}{[HA]} = -\phi(H_{-} - pK_{w} - \log[OH^{-}] + \log a_{H_2O}) + pK_{HA}$$

$$(2\text{-}36)$$

A plot of $H_{-} - \log([A^{-}]/[HA])$ against $(H_{-} - pK_{w} - \log[HO^{-}] + \log a_{H_2O})$, where $pK_{w} = 13.996$ at 25°C, and a_{H_2O} is the activity of water in the particular medium in which $[A^{-}]/[HA]$ is being measured, will give pK_{HA} as the intercept. Although this operation is more cumbersome than the simple extrapolation used in Fig. 2-3 and based on Eq. (2-35), it is less empirically based and, since it includes a water activity term, may well provide a more truly linear relationship and hence a more reliable extrapolation, although there appears to be little firm evidence at the moment in this regard.

An alternative extrapolative method that has found application in basic media derives from the studies of Marziano et al. in acidic media (340). Their method, which is described in Chapter 3, makes use of the indicator overlap principle but does not explicitly use an acidity function. The Bunnett–Olsen and Marziano–Cimino–Passerini methods give rather similar results when applied to weak carbon and nitrogen acids, although pK_{HA} values determined by the latter method tend to be slightly greater than those determined by the Bunnett–Olsen procedure.

5. Aqueous Acidity Constants for Nitrogen and Carbon Acids

Listed in Table 2-21 are a large number of nitrogen acids and their pK_{HA} values. Most of these have been determined by the use of acidity function and related extrapolative techniques using water and DMSO mixtures, as described earlier. For each such compound an average of these values has been taken, which tends to raise the acidity function pK_{HA} slightly, particularly for the weaker nitrogen acids. As we have seen there are indications that the acidity function values may be slightly low, and so the average values are probably better estimates of the intrinsic acid strengths of these compounds.

Listed in Table 2-22 are a large number of carbon acids together with their pK_{HA} values. An attempt has been made to make the list self-consistent by including only compounds whose pK_{HA} values can be referred to the aqueous state with some confidence.

Table 2-21

STRENGTHS OF NITROGEN ACIDS[a]

Compound	pK_{HA}	Ref.[b]	Compound	pK_{HA}	Ref.[b]
Anilines			Unsubstituted	27.7	4
2,4,6-Trinitro	12.2	1			
2,4-Diaza-3-chloro-6-nitro	13.3	2	Diphenylamines		
6-Bromo-2,4-dinitro	13.6	1	2,4,6,2',4',6'-Hexanitro	2.63	1
2-Aza-4,6-dinitro	13.8	2	2,4,6,2',4'-Pentanitro	6.72	1
2,4,6-Triaza	14.9	2	2,4,6,4'-Tetranitro	8.88	1
2,4-Dinitro	15.0	1	2,4,6,3'-Tetranitro	9.15	1
2,4-Diaza-3,5-dichloro	15.2	2	2,4,6-Trinitro	10.38	1
2,6-Dichloro-4-nitro	15.6	1	2,4,2',4'-Tetranitro	10.82	1
2-Aza-4-nitro	15.8	2	2,4,6-Trinitro-4'-amino	10.82	1
4-Aza-2-nitro	15.9	2	2,4,4'-Trinitro	12.3	3
2,5-Dichloro-4-nitro	16.0	1	2,4,3'-Trinitro	12.6	3
2,4-Diaza-3-chloro	16.4	2	2,4-Dinitro-4'-		
2-Aza-6-nitro	16.7	2	trifluoromethyl	12.9	3
2,6-Diaza-3,5-dichloro	16.7	2	2,4-Dinitro-3'-		
4-Chloro-2-nitro	16.9	3	trifluoromethyl	13.1	3
2,6-Di-*tert*-butyl-4-nitro	17.0	3	2,4-Dinitro-3'-chloro	13.2	3
2-Nitro	17.7	3	2,4-Dinitro	13.9	3
4-Nitro-N-triphenylmethyl	17.8	3	2,4-Dinitro-3'-methyl	14.0	3
2,6-Dimethyl-4-nitro	18.1	3	3,4'-Dinitro	14.5	3
2,6-Diaza-3-chloro	18.1	2	2,4-Dinitro-4'-amino	14.6	3
4-Nitro	18.2	3	4-Nitro-3'-trifluoromethyl	14.9	3
4-Nitro-N-methyl	18.2	3	4-Nitro-3'-chloro	15.0	3
4-Nitro-N-ethyl	18.2	3	4-Nitro	15.6	3
4-Nitro-N-isopropyl	18.3	3	4-Nitro-3'-methyl	15.6	3
2,4-Diaza	18.4	2	4-Nitro-4'-amino	16.4	3
4-Nitro-2-N-dimethyl	18.4	3	2-Nitro	18.0	3
4-Nitro-2-methyl	18.4	3	2,2'-Diaza	18.8	3
2,3,5,6-Tetrachloro	19.5	3	4-Methylsulfonyl	18.8	3
4-Nitro-N-*tert*-butyl	19.6	3	3-Nitro	19.4	3
2,6-Diaza	20.5	2	3,4'-Dichloro	19.7	3
2-Aza-4-chloro	21.8	2	3-Trifluoromethyl	20.5	3
4-Nitro-3,5-dimethyl	22.1	3	3-Chloro	20.7	3
2,3-Dichloro	22.1	3	4-Chloro	21.4	3
4-Aza	22.3	2	3-Methoxy	22.2	3
2,6-Dichloro	22.9	3	Unsubstituted	22.4	3
2,4-Dichloro	22.9	3	4-Methyl	23.1	3
4-Cyano	23.0	3	4-Methoxy	23.1	3
2,5-Dichloro	23.2	3			
3,4,5-Trichloro	23.4	3	Amides		
2-Aza	23.5	2	Saccharin	1.60	5
3,5-Dichloro	23.8	3	Nitrourea	4.16	5
3-Cyano	24.6	3	$C_6H_5SO_2NHC_6H_4$-4-NO_2	6.20	6
3,4-Dichloro	24.8	3	Phthalimide	8.30	5
4-Nitro-2,3,5,6-tetramethyl	25.4	3	$C_6H_5SO_2NHC_6H_5$	8.31	6
3-Trifluoromethyl	25.8	3	Hydantoin	9.12	7
3-Chloro	26.0	3	4-$O_2NC_6H_4SO_2NH_2$	9.14	6
			$CF_3CONHC_6H_5$	9.54	8

70

Table 2-21 (cont.)

Compound	pK_{HA}	Ref.[b]	Compound	pK_{HA}	Ref.[b]
6,7,8-Trimethyllumazine	9.90	9	Purine (diazabenzimidazole)	8.93	16
$C_6H_5SO_2NH_2$	10.0	6	6-Chloropurine[d]	7.82	16
4-$CH_3C_6H_4SO_2NH_2$	10.2	6	8-Phenylpurine[d]	8.09	16
N-Phenylthiobenzamide	10.6	10	8-Methylthiopurine[d]	7.67	16
5-Deaza-6,7,8-			Pyrazole	14.2	17
trimethyllumazine	11.4	11	1,2,3-Triazole	9.51	18
Thiobenzamide	12.8	10	1,2,4-Triazole	10.3	19
Thioacetamide	13.4	12	3,5-Dichloro-1,2,4-triazole	5.2	19
Acetamide	15.1	13	Benzotriazole	8.6	20
			Tetrazole	4.9	21
Miscellaneous heterocyclic			5-Chlorotetrazole	2.1	21
compounds[c]			1-Methylcytosine[d]	16.7	22
Imidazole	14.5	14	3-Methylcytosine[d]	13.4	22
2-Phenylimidazole	13.3	14	1-Methylisocytosine[d]	14.3	22
4- (or 5-)phenylimidazole	13.4	14	7-Methyladenine[d]	14.7	22
2,4- (or 2,5-)diphenylimidazole	12.5	14	9-Methyladenine[d]	16.7	22
Benzimidazole	12.9	15	1,7-Dimethylguanine[d]	15.0	22
5-Nitrobenzimidazole	10.8	15	1,9-Dimethylguanine[d]	14.6	22

[a] Standard state water.

[b] References: 1, R. Stewart and J. P. O'Donnell, Can. J. Chem. 42, 1681 (1964). 2, M. G. Harris and R. Stewart, Can. J. Chem. 55, 3800 (1977). 3, Data of D. Dolman and R. Stewart, Can. J. Chem. 45, 911 (1967) as recalculated by R. A. Cox and R. Stewart, J. Am. Chem. Soc. 98, 488 (1976) and averaged (see text). 4, Recalculated using the Hammett equation and pK_{HA} values in this table. 5, R. P. Bell and W. C. E. Higginson, Proc. R. Soc. London, Ser. A 197A, 141 (1949). 6, A. V. Willi, Helv. Chim. Acta 39, 46 (1956). 7, M. Zief and J. T. Edsall, J. Am. Chem. Soc. 59, 2245 (1937). 8, P. M. Mader, J. Am. Chem. Soc. 87, 3191 (1965). 9, W. Pfleiderer, J. W. Bunting, D. D. Perrin, and G. Nubel, Chem. Ber. 99, 3503 (1966). 10, W. Walter and R. F. Becker, Justus Liebigs Ann. Chem. 727, 71 (1969). 11, R. Stewart, R. Srinivasan, and S. J. Gumbley, Can. J. Chem. 59, 2755 (1981). 12, J. T. Edward and I. C. Wang, Can. J. Chem. 40, 399 (1962). 13, G. E. K. Branch and J. O. Clayton, J. Am. Chem. Soc. 50, 1680 (1928). 14, H. Walba and R. W. Isensee, J. Org. Chem. 16, 2789 (1961); 21, 702 (1956). 15, F. Terrier, F. Millot, and R. Schaal, Bull. Soc. Chim. Fr. p. 3002 (1969). 16, A. Albert and D. J. Brown, J. Chem. Soc. p. 2060 (1954). 17, G. Yagil, Tetrahedron 23, 2855 (1967). 18, D. P. Perrin, "Dissociation Constants of Organic Bases in Aqueous Solution." Butterworths, London, 1965. 19, A. Albert, "Heterocyclic Chemistry," 2nd ed. Athlone Press, London, 1968. 20, A. Albert, R. Goldacre, and J. Phillips, J. Chem. Soc. p. 2240 (1948). 21, E. Lieber, S. H. Patinken, and H. H. Tao, J. Am. Chem. Soc. 73, 1792 (1951). 22, M. G. Harris and R. Stewart, Can. J. Chem. 55, 3807 (1977). 23, J. R. Jones and S. E. Taylor, J. Chem. Res. (S) p. 154 (1980).

[c] For effect of temperature see footnote b, ref. 23.

[d] The structures and relevant numbering of one tautomer of the parent compounds are as follows:

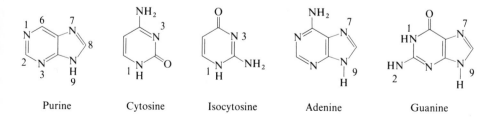

Purine   Cytosine   Isocytosine   Adenine   Guanine

Table 2-22

STRENGTHS OF CARBON ACIDS[a]

Compound	pK$_{HA}$	Ref.[b]
Hydrocarbons		

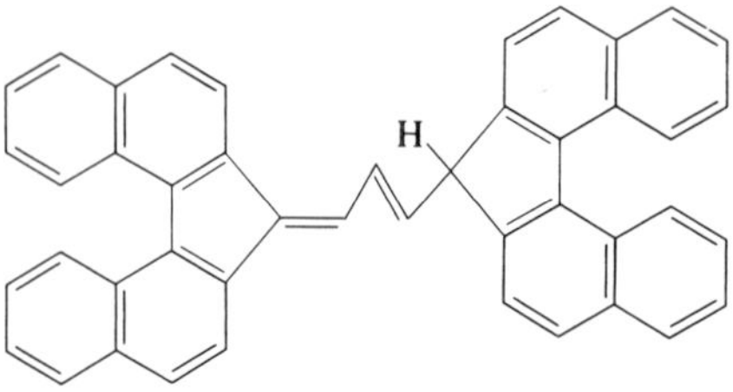	8.2	1
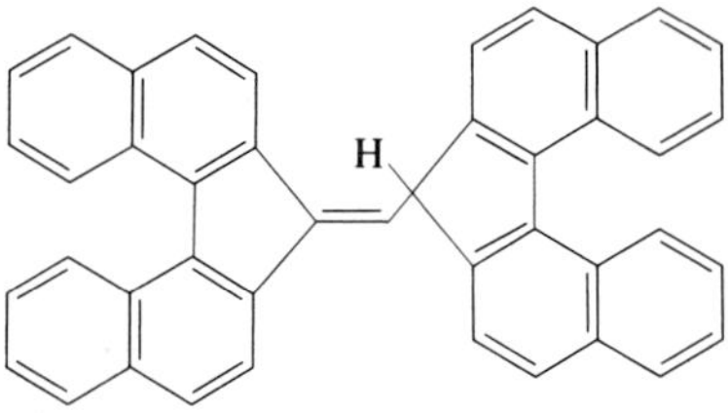	9.1	1
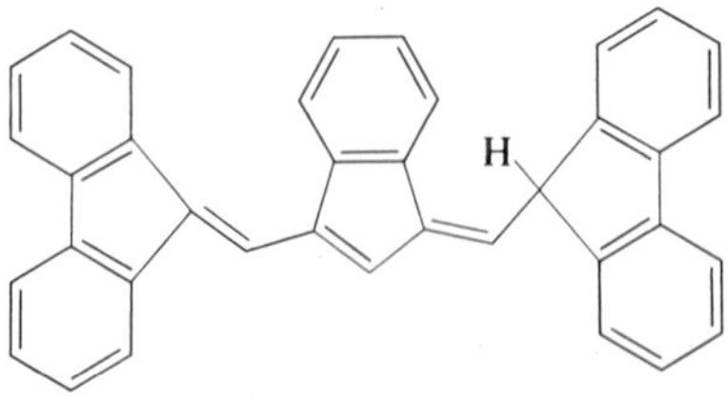	10.0	1
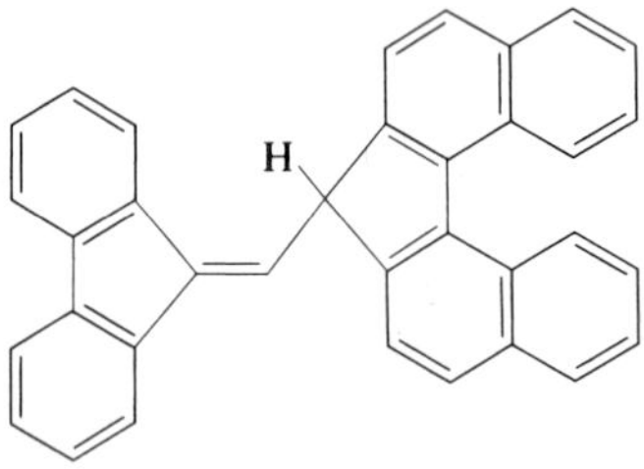	11.2	1
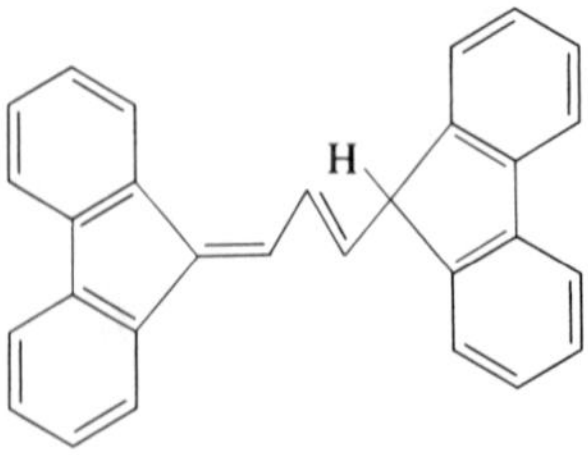	12.3	1

Table 2-22 (*cont.*)

Compound	pK_{HA}	Ref.[b]

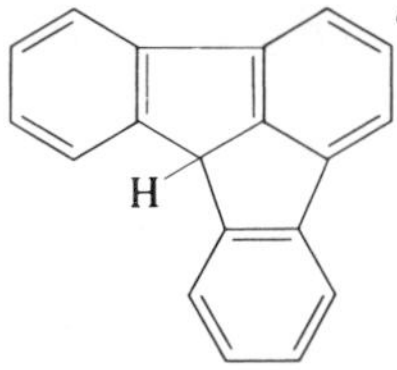	14.0	1
1,4-Diphenyl-1,3-cyclopentadiene	15.4[d]	2
Cyclopentadiene	16.0	3
9-Phenylfluorene[e]	18.5[f]	4, 5

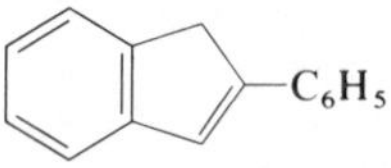	18.6[d]	6

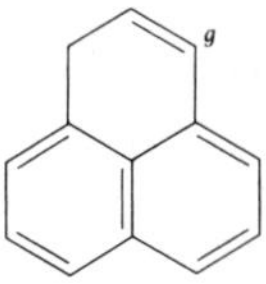	19.5	7

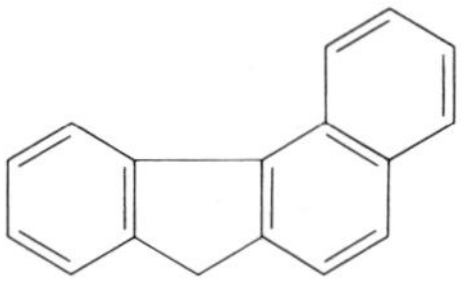	19.7	8
Phenylacetylene	20.0	9
9-Methylfluorene[e]	22.1	7, 8
Fluorene	22.2	7, 8
9-Ethylfluorene[e]	22.2	7, 8
9-Isopropylfluorene[e]	23.2	7, 8
9-*tert*-Butylfluorene[e]	23.9	7, 8
1,2,3-Triphenylpropene	26.6	10
4-Phenyltriphenylmethane	30.2	11
Triphenylmethane	31.5	12
Diphenylmethane	33.4	12
Cumene	39.1	13
Toluene	~41	12
Methane	~58	14
Cyclopropene	~61	15
Isobutane	~71	16
Carbonyl compounds		
$NO_2CHClCO_2C_2H_5$	4.16	17
$CH_2(CHO)_2$	5.0	18
$NO_2CH_2CO_2C_2H_5$	5.75	17
$(CH_3CO)_3CH$	5.86	19
CH_3COCH_2CHO	5.92	20
$NO_2CHFCO_2C_2H_5$	6.28	17
$(CH_3O_2C)_3CH$	7.8	21

(*cont.*)

Table 2-22 (*cont.*)

Compound	pK_{HA}	Ref.[b]
$(CH_3CO)_2CH_2$	8.84	22
$(C_6H_5CO)_2CH_2$	8.95	23
$(C_2H_5CO)_2CH_2$	9.55	22
$(iPrCO)_2CH_2$	9.82	22
$CH_3COCH_2CO_2C_2H_5$	10.7	20
$(CH_3CO)_2CHCH_3$	11.0	24
$(tBuCO)_2CH_2$	11.57	22
$CH_3COCH(C_2H_5)CO_2C_2H_5$	12.7	24
$CH_2(CO_2C_2H_5)_2$	13.3	24
$C_2H_5CH(CO_2C_2H_5)_2$	15	24
$(CH_3)_2CHCHO$	15.5	25
$C_6H_5COCH(CH_3)_2$	16.5	26
CH_3CHO	16.5	27
$C_6H_5COCH_3$	17.7, 18.6	27, 28
$(CH_3)_2C{=}O$	19.0	27
$(C_2H_5)_2C{=}O$	19.9	27
Other compounds[h]		
Tris(pentafluorophenyl)methane	15.8	29
Bis(pentafluorophenyl)methane	21.3	29
Chloroform	24	30
Phenyl methyl sulfone	26.7	31
9-Phenylxanthene	28.5	11
Dimethyl sulfoxide	33	32, 33

[a] Standard state water.
[b] References: 1, R. Kuhn and D. Rewicki, *Justus Liebigs Ann. Chem.* **706**, 250 (1969). 2, L. G. Greifenstein, J. B. Lambert, R. J. Nienhuis, G. E. Drucker, and G. A. Pagani, *J. Am. Chem. Soc.* **103**, 7753 (1981). 3, A. Streitwieser, Jr. and L. L. Nebenzahl, *J. Am. Chem. Soc.* **98**, 2188 (1976). 4, C. H. Langford and R. L. Burwell, *J. Am. Chem. Soc.* **82**, 1503 (1960). 5, K. Bowden and R. Stewart, *Tetrahedron* **21**, 261 (1965). 6, L. G. Greifenstein, J. B. Lambert, R. J. Nienhuis, H. E. Fried, and G. A. Pagani, *J. Org. Chem.* **46**, 5125 (1981). 7, R. A. Cox and R. Stewart, *J. Am. Chem. Soc.* **98**, 488 (1976). 8, K. Bowden and A. F. Cockerill, *J. Chem. Soc. B* p. 173, (1970). 9, A. C. Lin, Y. Chiang, D. B. Dahlberg, and A. J. Kresge, *J. Am. Chem. Soc.* **105**, 5380 (1983). 10, A. Streitwieser, Jr., E. Juaristi, and L. L. Nebenzahl, *in* "Comprehensive Carbanion Chemistry" (E. Buncel and T. Durst, eds.), Elsevier, Amsterdam, 1980. 11, A. Streitwieser, Jr., E. Ciuffarin, and J. H. Hammons, *J. Am. Chem. Soc.* **89**, 63 (1967). 12, A. Streitwieser, Jr., J. R. Murdoch, G. Hafelinger, and C. J. Chang, *J. Am. Chem. Soc.* **95**, 4248 (1973). 13, R. R. Fraser, M. Bresse, and T. S. Mansour, *J. Chem. Soc., Chem. Commun.* p. 620 (1983). 14, B. Jaun, J. Schwarz, and R. Breslow, *J. Am. Chem. Soc.* **102**, 5741 (1980). 15, M. R. Wasielewski and R. Breslow, *J. Am. Chem. Soc.* **98**, 4222 (1976). 16, R. Breslow and R. Goodin, *J. Am. Chem. Soc.* **98**, 6076 (1976). 17, H. G. Adolph and M. J. Kamlet, *J. Am. Chem. Soc.* **88**, 4761 (1966). 18, G. Schwarzenbach and E. Felder, *Helv. Chim. Acta* **27**, 1701 (1944). 19, See ref. 20. 20. R. G. Pearson and R. L. Dillon, *J. Am. Chem. Soc.* **75**, 2439 (1953). 21, D. J. Barnes and R. P. Bell, *Proc. R. Soc. London*

The most acidic carbon acids are the cyanocarbons, which have been discussed in Section IV,B. The most acidic hydrocarbon is **58**, which has a

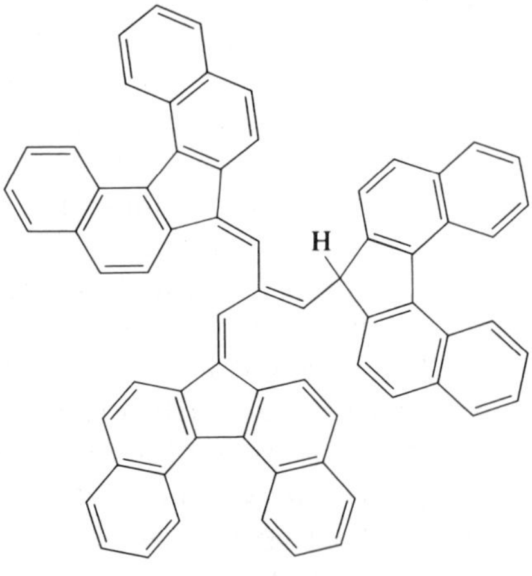

58

pK_{HA} below 7. This compound and others like it are, of course, insoluble in water and so Kuhn and Rewicki (*341*), who studied this family of compounds, used DMSO containing acetate buffers or tripropylamine to measure their extents of ionization. The link to aqueous values was made by using 9-

---

318, 421 (1970). 22, H. Koshimura and T. Okuba, *Anal. Chim. Acta* **49**, 67 (1970). 23, M. Laloi-Diard and M. Rubenstein, *Bull. Soc. Chim. Fr.* p. 310 (1965). 24, R. G. Pearson and J. M. Mills, *J. Am. Chem. Soc.* **72**, 1692 (1950). 25, Y. Chiang, A. J. Kresge, and P. A. Walsh, *J. Am. Chem. Soc.* **104** 6122 (1982). 26, A. J. Kresge, personal communication. 27, J. P. Guthrie, *Can. J. Chem.* **57**, 1177 (1979) and ref. *307*. 28, J. Toullec and C. Vervy-Doussin in ref. *308*. 29, R. Filler and C.-S. Wang, *Chem. Commun.* p. 287 (1968). 30, Z. Margolin and F. A. Long, *J. Am. Chem. Soc.* **94**, 5108 (1972); **95**, 2757 (1973). 31, See ref. *239*, p. 339. 32, E. C. Steiner and J. M. Gilbert, *J. Am. Chem. Soc.* **87**, 382 (1965). 33, R. Stewart and J. R. Jones, *J. Am. Chem. Soc.* **89**, 5069 (1967).

[c] Fluoradene.

[d] By comparison with nitrogen indicators and the H_- scale.

[e] The structure and pertinent numbering of fluorene is as follows:

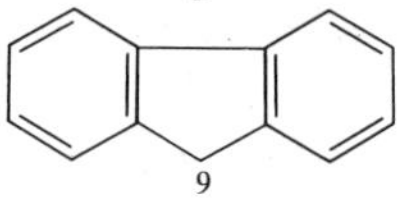

9

[f] Acidity function pK; extrapolative methods give significantly higher values for 9-arylfluorenes (*319*).

[g] Phenalene.

[h] For nitro- and cyanoalkanes, and arenes, see Tables 2-18, 2-19, Section IV,B, and refs. *227* and *349*.

cyanofluorene, which has a pK_{HA} of 11.4. For those acids that produce highly delocalized anions we have seen that their acidity constants in water and in DMSO differ by a rather small and generally fixed amount, and we can therefore assume that Kuhn and Rewicki's values for the polycyclic hydrocarbons listed in Table 2-22 do, in fact, refer to the aqueous state.

Those hydrocarbons with pK_{HA} values between 18 and 22, mostly fluorene derivatives, were determined by acidity function and related extrapolative techniques, and the values listed in Table 2-22 for such compounds are average values, as is the case with the nitrogen acids of Table 2-21. It is highly significant that these numbers are very close to those obtained by Streitwieser *et al.* for the ion-product pK values in cyclohexylamine [and, indeed, to the pK_{HA}^{DMSO} values obtained by Bordwell *et al.* for carbon acids that give delocalized anions, some of which appear to be somewhat stronger in DMSO than in water, and others somewhat weaker (229,342)]. This correspondence allows us to take the ion-product measurements, which can be extended to acids as weak as toluene, as giving reasonable estimates of the acidity constants of these compounds in water. A value near 41 is thus obtained for the pK_{HA} of toluene, with propene being approximately two pK units weaker still (343).

There is only one group of hydrocarbons of any appreciable acid strength that produces anions with localized charge, that being the alkynes (343a). Estimates of the acid strength of phenylacetylene based on its rate of ionization show it to be considerably stronger in water ($pK_{HA} = 20.0$) (344) than in DMSO ($pK_{HA}^{DMSO} = 28.7$) (345) or in cyclohexylamine (ion-pair p$K = 23.2$) (346). These results are in agreement with the notion that the strength of acids that give anions with localized charge should be strongly medium dependent. Furthermore, the proton transfer from phenylacetylene to water appears to be rapid and reversible, with separation of the ion pairs being the rate-controlling step. That is, phenylacetylene, and presumably most other carbon acids that give charge-localized anions, behave more like normal acids (13) than like pseudoacids (295).

The ion-pair pK values (per hydrogen) of a number of acids that produce charge-localized anions have been reported by Streitwieser and his group; these include pentafluorobenzene (25.8), pentachlorobenzene (29.9), thiazole (29.5), 1,3,5-trichlorobenzene (33.6), benzofuran (36.8), thiophene (38.2), and benzene (43) (239, 347). It is not clear how closely these values correspond to aqueous-state values, but it may be significant that a number of these values are fairly close to those found by Reutov and his group using a polarographic method and the high-dielectric solvent dimethylformamide (274).

It is interesting that the ion-pair pK_1 and pK_2 values determined in cyclohexylamine–cesium cyclohexylamide of the biindenyl and bifluorenyl compounds **59** and **60** are very close together, indicating the considerable

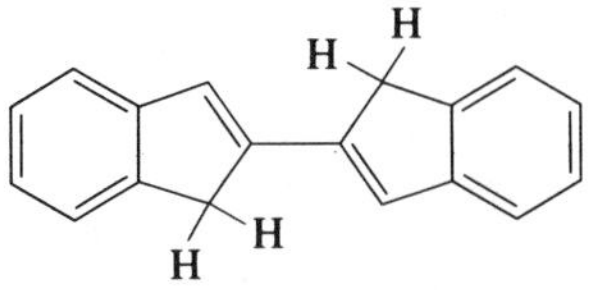

59	**60**
$pK_1 = 19.8$	$pK_1 = 20.5$
$pK_2 = 20.3$	$pK_2 = 20.7$

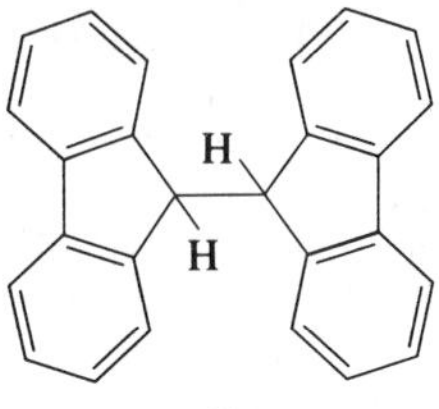

degree of stabilization that is conferred by an ion-triplet structure dicarb-anion plus two cations) (*353*).

The acidities of a number of monosubstituted benzenes have been determined in tetrahydrofuran by Fraser *et al.* and linked to Streitwieser's scale through 2-methyl-1,3-dithiane (**61**, $pK_{HA} = 37.8$) and tetramethylpiper-idine ($pK_{HA} = 37.3$) (*354, 355*).

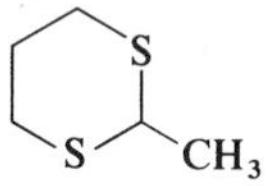

61

The gas phase acidities of a large number of carbon acids have been measured, and this subject has been reviewed by Pellerite and Brauman (*348*). For hydrocarbons that form large, highly delocalized carbanions the relative order of acid strength is found to be the same in the gas phase as in polar aprotic solvents such as DMSO (*350–352*).

References

1. R. D. Howells and J. D. McCown, *Chem. Rev.* **77**, 69 (1977).
2. H. J. Bakoss, R. J. Ranson, R. M. G. Roberts, and A. R. Sadri, *Tetrahedron* **38**, 623 (1982).
3. T. Gramstad, *Tidsskr. Kjemi, Bergves. Metall.* **19**, 62 (1959).
4. A. Engelbrecht and B. M. Rode, *Monatsh. Chem.* **103**, 1315 (1972).
5. I. M. Kolthoff and A. W. Willman, *J. Am. Chem. Soc.* **56**, 1007 (1934).
6. D. G. Russell and J. B. Senior, *Can. J. Chem.* **52**, 2975 (1974).
7. R. J. Gillespie, T. E. Peel, and E. A. Robinson, *J. Am. Chem. Soc.* **93**, 5083 (1971).
8. J. B. Spencer and J.-O. Lundgren, *Acta Crystallogr., Sect. B* **29B**, 1923 (1973).
9. R. Stewart and J. P. O'Donnell, *Can. J. Chem.* **42**, 1694 (1964).
10. O. W. Webster, *J. Am. Chem. Soc.* **87**, 3046 (1966).
11. J. R. Jones, "The Ionization of Carbon Acids," p. 65. Academic Press, London, 1973.
12. M. Leuchs and G. Zundel, *Can. J. Chem.* **60**, 2118 (1982).

13. M. Eigen, *Angew. Chem., Int. Ed. Engl.* **3,** 1 (1964).

14. A. K. Covington, J. G. Freeman, and T. H. Lilley, *J. Phys. Chem.* **74,** 3773 (1970).

15. R. I. Gelb and L. M. Schwartz, *Anal. Chem.* **44,** 554 (1972).

16. O. D. Bonner and P. R. Prichard, *J. Solution Chem.* **8,** 113 (1979).

17. R. I. Gelb, L. M. Schwartz, and D. A. Laufer, *J. Am. Chem. Soc.* **103,** 5664 (1981).

18. L. M. Schwartz and L. O. Howard, *J. Phys. Chem.* **74,** 4374 (1970).

19. L. M. Schwartz and L. O. Howard, *J. Phys. Chem.* **75,** 1798 (1971).

20. J. H. R. Clarke and L. A. Woodward, *Trans. Faraday Soc.* **62,** 2226 (1966).

21. A. K. Covington and R. Thompson, *J. Solution Chem.* **3,** 603 (1974).

22. J. P. Guthrie, *Can. J. Chem.* **56,** 2342 (1978).

23. R. A. Cox, U. L. Haldna, K. L. Idler, and K. Yates, *Can. J. Chem.* **59,** 2591 (1981).

23a. P. A. Giguère and S. Turrell, *J. Am. Chem. Soc.* **102,** 5473 (1980).

24. I. M. Kolthoff, S. Bruckenstein, and M. K. Chantooni, *J. Am. Chem. Soc.* **83,** 3927 (1961).

25. I. M. Kolthoff and M. K. Chantooni, *J. Phys. Chem.* **70,** 856 (1966).

26. J. F. Coetzee, *in* "Progress in Physical Organic Chemistry" (S. G. Cohen, A Streitwieser, and R. W. Taft, eds.), Vol. IV, p. 45. Wiley (Interscience), New York, 1967.

27. J. Bessière, *Bull. Soc. Chim. Fr.* p. 3353 (1969).

28. J. Barr, R. J. Gillespie, and E. A. Robinson, *Can. J. Chem.* **39,** 1266 (1961).

29. R. J. Gillespie, *J. Chem. Soc.* p. 2537 (1950).

30. R. L. Benoit, C. Buisson, and G. Choux, *Can. J. Chem.* **48,** 2353 (1970).

31. R. L. Benoit and C. Buisson, *Electrochim. Acta* **18,** 105 (1973).

32. M. L'Her and J. Courtot-Coupez, *J. Electroanal. Chem.* **48,** 265 (1973).

32a. N. A. Baranov, G. B. Manukhina, and N. A. Vlasov, *Zh. Fiz. Khim.* **53,** 2634 (1979); *Chem. Abstr.* **92,** 93733 (1980).

33. G. Schwarzenbach, *Z. Phys. Chem.* **176,** 133 (1936).

34. A. Kossiakoff and D. Harker, *J. Am. Chem. Soc.* **60,** 2047 (1938).

35. L. Pauling, "General Chemistry," p. 394. Freeman, San Francisco, California, 1947.

36. A. J. Kresge and Y. C. Tang, *J. Org. Chem.* **42,** 757 (1977).

37. R. W. Taft, *in* "Steric Effects in Organic Chemistry" (M. S. Newman, ed.), Chap. 13. Wiley, New York, 1956.

38. C. D. Johnson, "The Hammett Equation." Cambridge Univ. Press, 1973.

39. H. H. Jaffé, L. D. Freedman, and G. O. Doak, *J. Am. Chem. Soc.* **75,** 2209 (1953).

40. D. J. Martin and C. E. Griffin, *J. Organomet. Chem.* **1,** 292 (1964).

41. L. D. Freedman and G. O. Doak, *Chem. Rev.* **57,** 479 (1957).

42. F. G. Ryss and V. B. Tulchinskii, *Russ. J. Inorg. Chem.* **6,** 947 (1961).

43. C. A. Bunton, D. R. Llewellyn, K. G. Oldham, and C. A. Vernon, *J. Chem. Soc.* p. 3574 (1958).

44. I. M. Kolthoff and S. Bruckenstein, *J. Am. Chem. Soc.* **78,** 1 (1956).

45. S. Brunkenstein and I. M. Kolthoff, *J. Am. Chem. Soc.* **78,** 2974 (1956).

46. J. Bessière, *Bull. Soc. Chim. Fr.* p. 3356 (1969).

47. J. Bessière, *Anal. Chem. Acta* **52,** 55 (1970).

48. A. I. Razumov and S. D. Khen, *Zh. Obsheh. Khim.* **26,** 2233 (1956).

49. W. D. Kumler and J. J. Eiler, *J. Am. Chem. Soc.* **65,** 2355 (1943).

50. P. C. Crofts and G. M. Kosolapoff, *J. Am. Chem. Soc.* **75,** 3379 (1953).

51. H. Cerfontain, A. Koeberg-Telder, and C. Kruk, *Tetrahedron Lett.* p. 3639 (1975).

52. V. A. Kozlov, B. D. Berezin and I. A. Popkova, *Zh. Fiz. Khim.* **55,** 1481 (1981).

53. P. K. Maarsen, R. Bregman, and H. Cerfontain, *Tetrahedron* **30,** 1211 (1975).

54. H. Cerfontain and B. W. Schnitzer, *Rec. Trav. Chim. Pays-Bas* **91,** 199 (1972).

55. D. D. Perrin, *Pure Appl. Chem.* **20,** 133 (1960); this paper in its entirety has been published as "Dissociation Constants of Inorganic Acids and Bases in Aqueous Solution." Butterworths, London, 1969.

56. D. C. Hodgkin and P. Kruss, *Can. J. Chem.* **51**, 2297 (1973).
57. A. A. Kryuchkov, A. G. Kozachenko, E. I. Matrosov, and M. I. Kabachnik, *Izv. Akad. Nauk SSR, Ser. Khim.* p. 1985 (1978).
58. C. A. Bunton, D. R. Llewellyn, K. G. Oldham, and C. A. Vernon, *J. Chem. Soc.* p. 3574 (1958).
59. D. F. Peppard, G. W. Mason, and C. M. Andrejasish, *J. Inorg. Nucl. Chem.* **27**, 697 (1965).
60. D. D. Perrin, B. Dempsey, and E. P. Serjeant, "pK_a Prediction for Organic Acids and Bases." Chapman and Hall, London, 1981.
61. J. N. Phillips, *Aust. J. Chem.* **14**, 183 (1961).
62. R. P. Bell, "The Proton in Chemistry," 2nd ed. p. 40 and p. 92. Cornell Univ. Press, Ithaca, New York, 1973.
63. D. F. DeTar, *J. Org. Chem.* **45**, 5166 (1980) and ref. therein.
63a. R. W. Taft, E. Price, I. R. Fox, I. C. Lewis, K. K. Andersen, and G. T. Davis, *J. Am. Chem. Soc.* **85**, 709 (1963).
64. O. Exner, *in* "Correlation Analysis in Organic Chemistry" (N.B. Chapman and J. Shorter, eds.), p. 471. Plenum, New York, 1978.
65. A. Kossiakoff and D. Harker, *J. Am. Chem. Soc.* **60**, 2047 (1938).
66. L. Pauling, "The Nature of the Chemical Bond," p. 557. Cornell Univ. Press, Ithaca, New York, 1939.
67. J. E. Ricci, *J. Am. Chem. Soc.* **70**, 109 (1948).
68. R. P. Bell, "Acids and Bases." Methuen, London, 1952.
69. R. J. Gillespie, *J. Chem. Soc.* p. 2537 (1950).
70. F. Wudl, D. A. Lightner, and D. J. Cram, *J. Am. Chem. Soc.* **89**, 4099 (1967).
71. C. D. Ritchie, J. D. Saltiel, and E. S. Lewis, *J. Am. Chem. Soc.* **83**, 4601 (1961).
72. P. Rumpf and J. Sadet, *Bull. Soc. Chim. Fr.* p. 447 (1958).
73. D. Veltwisch, E. Janata, and K.-D Asmus, *J. Chem. Soc., Perkin Trans. 2* p. 146 (1980).
74. D. DeFilippo and F. Momicchioli, *Tetrahedron* **25**, 5733 (1969).
75. R. E. Penn, E. Block, and L. K. Revelle, *J. Am. Chem. Soc.* **100**, 3622 (1978).
75a. See, however, N. Nakamura, *J. Am. Chem. Soc.* **105**, 7172 (1983).
76. J. L. Kice, *in* "Advances in Physical Organic Chemistry," (V. Gold and D. Bethell eds.), Vol. 17, p. 65. Academic Press, London, 1980.
77. A. G. Cook and G. W. Mason, *J. Org. Chem.* **37**, 3342 (1972).
78. P. Lesfauries and P. Rumpf, *C. R. Hebd. Seances Acad. Sci.* **228**, 1018 (1949).
79. E. M. Evleth, L. D. Freedman, and R. I. Wagner *J. Org. Chem.* **27**, 2192 (1962).
80. M. L. Kilpatrick, *J. Am. Chem. Soc.* **71**, 2607 (1949).
81. J. D. McCullough and E. S. Gould, *J. Am. Chem. Soc.* **71**, 674 (1949).
82. O. D. Bonner and P. R. Prichard, *J. Solution Chem.* **8**, 113 (1979).
83. P. D. Bonner, H. B. Flora, and H. W. Aitken, *J. Phys. Chem.* **75**, 2492 (1971).
84. J. Hine and W. C. Bailey, *J. Org. Chem.* **26**, 2098 (1961).
85. J. L. Kurz and J. M. Farrar, *J. Am. Chem. Soc.* **91**, 6057 (1969).
86. A. L. Henne and C. J. Fox, *J. Am. Chem. Soc.* **73**, 2323 (1951).
87. R. A. Robinson and R. H. Stokes, "Electrolyte Solutions," 2nd ed., p. 524. Butterworths, London, 1959.
88. H. von Halban and J. Brull, *Helv. Chim. Acta* **27**, 1719 (1944).
89. Ref. *62*, pp. 98–101.
90. A. J. Hoefnagel, M. A. Hoefnagel, and B. M. Wepster, *J. Org. Chem.* **43**, 4720 (1978).
91. A. J. Hoefnagel, and B. M. Wepster, *J. Org. Chem.* **47**, 2318 (1982).
92. L. Eberson, *in* "The Chemistry of Carboxylic Acids and Esters" (S. Patai, ed.), Chap. 6. Interscience, London, 1969.
93. J. Hine, "Structural Effects on Equilibria in Organic Chemistry," p. 85. Wiley (Interscience), New York, 1975.
94. J. Steigman and D. Sussman, *J. Am Chem. Soc.* **89**, 6406 (1967).

95. J. F. J. Dippy, S. R. C. Hughes, and J. W. Laxton, *J. Chem. Soc.* p. 2995 (1956).

96. M. T. Ryan and K. J. Berman. *Spectrochim. Acta, Part A* **25A,** 1155 (1969).

97. J. Chatt and A. A. Williams, *J. Chem. Soc.* p. 4403 (1954); p. 688 (1956).

98. J. M. Wilson, A. G. Briggs, J. E. Sawbridge, P. Tickle, and J. J. Zuckerman, *J. Chem. Soc. A* p. 1024 (1970).

99. M. Bos and E. A. M. F. Dahmen, *Anal. Chim. Acta* **55,** 285 (1971).

100. Ref. *62,* p. 56.

101. J. H. Elliott and M. Kilpatrick, *J. Phys. Chem.* **45,** 454, 466 (1941).

102. J. B. Cumming and P. Kebarle, *Can. J. Chem.* **56,** 1 (1978).

103. J. T. Edward, *J. Chem. Educ.* **59,** 354 (1982).

104. R. W. Gurney, "Ionic Processes in Solution." McGraw-Hill, New York, 1953.

105. M. J. Blandamer, J. Burgess, P. P. Duce, R. E. Robertson, and J. W. M. Scott, *Can. J. Chem.* **59,** 2845 (1981).

106. Thermodynamic parameters for the dissociation constants of a large number of acids are listed by R. M. Izatt and J. J. Christensen, *in* "Handbook of Biochemistry" (H. A. Sober, ed.), Chem. Rubber Publ. Co., Cleveland, Ohio, 1970, and J. J. Christensen, L. D. Hansen, and R. M. Izatt, "Handbook of Proton Ionization Heats and Related Thermodynamic Quantities." Wiley, New York, 1976.

107. J. M. Wilson, N. E. Gore, J. E. Sawbridge, and F. Cardenas-Cruz, *J. Chem. Soc. B* p. 852 (1967).

108. E. C. W. Clarke and D. N. Glew, *Trans. Faraday Soc.* **62,** 539 (1966).

109. M. J. Blandamer, J. W. M. Scott, and R. E. Robertson, *J. Chem. Soc., Perkin Trans. 2* p. 447 (1981).

110. See, however, Chapter 3, ref. *33a.*

111. See, for example, ref. *62,* p. 83.

112. J. Jones and F. G. Soper, *J. Chem. Soc.* p. 133 (1936).

113. D. H. McDaniel and H. C. Brown, Science **118,** 370 (1953).

114. L. Hunter, *Chem. Ind.* p. 155 (1953).

115. N. Bjerrum, *Z. Phys. Chem.* **106,** 219 (1923).

116. R. Gase and C. K. Ingold, *J. Chem. Soc.* p. 1691 (1929).

117. E. Eucken, *Angew. Chem.* **45,** 203 (1932).

118. J. G. Kirkwood and F. H. Westheimer, *J. Chem. Phys.* **6,** 506, 513 (1938).

119. A. J. Hoefnagel and B. M. Wepster, *J. Org. Chem.* **47,** 2318 (1982) and refs. therein.

120. E. J. King, "Acid–Base Equilibria," Chap. 7. Pergamon, London, 1965.

121. F. H. Westheimer and O. T. Benfey, *J. Am. Chem. Soc.* **78,** 5309 (1956).

122. J. Jager, T. Graafland, H. Schenk, A. J. Kirby, and J. B. F. N. Engberts, *J. Am. Chem. Soc.* **106,** 139 (1984) and refs. therein.

123. J. T. Edward, E. Cooke, and T. C. Paradellis, *Can. J. Chem.* **60,** 2546 (1982).

124. L. Eberson, *Acta Chem. Scand.* **13,** 211 (1959).

125. L. L. McCoy and G. W. Nachtigal, *J. Am. Chem. Soc.* **85,** 1321 (1963).

126. H. C. Brown, D. H. McDaniel, and O. Hafliger, *in* "Determination of Organic Structures by Physical Methods" (E. A. Braude and F. C. Nachod, eds.), Vol. 1, Chap. 14. Academic Press, New York.

127. P. K. Glasoe and L. Eberson, *J. Phys. Chem.* **68,** 1560 (1964).

128. T. Watanabe, Y. Otsuji, and M. Hamada, *Bull. Chem. Soc. Jpn.* **38,** 1247 (1965).

129. L. L. McCoy, *J. Am. Chem. Soc.* **89,** 1673 (1967).

130. S. J. Gumbley and R. Stewart, *J. Chem. Soc., Perkin Trans. 2* p. 529 (1984); see also ref. *131.*

131. G. Kohnstam and D. L. H. Williams, *in* "The Chemistry of Carboxylic Acids and Esters" (S. Patai, ed.), Chap. 16. Interscience, London, 1969.

132. L. A. Cohen and W. M. Jones, *J. Am. Chem. Soc.* **85,** 3397 (1963).

133. C. H. Rochester, *J. Chem. Soc.* p. 4603 (1965).

134. A. Albagli, A. Buckley, A. M. Last, and R. Stewart, *J. Am. Chem. Soc.* **95,** 4711 (1973).

135. P. J. Pearce and R. J. J. Simpkins, *Can. J. Chem.* **46,** 241 (1968).

136. A. I. Biggs and R. A. Robinson, *J. Chem. Soc.* p. 388 (1961).

137. G. W. Wheland, R. M. Brownell, and E. C. Mayo, *J. Am. Chem. Soc.* **70,** 2493 (1948).

138. G. Kortum, W. Vogel, and K. Andrussow, "Dissociation constants of Organic Acids in Aqueous Solution." Butterworths, London, 1961.

139. B. Nowak and Z. Pawlak, *J. Chem. Soc. Faraday Trans. 1* **78,** 2693 (1982).

140. G. Schwarzenbach and E. Rudin, *Helv. Chim. Acta.* **22,** 360 (1939).

141. G. H. Parsons and C. H. Rochester, *J. Chem. Soc. Faraday Trans. 1* p. 1058 (1975).

142. E. Grunwald and E. Price, *J. Am. Chem. Soc.* **86,** 4517 (1964).

143. M. Fujio, R. T. McIver, and R. W. Taft, *J. Am. Chem. Soc.* **103,** 4017 (1981).

144. T. B. McMahon and P. Kebarle, *J. Am. Chem. Soc.* **99,** 2222 (1977).

145. L. P. Fernandez and L. G. Hepler, *J. Am. Chem. Soc.* **81,** 1783 (1959).

146. L. G. Hepler, *J. Am. Chem. Soc.* **85,** 3089 (1963).

147. D. T. Y. Chen and K. J. Laidler, *Trans. Faraday Soc.* **58,** 480 (1962).

148. C. H. Rochester, *in* "The Chemistry of the Hydroxyl Group" (S. Patai, ed.), Part 1, Chap. 7. Interscience, London, 1971.

149. H. Hart, *Chem. Rev.* **79,** 515 (1979); see also S. E. Biali and Z. Rapaport, *J. Am. Chem. Soc.* **106,** 477 (1984) and refs. therein.

150. L. M. Schwartz, R. I. Gelb, and D. A. Laufer, *in* "Oxocarbons" (R. West, ed.), Chap. 3. Academic Press, New York, 1980.

151. Y. Chiang, A. J. Kresge, and P. A. Walsh, *J. Am. Chem. Soc.* **104,** 6122 (1982).

151a. Y. Chiang, A. J. Kresge., Y. S. Tang, and J. Wirz, *J. Am. Chem. Soc.* **106,** 460 (1984).

152. E. Taphui and W. P. Jencks, *J. Am. Chem. Soc.* **104,** 5758 (1982).

153. P. Haspra, A Sutter, and J. Wirz, *Angew. Chem., Int. Ed. Engl.* **18,** 617 (1979); see also ref. *154.*

154. M. Novak and G. M. Louden, *J. Org. Chem.* **42,** 2494 (1977).

155. A. Beauchamp and R. L. Benoit, *Can. J. Chem.* **44,** 1607 (1966).

155a. S.-F. Tan and K.-P. Ang, *J. Chem. Soc., Perkin Trans. 2* p. 471 (1983).

156. D. L. Campbell and T. Moeller, *J. Inorg. Nucl. Chem.* **31,** 1077 (1969).

157. R. I. Gelb, L. M. Schwartz, and D. A. Laufer, *J. Phys. Chem.* **82,** 1985 (1978); see also refs. *158–160.*

158. D. Alexandersson and N.-G. Vannerberg, *Acta Chem. Scand.* **26,** 1909 (1972).

159. E. Patton and R. West, *J. Phys. Chem.* **74,** 2512 (1970).

160. M. F. Fleury and G. Molle, *C. R. Hebd. Seances Acad. Sci.* **273,** 605 (1971).

161. M. M. T. Khan, and A. E. Martell, *J. Am. Chem. Soc.* **91,** 4668 (1969).

162. T. W. Birch and L. J. Harris, *Biochem. J.* **27,** 595 (1933).

163. G. Schwarzenbach and H. Suter, *Helv. Chim. Acta* **24,** 617 (1941).

164. R. I. Gelb and L. M. Schwartz, *J. Chem. Soc., Perkin Trans. 2* p. 930 (1976).

165. L. M. Schwartz and L. O. Howard, *J. Phys. Chem.* **75,** 1798 (1971).

166. R. I. Gelb, *Anal. Chem.* **43,** 1110 (1971).

167. A. Beauchamp and R. L. Benoit, *Bull. Soc. Chim. Fr.* p. 672 (1967).

168. R. H. Boyd, *J. Phys. Chem.* **67,** 737 (1963).

169. J. P. Guthrie and P. A. Cullimore, *Can. J. Chem.* **57,** 240 (1979).

169a. J. P. Guthrie, *Acc. Chem. Res.* **16,** 122 (1983).

170. L. M. Schwartz, R. I. Gelb, and J. O. Yardley, *J. Phys. Chem.* **79,** 2246 (1975).

171. J. W. Larson and L. G. Hepler, *in* "Solute–Solvent Interactions" (J. F. Coetzee and C. D. Ritchie, eds.), Chap. 1. Dekker, New York, 1969.

171a. P. J. Krueger, "The Chemistry of the Hydrazo, Azo, and Azoxy Groups" (S. Patai, ed.), Part 1, Chap. 7, Wiley, London, 1975, and refs. therein.

172. J. Murto, *Acta Chem. Scand.* **18,** 1043 (1964).

173. W. P. Jencks, S. R. Brant, J. R. Gandler, G. Fendrich, and C. Nakamura, *J. Am. Chem. Soc.* **104,** 7045 (1982).

174. R. Stewart and R. Van der Linden, *Can. J. Chem.* **38,** 399 (1960).

175. P. Ballinger and F. A. Long, *J. Am. Chem. Soc.* **82,** 795 (1960).

176. P. Ballinger and F. A. Long, *J. Am. Chem. Soc.* **81,** 1050 (1959).

177. F. A. Long and P. Ballinger, in "Electrolytes" (B. Pesce, ed.), p. 152. Pergamon, Oxford, 1962.

178. J. Hine and M. Hine, *J. Am. Chem. Soc.* **74,** 5266 (1952).

179. R. P. Bell and D. P. Onwood, *Trans. Faraday Soc.* **58,** 1557 (1962).

180. R. M. Izatt, J. H. Rytting, L. D. Hansen, and J. J. Christensen, *J. Am. Chem. Soc.* **88,** 2641 (1966).

181. See ref. *174.*

182. W. J. Middleton and R. V. Lindsey, *J. Am. Chem. Soc.* **86,** 4948 (1964).

183. A. C. Satterthwait and W. P. Jencks, *J. Am. Chem. Soc.* **96,** 7031 (1974).

184. I. L. Krunyants, A. V. Fokin, B. K. Dyatkin, and V. A. Komarow, *Izv. Akad. Nauk USSR, Chem. Ser.* p. 1425 (1964).

185. B. L. Dyatkin, E. P. Mochalina, and I. L. Knunyants, *Tetrahedron* **21,** 2991 (1965).

185a. W. Reeve, C. M. Erikson, and P. F. Aluotto, *Can. J. Chem.* **57,** 2747 (1979).

186. T. N. Hall, *Tetrahedron,* **19,** Suppl. 1, 115 (1963).

187. B. C. Hendricks and W. H. Steinbach, *J. Phys. Chem.* **42,** 335 (1938).

188. G. Kilde and W. F. K. Wynne-Jones, *Trans. Faraday Soc.* **49,** 243 (1953).

189. P. A. Levene, H. S. Simms, and L. W. Bass, *J. Biol. Chem.* **70,** 243 (1926).

190. A. Albert, "Synthetic Procedures in Nucleic Acid Chemistry" (W. W. Zorbach and R. S. Tipson, eds.), Vol. 2, Chap. 1. Wiley (Interscience), New York, 1973.

191. R. A. Alberty, R. M. Smith, and R. Back, *J. Biol. Chem.* **193,** 425 (1951).

192. A. Albert, *Biochem. J.* **54,** 646 (1953).

193. A. J. Everett and G. J. Minkoff, *Trans. Faraday Soc.* **49,** 410 (1953).

194. R. P. Bell, "The Proton in Chemistry," 1st ed., p. 159. Cornell Univ. Press, Ithaca, New York, 1959; see also 2nd ed., 1973, pp. 198–199.

195. P. A. Bernstein, F. A. Hohorst, and D. D. DesMarteau, *J. Am. Chem. Soc.* **93** 3882 (1971).

196. P. H. Ogden and G. C. Nicholson, *Tetrahedron Lett.* p. 3553 (1968).

197. R. J. Irving, L. Melander, and I. Wadso, *Acta Chem. Scand.* **18,** 769 (1964).

198. P. De Maria, A. Fini, and F. M. Hall, *J. Chem. Soc., Perkin Trans 2* p. 1969 (1973).

199. N. Sheppard, *Trans. Faraday Soc.* **45,** 693 (1949).

200. R. Engler and G. Gattow, *Z. Anorg. Allg. Chem.* **389,** 151 (1972).

201. A. P. Gavrish, *Ukr. Khim. Zh. (Ukr. Ed.)* **29,** 900 (1963).

202. M. M. Kreevoy, B. E. Eichinger, F. E. Stary, E. A. Katz, and J. H. Sellstedt, *J. Org. Chem.* **29,** 1641 (1964).

203. D. L. Leussing and G. S. Alberts, *J. Am. Chem. Soc.* **82,** 4458 (1961).

204. P. J. Antikainen and K. Tevanari, *Suom. Kemistihil. B* **42B,** 178 (1969).

205. M. M. Kreevoy, E. T. Harper, R. E. Duvall, H. S. Wilgus, and L. T. Ditsch, *J. Am. Chem. Soc.* **82,** 4899 (1960).

206. D. L. Yabroff, *Ind. Eng. Chem.* **32,** 257 (1940).

207. P. De Maria, A. Fini, and F. M. Hall, *J. Chem. Soc., Perkin Trans. 2* p. 1443 (1974).

208. H. A. Pohl, *J. Chem. Eng. Data* **6,** 515 (1961).

209. R. H. Wright and O. Maass, *Can. J. Res.* **6,** 588 (1932).

210. P. De Maria, A. Fini, and F. M. Hall, *J. Chem. Soc., Perkin Trans. 2* p. 149 (1977).

211. E. Jansons, *Russ. Chem. Rev.* **45,** 1035 (1976).

212. J. Voss, "The Chemistry of Functional Groups Supplement B, The Chemistry of Acid Derivatives Part 2" (S. Patai, ed.), Chap. 18. Wiley, Chichester, 1979.

213. M. J. Jannsen, "The Chemistry of Carboxylic Acids and Esters" (S. Patai, ed.), Chap. 15. Wiley, London, 1969.

214. M. A. Bernard, M. M. Borel, and G. Dupriez, *Rev. Chim. Miner.* **12**, 181 (1975); *Chem. Abstr.* **83**, 157060 (1975).

215. J. Hipkin and D. P. N. Satchell, *Tetrahedron* **21**, 835 (1965).

216. J. E. Bartmess, J. A. Scott, and R. T. McIver, *J. Am. Chem. Soc.* **101**, 6046 (1979).

217. G. I. McKay, A. B. Rakshit, and D. K. Boehme, *Can. J. Chem.* **60**, 2594 (1982).

218. R. W. Taft, M. Taagepera, J. L. M. Abboud, J. F. Wolf, D. J. Defrees, W. J. Hehre, J. E. Bartmess, and R. T. McIver, *J. Am. Chem. Soc.* **100**, 7765 (1978).

219. G. Boand, R. Houriet, and T. Gaumann, *J. Am. Chem. Soc.* **105**, 2203 (1983).

220. J. I. Brauman and L. K. Blair, *J. Am. Chem. Soc.* **92**, 5986 (1970).

221. R. W. Taft, *in* "Proton Transfer Reactions" (E. F. Caldin and V. Gold, eds.), Chap. 2. Chapman and Hall, London, 1975.

222. See, however, G. Caldwell *et al.*, *J. Am. Chem. Soc.* **107**, 80 (1985).

223. W. J. Middleton, E. L. Little, D. D. Coffman, and V. A. Engelhardt, *J. Am. Chem. Soc.* **80**, 2795 (1958).

224. R. H. Boyd, *J. Am. Chem. Soc.* **83**, 4288 (1961).

225. F. G. Bordwell, R. H. Imes, and E. C. Steiner, *J. Am. Chem. Soc.* **89**, 3905, (1967).

226. F. G. Bordwell and D. Algrim, *J. Org. Chem.* **41**, 2507 (1976).

227. J. R. Jones, "The Ionization of Carbon Acids," pp. 74–76. Academic Press, London, 1973.

228. F. G. Bordwell, D. Algrim, and N. R. Vanier, *J. Org. Chem.* **42**, 1817 (1977).

229. F. G. Bordwell, *Pure Appl. Chem.* **49**, 963 (1977).

230. J.-C. Hallé, F. Terrier, and R. Gaboriaud, *Bull. Soc. Chim. Fr.* p. 1231 (1973).

231. K. Bowden, A. Buckley, and R. Stewart, *J. Am. Chem. Soc.* **88**, 947 (1966).

232. F. G. Bordwell, G. E. Drucker, and H. E. Fried, *J. Org. Chem.* **46**, 632 (1981).

232a. See also M. I. Terekhova, E. S. Petrov, O. P. Shkurko, M. A. Miklaleva, V. P. Mamaev, and A. J. Shatenshtein, *J. Org. Chem. USSR (Engl. Ed.)* **19**, 405 (1983).

233. R. B. Homer and C. D. Johnson, *in* "Chemistry of Amides" (J. Zabicky, ed.), p. 240. Interscience, London, 1970.

233a. E. B. Skibo and T. C. Bruice, *J. Am. Chem. Soc.* **105**, 3304 (1983).

234. Ref. *62*, p. 87.

235. D. J. Cram, "Fundamentals of Carbanion Chemistry," Chap. 1. Academic Press, New York, 1965.

236. T. H. Lowry and K. S. Richardson "Mechanism and Theory of Organic Chemistry," 2nd ed., Chap. 3, Harper and Row, New York, 1981.

237. G. Schwarzenbach, *Z. Phys. Chem. A* **176A**, 133 (1936).

238. Ref. *194*; 1st ed., p. 87; 2nd ed., p. 86.

239. A. Streitwieser, Jr., E. Juaristi, and L. L. Nebenzahl, "Comprehensive Carbanion Chemistry" (E. Buncel and T. Durst, eds.) Chap. 7. Elsevier, Amsterdam 1980.

240. S. A. Chaudhri and K.-D. Asmus, *J. Chem. Soc. Faraday Trans. 1*, **68**, 385 (1972).

241. B. H. J. Bielski and A. O. Allen, *J. Phys. Chem.* **71**, 4544 (1967).

242. R. G. Pearson and R. L. Dillon, *J. Am. Chem. Soc.* **75**, 2439 (1953).

243. J. P. Schaefer and J. J. Bloomfield, *Org. React. (N.Y.)* **15**, 1 (1967).

244. K. F. Bonhoeffer, K. H. Gieb, and O. Reitz, *J. Chem. Phys.* **7**, 664 (1939).

245. D. Algrim, J. E. Bares, J. C. Branca, and F. G. Bordwell, *J. Org. Chem.* **43**, 5024 (1978).

246. S. Trofimenko, *J. Org. Chem.* **28**, 217 (1963).

247. J. K. Williams, E. L. Martin, and W. A. Sheppard, *J. Org. Chem.* **31**, 919 (1966).

248. R. H. Boyd, *J. Am. Chem. Soc.* **83**, 4288 (1961).

249. See, for example, K. E. Edgecombe and R. J. Boyd, *Can. J. Chem.* **61**, 45 (1983).

250. E. Ciganek, W. J. Linn, and O. W. Webster, *in* "Chemistry of the Cyano Group" (Z. Rapoport, ed.), Chap. 9. Interscience, London, 1970.

251. O. W. Webster, *J. Am. Chem. Soc.* **88**, 3046 (1966).

251a. T. Okuyama, Y. Ikenouchi, and T. Fueno, *J. Am. Chem. Soc.* **100**, 6162 (1978).

252. E. D. Laganis and D. M. Lemal, *J. Am. Chem. Soc.* **102**, 6633 (1980).

252a. O. S. Tee and N. R. Iyengar, *J. Am. Chem. Soc.* **107**, 455 (1985).

253. J. B. Conant and G. W. Wheland, *J. Am. Chem. Soc.* **54**, 1912 (1932).

254. W. K. McEwen, *J. Am. Chem. Soc.* **58**, 1124 (1936).

255. F. G. Bordwell, J. C. Branca, D. L. Hughes, and W. N. Olmstead. *J. Org. Chem.* **45**, 3305 (1980).

256. E. S. Petrov, M. I. Terekhova, and A. I. Shatenshtein, *Zh. Obshch. Khim.* **44**, 1118 (1974).

257. M. M. Kreevoy and E. H. Baughman, *J. Am. Chem. Soc.* **95**, 8178 (1973).

258. A. Streitwieser, Jr., E. Ciuffarin, and J. H. Hammons, *J. Am. Chem. Soc.* **89**, 63 (1967).

259. A. Streitwieser, Jr., C. J. Chang, and D. M. E. Reuben, *J. Am. Chem. Soc.* **94**, 5730 (1972).

260. C. H. Langford and R. W. Burwell, *J. Am. Chem. Soc.* **82**, 1503 (1960).

261. K. Bowden and R. Stewart, *Tetrahedron* **21**, 261 (1965).

262. V. M. Vlasov, M. I. Terekhova, E. S. Petrov, V. D. Sutula, and A. I. Shatenshtein, *J. Org. Chem. USSR (Engl. Ed.)* **18**, 1461 (1982) and refs. therein.

262a. I. O. Shapiro, M. I. Terekhova, Y. I. Ranneva, E. S. Petrov, and A. I. Shatenshtein, *J. Gen. Chem. USSR (Engl. Transl.)* **53**, 1245 (1983).

263. R. Kuhn and D. Rewicki, *Ann. Chem.* **704**, 9 (1967); **706**, 250 (1967).

264. J. H. T. Takemoto and J. J. Lagowski, *J. Am. Chem. Soc.* **91**, 3785 (1969).

265. W. L. Jolly and T. Birchall, *J. Am. Chem. Soc.* **88**, 5439 (1966).

266. R. R. Fraser, A. Baignée, M. Bresse, and K. Hata, *Tetrahedron Lett.* p. 4195 (1982).

267. R. A. Olofson and C. M. Dougherty, *J. Am. Chem. Soc.* **95**, 582 (1973).

268. C. A. Brown, *J. Chem. Soc., Chem. Commun.* p. 222 (1975).

269. E. M. Arnett, K. G. Venkatasubramaniam, R. T. McIver, E. K. Fukuda, F. G. Bordwell, and R. D. Press, *J. Am. Chem. Soc.* **104**, 325 (1982).

270. E. M. Arnett and K. G. Venkatasubramaniam, *Tetrahedron Lett.* p. 987 (1981).

271. E. S. Petrov, M. I. Terekhova, and A. I. Shatenshtein, *Zh. Obsch. Khim.* **44**, 1118 (1974).

272. D. E. Applequist and D. F. O'Brien, *J. Am. Chem. Soc.* **85**, 743 (1963).

273. R. M. Salinger and R. E. Dessy, *Tetrahedron Lett.* p. 729 (1963).

273a. See also R. R. Fraser, M. Bresse, N. Chuaqui-Offermanns, K. N. Houk, and N. G. Rondan, *Can. J. Chem.* **61**, 2729 (1983).

274. O. A. Reutov, I. P. Beletskaya, and K. P. Butin, "CH-Acids." Pergamon, Oxford, 1978 (transl. T. R. Compton).

274a. See also M. Schlosser and S. Strunk, *Tetrahedron Lett.* p. 741 (1984).

275. I. M. Kolthoff and T. B. Reddy, *Inorg. Chem.* **1**, 189 (1962).

276. C. D. Ritchie and R. E. Uschold, *J. Am. Chem. Soc.* **89**, 1721 (1967); **90**, 2821 (1968).

277. C. D. Ritchie *in* "Solute-Solvent Interactions" (J. F. Coetzee and C. D. Ritchie, eds.), p. 231. Dekker, New York, 1969.

278. A. Streitwieser, Jr., and C. Perrin, *J. Am. Chem. Soc.* **86**, 4938 (1964).

279. R. E. Dessy, W. Kitching, T. Psarras, R. Salinger, A. Chen, and T. Chivers, *J. Am. Chem. Soc.* **88**, 460 (1966).

280. G. Nisli, D. Barnes, and P. Zuman, *J. Chem. Soc. B* p. 764 (1970).

281. K. P. Butin, I. P. Beletskaya, and O. A. Reutov, *Elektrochimiya* **2**, 635 (1966).

282. O. A. Reutov, K. P. Butin, and I. P. Beletskaya, *Bull. Inst. Polytech. Iasi, II* **16**, 33 (1970).

283. R. Breslow and K. Balasubramanian, *J. Am. Chem. Soc.* **91**, 5182 (1969).

284. R. Breslow and W. Chu. *J. Am. Chem. Soc.* **92**, 2165 (1970).

285. R. Breslow and W. Chu, *J. Am. Chem. Soc.* **95**, 411 (1973).

286. C. H. Rochester, "Acidity Functions." Academic Press, London, 1970.

287. R. Breslow and R. Goodin, *J. Am. Chem. Soc.* **98,** 6077 (1976).

288. R. Breslow and J. L. Grant, *J. Am. Chem. Soc.* **99,** 7745 (1977).

289. B. Jaun, J. Schwarz, and R. Breslow, *J. Am. Chem. Soc.* **102,** 5741 (1980).

290. R. Breslow, *Pure Appl. Chem.* **40,** 493 (1974).

291. E. M. Arnett, T. C. Moriarity, L. E. Small, J. P. Rudolph, and R. P. Quirk, *J. Am. Chem. Soc.* **95,** 1492 (1973).

291a. E. M. Arnett and K. G. Venkatasubramaniam, *J. Org. Chem.* **48,** 1569 (1983).

292. C. D. Ritchie, *J. Am. Chem. Soc.* **91,** 6749 (1969).

293. A. J. Kresge, *Acc. Chem. Res.* **8,** 354 (1978).

294. H. F. Koch, J. G. Koch, N. H. Koch, and A. S. Koch, *J. Am. Chem. Soc.* **105,** 2388 (1983).

295. A. Hantzsch, *Ber.* **32,** 575 (1899).

296. G. N. Lewis and G. T. Seaborg, *J. Am. Chem. Soc.* **61,** 1886 (1939).

297. R. P. Bell, *Trans. Faraday Soc.* **39,** 253 (1943).

298. J. R. Jones, R. E. Marks, and S. C. Subba Rao, *Trans. Faraday Soc.* p. 111 (1963).

299. J. P. Guthrie, *Can. J. Chem.* **57,** 1177 (1979).

300. J.-E. Dubois, M. El-Alaoui, and J. Toullec, *J. Am. Chem. Soc.* **103,** 5393 (1981).

301. A. I. Shatenshtein, *in* "Advances in Physical Organic Chemistry" (V. Gold, ed.), Vol. 1, p. 156. Academic Press, London, 1963.

302. A. Streitwieser, Jr., R. A. Caldwell, and M. R. Granger, *J. Am. Chem. Soc.* **86,** 3578 (1964).

303. A. Streitwieser, Jr., W. C. Langworthy, and J. I. Brauman, *J. Am. Chem. Soc.* **85,** 1761 (1963).

304. A. J. Kresge and M. F. Powell, *Int. J. Chem. Kinet.* **14,** 19 (1982).

305. A. F. Cockerill, D. W. Earls, J. R. Jones, and T. G. Rumney, *J. Am. Chem. Soc.* **96,** 575 (1974).

306. Y. Chiang, A. J. Kresge, and P. A. Walsh, *J. Am. Chem. Soc.* **104,** 6122 (1982).

307. J. P. Guthrie, J. Cossar, and A. Klym, *J. Am. Chem. Soc.* **106,** 1351 (1984).

307a. J. P. Guthrie, J. Cossar, and A. Klym, *J. Am. Chem. Soc.* **104,** 895 (1982).

307b. E. D. Stroud, D. J. Fife, and G. G. Smith, *J. Org. Chem.* **48,** 5368 (1983).

308. J. Toullec, *in* "Advances in Physical Organic Chemistry" (V. Gold and D. Bethell, eds.), Vol. 18, p. 1. Academic Press, London, 1982.

309. L. P. Hammett and A. J. Deyrup, *J. Am. Chem. Soc.* **54,** 2721 (1932).

310. L. P. Hammett and A. J. Deyrup, *J. Am. Chem. Soc.* **55,** 1900 (1933).

311. L. P. Hammett, "Physical Organic Chemistry," 2nd ed., Chap. 9. McGraw-Hill, New York, 1970.

312. R. H. Boyd *in* "Solute–Solvent Interactions" (J. F. Coetzee and C. D. Ritchie, eds.), Chap. 3. Dekker, New York, 1969.

313. K. Bowden, *Chem. Rev.* **66,** 119 (1966).

314. R. A. Côx and K. Yates, *Can. J. Chem.* **61,** 2225 (1983).

315. C. H. Rochester, *J. Chem. Soc. Q. Rev.* p. 511 (1966).

316. G. Schwarzenbach and R. Sulzberger, *Helv. Chim. Acta* **27**, 348 (1944).

317. J. T. Edward and I. C. Wang, *Can. J. Chem.* **40,** 399 (1962).

318. G. Yagil, *J. Phys. Chem.* **71,** 1034 (1967).

319. R. A. Cox and R. Stewart, *J. Am. Chem. Soc.* **98,** 488 (1976).

320. C. Perrin, *J. Am. Chem. Soc.* **86,** 256 (1964).

321. R. Schaal, *C. R. Hebd. Seances Acad. Sci.* **239,** 1036 (1954).

322. N. C. Deno, *J. Am. Chem. Soc.* **74,** 2039 (1952).

323. R. Schaal and P. Favier, *Bull. Soc. Chim. Fr.* p. 2011 (1959).

324. C. H. Langford and R. L. Burwell, *J. Am. Chem. Soc.* **82,** 1503 (1960).

325. R. Stewart and J. P. O'Donnell, *J. Am. Chem. Soc.* **84,** 493 (1962).

326. R. Stewart and J. P. O'Donnell, *Can. J. Chem.* **42,** 1681 (1964).

327. D. Dolman and R. Stewart, *Can. J. Chem.* **45,** 911 (1967).

328. F. G. Bordwell, personal communication.
329. J. C. Hallé, R. Gaboriaud, and R. Schaal, *Bull. Soc. Chim. Fr.* p. 1851 (1969).
330. J. C. Hallé, F. Terrier, and R. Schaal, *Bull. Soc. Chim. Fr.* p. 4569 (1969); p. 1231 (1973).
331. J. C. Hallé, *Bull. Soc. Chim. Fr.* p. 1553 (1973).
332. J. T. Edward, M. Sjöström, and S. Wold, *Can. J. Chem.* **59,** 2350 (1981).
333. A. Albagli, A. Buckley, A. M. Last, and R. Stewart, *J. Am. Chem. Soc.* **95,** 4711 (1973).
334. K. Bowden and A. F. Cockerill, *J. Chem. Soc. B* p. 173 (1970).
335. R. Stewart and M. G. Harris, *Can. J. Chem.* **55,** 3807 (1977).
336. R. A. Cox, R. Stewart, M. J. Cook, A. R. Katritzky, and R. D. Tack, *Can. J. Chem.* **54,** 900 (1976).
337. J. Janata and R. D. Holtby-Brown, *J. Chem. Soc., Perkin Trans. 2* p. 991 (1973).
338. K. Bowden, A. Buckley, and R. Stewart, *J. Am. Chem. Soc.* **88,** 947 (1967).
339. J. F. Bunnett and F. P. Olsen, *Can. J. Chem.* **44,** 1899 (1966).
340. N. C. Marziano, G. M. Cimino, and R. C. Passerini, *J. Chem. Soc., Perkin Trans. 2* p. 1915 (1973).
341. See ref. *263.*
342. L. G. Greifenstein, J. B. Lambert, R. J. Nienhuis, G. E. Drucker, and G. A. Pagini, *J. Am. Chem. Soc.* **103,** 7753 (1981).
343. F. D. W. Boerth and A. Streitwieser, Jr., *J. Am. Chem. Soc.* **103,** 6443 (1981).
343a. A. C. Hopkinson, "The Chemistry of the Carbon–Carbon Triple Bond" (S. Patai, ed.), Chap. 4. Wiley, Chichester, 1978.
344. A. C. Lin, Y. Chiang, D. B. Dahlberg, and A. J. Kresge, *J. Am. Chem. Soc.* **105,** 5380 (1983).
345. F. G. Bordwell, D. Algrim, and H. E. Fried, *J. Chem. Soc., Perkin Trans 2* p. 726 (1979).
346. A. Streitwieser, Jr., and D. M. E. Reuben, *J. Am. Chem. Soc.* **93,** 1794 (1971).
347. A. Streitwieser, and P. J. Scannon, *J. Am. Chem. Soc.* **95,** 6273 (1973).
348. M. J. Pellerite and J. I. Brauman, "Comprehensive Carbanion Chemistry" (E. Buncel and T. Durst, eds.), Chap. 2. Elsevier, Amsterdam, 1980.
349. V. I. Slovetskii, L. V. Okhobystina, A. A. Fainzilberg, A. I. Ivanov, L. J. Biryukova, and S. S. Novikoff, *Izv. Akad. Nauk SSR,* p. 2063 (1965).
350. F. G. Bordwell, J. E. Bartmess, G. E. Drucker, Z. Margolin, and W. S. Mathews, *J. Am. Chem. Soc.* **97,** 3226 (1975).
351. R. W. Taft, *Prog. Phys. Org. Chem.* **14,** 247 (1983).
352. M. Mishima, I. Koppel and R. W. Taft, personal communication.
353. A. Streitwieser Jr. and J. T. Swanson, *J. Am. Chem. Soc.* **105,** 2502 (1983).
354. R. R. Fraser, M. Bresse, and T. S. Mansour, *J. Am. Chem. Soc.* **105,** 7790 (1983).
355. R. R. Fraser, M. Bresse, and T. S. Mansour, *Chem. Commun.* p. 620 (1983).

3

Strengths of Neutral Organic Bases

I. Introduction

All of the functional groups of organic chemistry are to some extent basic since all of them possess at least one pair of electrons that can form a bond to the proton. Although the most basic single functional group is probably the amidino unit (**1**), the most basic organic molecules are polyfunctional amines. These include the 1,8-diaminonaphthalenes (**2**) (*1, 2*) and macrobicyclic amines of general structure **3** (*3–5*), some of which remain protonated in concentrated aqueous potassium hydroxide. Moreover, they transfer protons sluggishly (Section III,C).

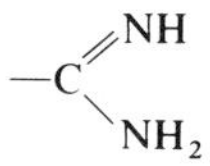

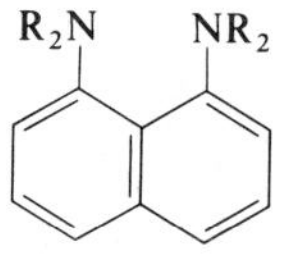

The least basic functional groups include nitro, cyano, and halo, which require extremely acidic conditions (e.g., 100% sulfuric acid) to effect significant degrees of protonation (*6–8*).

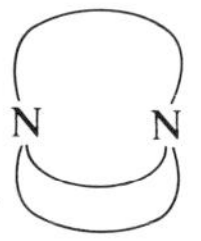

As described in Chapter 1 we shall refer to the strengths of bases in water in terms of the acid strengths of their conjugate acids, using the notation K_{BH^+} or pK_{BH^+} for the purpose. In these terms the base strengths of organic molecules range from a pK_{BH^+} near 16 for some compounds of type **2** and 18 or more for some of type **3** to values of -10 or less for many nitriles and nitro compounds, with haloalkanes being less basic still. Alkanes lack an available pair of electrons and are thus not formally bases in the Brønsted sense. (Alkenes and alkynes do not have an unshared pair of electrons but do have a pair of electrons that is surplus to the minimum bonding requirements of the molecules.) Alkanes undergo protonation in the gas phase to form pentacoordinated species, but the extent to which this occurs in the condensed phase is a matter of controversy that will be dealt with in Section VI.

The various techniques that have been used to determine the basicity of organic molecules were described in Arnett's monumental 1963 review (9).

II. Amidines and Guanidines

Amidines (4), are some two pK units and guanidines (5) some three to four pK units more basic than aliphatic amines [Eqs. (3-1) and (3-2)].

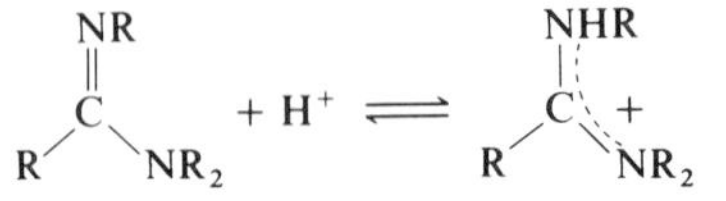

$$
\begin{array}{ccc}
\underset{\substack{\| \\ \text{C} \\ R \qquad NR_2}}{NR} & + H^+ \rightleftharpoons & \underset{\substack{| \\ \text{C} \\ R \qquad NR_2}}{NHR} +
\end{array}
\qquad (3\text{-}1)
$$

4

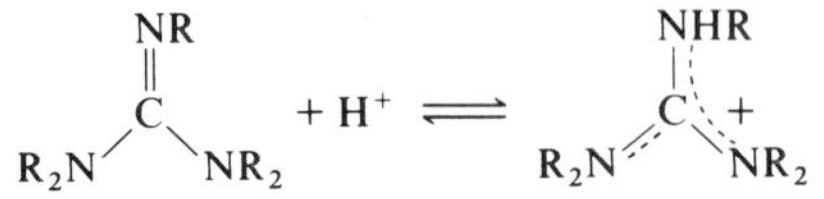

$$
\begin{array}{ccc}
\underset{\substack{\| \\ \text{C} \\ R_2N \qquad NR_2}}{NR} & + H^+ \rightleftharpoons & \underset{\substack{| \\ \text{C} \\ R_2N \qquad NR_2}}{NHR} +
\end{array}
\qquad (3\text{-}2)
$$

5

Guanidines, with pK_{BH^+} values near 14, are virtually completely ionized in dilute aqueous solution. Pentamethylguanidine ($pK_{BH^+} = 13.8$), for example, occupies a position on the basic side that is comparable to that which trifluoroacetic acid ($pK_{BH^+} \approx 0$) occupies on the acidic side. Guanidine, the parent compound, is a crystalline, hygroscopic substance melting at 50°C whose pK_{BH^+} is 13.6. In view of its high basicity and ready miscibility with water it is surprising that it was not exploited in the early development of the H_- scale, rather than hydrazine or ethylenediamine, which were used for this purpose (Chapter 2, Section IV,C,4).

Polyalkylation of guanidine causes fairly modest increases in base strength (Table 3-1). In this respect the most effective distribution of alkyl groups is one that places an alkyl group at each nitrogen atom.

The importance to synthetic organic chemistry of having available strong bases of low nucleophilicity led Barton *et al.* (10) to prepare and utilize the species 6 (Table 3-1), whose pK_{BH^+} in water is slightly greater than 14, judging from its measured pK in aqueous ethanol. Hindered amidines such as 7 have been used for the same purpose (10a).

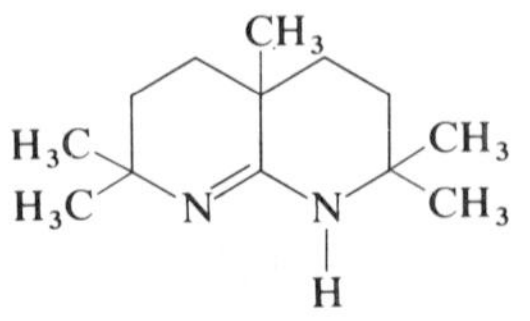

7

Table 3-1

BASICITY CONSTANTS OF GUANIDINES AND AMIDINES IN WATER AT 25°C

Structure	pK_{BH^+}	Ref.	Structure	pK_{BH^+}	Ref.
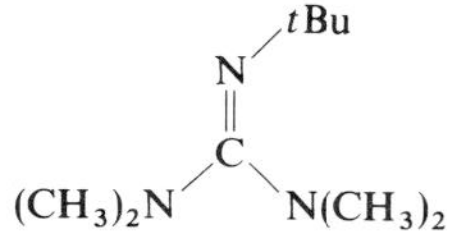 $(CH_3)_2N\!-\!C(=\!N\!-\!tBu)\!-\!N(CH_3)_2$ **6**	~14.3	10	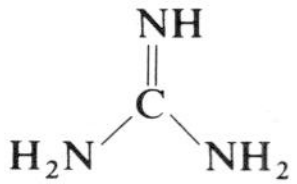 $H_2N\!-\!C(=\!NH)\!-\!NH_2$	13.6	11
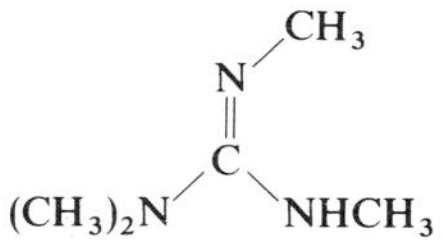 $(CH_3)_2N\!-\!C(=\!N\!-\!CH_3)\!-\!NHCH_3$	13.9	11	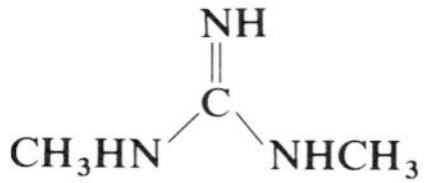 $CH_3HN\!-\!C(=\!NH)\!-\!NHCH_3$	13.6	11
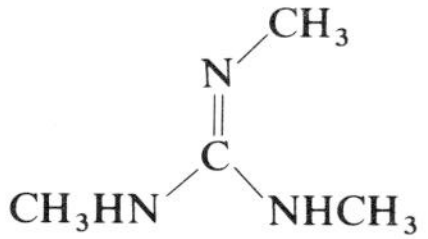 $CH_3HN\!-\!C(=\!N\!-\!CH_3)\!-\!NHCH_3$	13.9	11	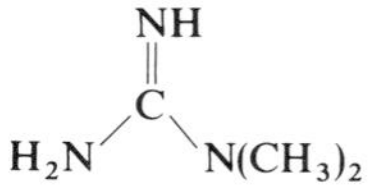 $H_2N\!-\!C(=\!NH)\!-\!N(CH_3)_2$	13.4	11
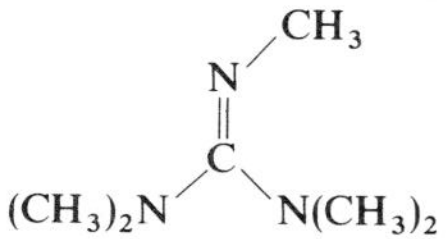 $(CH_3)_2N\!-\!C(=\!N\!-\!CH_3)\!-\!N(CH_3)_2$	13.8	11	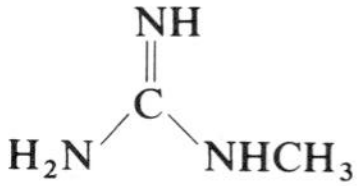 $H_2N\!-\!C(=\!NH)\!-\!NHCH_3$	13.4	11
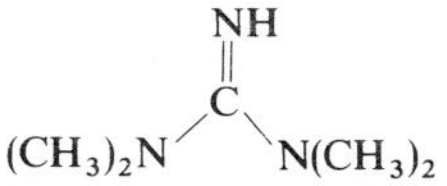 $(CH_3)_2N\!-\!C(=\!NH)\!-\!N(CH_3)_2$	13.6	9, 11	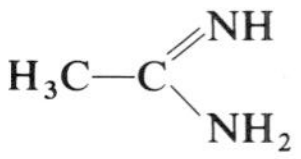 $H_3C\!-\!C(=\!NH)\!-\!NH_2$	12.4	12
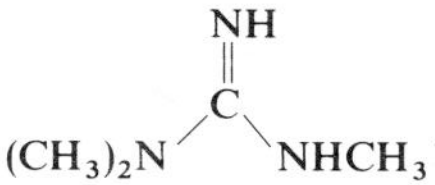 $(CH_3)_2N\!-\!C(=\!NH)\!-\!NHCH_3$	13.6	11	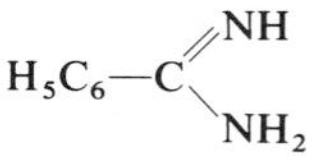 $H_5C_6\!-\!C(=\!NC_6H_5)\!-\!NH_2$	11.6 (20°C)	13
			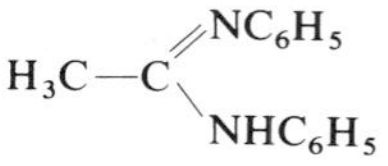 $H_3C\!-\!C(=\!NH)\!-\!NHC_6H_5$	8.3	12

III. Amines

Monoamino aliphatic compounds have base strengths that range from pK_{BH^+} values near 11 to values of less than zero (Table 3-2). Monoamino aromatic compounds span the range between about 6 and -10 (Tables 3-3 and 3-4).

It is not difficult to measure the strengths of most aliphatic and aromatic amines. Aliphatic amines are generally strong enough that glass electrode measurements at half-neutralization give pK_{BH^+} directly. Arylamines are too weak for this technique to give accurate results but, fortunately, their degree of protonation can be determined spectrophotometrically and in most cases these results can be combined with a glass electrode measurement to give pK_{BH^+}. Arylamines containing strongly electron-withdrawing groups are

Table 3-2

pK_{BH^+} VALUES OF SOME ALIPHATIC MONOAMINES IN WATER AT 25°C[a]

Compound	pK_{BH^+}	Compound	pK_{BH^+}
$(C_2H_5)_2NH$	11.04	$(H_2C{=}CHCH_2)_2NH$	9.29
$[CH_3(CH_2)_5]_2NH$	11.01	$Br(CH_2)_3NH_2$	8.9
$(CH_3)_3SiCH_2CH_2NH_2$	10.97	$(HOCH_2CH_2)_2NH$	8.88
$(C_2H_5)_3N$	10.75	$BrCH_2CH_2NH_2$	8.5
$(CH_3)_3Si(CH_2)_3NH_2$	10.73	$(H_2C{=}CHCH_2)_3N$	8.31
$(CH_3)_2NH$	10.73	$HC{\equiv}CCH_2NH_2$	8.15
$(CH_3)_3CNH_2$	10.68	$(HOCH_2)_3CNH_2$	8.08
$c\text{-}C_6H_{11}NH_2$	10.68	$(HOCH_2CH_2)_3N$	7.76
$CH_3(CH_2)_xNH_2$	10.60–10.70	$CNCH_2CH_2NH_2$	7.7
$(x = 0\text{--}21)$		$(HC{\equiv}CCH_2)_2NH$	6.10
$(CH_3)_2CHNH_2$	10.60	$CF_3CH_2NH_2$	5.7
$HO(CH_2)_5NH_2$	10.43	$CCl_3CH_2NH_2$	5.4
$HO(CH_2)_4NH_2$	10.3	$(CNCH_2CH_2)_2NH$	5.26
$(CH_3CH_2CH_2)_3N$	10.26	$(ClCH_2CH_2)_3N$	4.37
$HO(CH_2)_3NH_2$	9.96		
$(CH_3CH_2CH_2CH_2)_3N$	9.93		
$(CH_3)_3N$	9.75		4.05^b
$CCl_3(CH_2)_3NH_2$	9.9		
$Br(CH_2)_5NH_2$	9.6		
$HOCH_2CH_2NH_2$	9.50	$(HC{\equiv}CCH_2)_3N$	3.09
$H_2C{=}CHCH_2NH_2$	9.49	$(CF_3)_2CHNH_2$	1.22^b
$CH_3OCH_2CH_2NH_2$	9.44	$(CNCH_2CH_2)_3N$	1.1
$Cl_2C{=}CHCH_2NH_2$	9.4	$(C_2H_5)_2NCl$	1.0
$CH_3SCH_2CH_2NH_2$	9.33	$(C_2H_5)_2NCN$	<0

[a] Ref. 26.

[b] R. D. Roberts, H. E. Ferran, Jr., M. J. Gula, and T. A. Spencer, *J. Am. Chem. Soc.* **102**, 7054 (1980).

Table 3-3

pK_{BH^+} OF MONOSUBSTITUTED ANILINES IN WATER AT 25°C[a]

Aniline	Ortho	Meta	Para
Unsubstituted	4.60	4.60	4.60
Amino	4.7	5.0	6.2
Bromo	2.53	3.56	3.86
tert-Butyl	3.78	4.66	4.95
Carbomethoxy	2.2	3.5	2.4
Chloro	2.64	3.52	4.00
Cyano	0.95	2.80	1.75
Ethyl	4.37	4.70	5.00
Fluoro	3.20	3.59	4.65
Hydroxy	4.7	4.3	5.6
Iodo	2.60	3.61	3.78
Methyl	4.45	4.71	5.08
Methoxy	4.52	4.23	5.34
Methylthio	—	4.00	4.35
Nitro	−0.29[b]	2.50[b]	0.99[b]
Trifluoromethoxy	2.44	3.25	3.82
Trifluoromethylsulfonyl	—	1.8	0

[a] Values taken from Perrin (26), except where noted.

[b] Reference 17. See also ref. 87, where almost identical values are given for the ortho and para compounds.

Table 3-4

pK_{BH^+} VALUES OF PRIMARY NITROANILINES IN WATER AT 25°C[a]

Substituent	pK_{BH^+}	Substituent	pK_{BH^+}
3-Nitro	2.50	2,6-Dinitro-4-ethoxy	−3.79
4-Nitro	0.99	2,4-Dinitro	−4.31[b]
2-Nitro-4-methyl	0.45	2,6-Dinitro-4-methyl	−4.36
2-Nitro	−0.30	2,4-Dinitro	−4.38
2-Nitro-4-fluoro	−0.44	2,6-Dinitro-4-fluoro	−5.41
2-Nitro-5-hydroxy	−0.61	2,6-Dinitro	−5.43
2-Nitro-4-chloro	−1.04	2,6-Dinitro-4-chloro	−6.16
2-Nitro-4-bromo	−1.20	2,4-Dinitro-6-bromo	−6.71
2-Nitro-5-chloro	−1.51	2,6-Dinitro-4-carboxy	−8.26
4-Nitro-2,5-dichloro	−1.82	2,4,6-Trinitro-3-methyl	−8.37
2-Nitro-6-chloro	−2.46	2,6-Dinitro-4-acetyl	−8.79
2-Nitro-4-carboxy	−2.67	2,4,6-Trinitro-3-bromo	−9.34[c]
4-Nitro-2,6-dichloro	−3.24	2,4,6-Trinitro-3-chloro	−9.71
2-Nitro-4,6-dichloro	−3.29	2,6-Dinitro-4-methylsulfonyl	−9.75
		2,4,6-Trinitroaniline	−10.04[d]

[a] The values are taken from the compilations of Rochester (16, p. 67) and Kuznetsov and Gidaspo (27), who drew heavily on the work of Jorgenson and Hartter (18), Ryabova et al. (28), Yates and Wai (29), and Johnson et al. (30). Where discrepancies (usually small) were found, an average was taken.

[b] Reference 82.

[c] Reference 30.

[d] An extrapolative procedure (30a) gives −10.35.

virtually unprotonated in aqueous solution, and they were used by Hammett in the 1930s in his development of the acidity function, a concept that allowed the acidity of strongly acidic systems to be quantitatively explored for the first time (*14–17*).

Hammett and Deyrup chose 4-nitroaniline as the anchor compound since its extent of protonation [Eq. (3-3)] can be determined in what is essentially

$$O_2N-\langle\!\!\!\bigcirc\!\!\!\rangle-\overset{+}{N}H_3 \;\rightleftharpoons\; O_2N-\langle\!\!\!\bigcirc\!\!\!\rangle-NH_2 + H^+ \qquad (3\text{-}3)$$

aqueous solution. This compound is $\sim 10\%$ protonated at pH 2.0 but, because it is difficult to measure accurately small degrees of protonation, the pK_{BH^+} is obtained by extrapolating plots of $\log([BH^+]/[B]) - \log[H^+]$ against concentration of acid to infinite dilution (*17*). When this is done for a variety of mixed acids, an average pK_{BH^+} of 0.99 is obtained (*17a*), a value slightly different from that originally used by Hammett and Deyrup.

By then measuring the extent of protonation of 4-nitroaniline in more concentrated acid, where activity coefficients will no longer be unity, the protonating capacity of the medium, designated H_0, can be determined. (4-Nitroaniline is 90% protonated in 0.7 M sulfuric acid.) In dilute aqueous solution, H_0, which is given by Eqs. (3-4) and (3-5), becomes equal to pH and

$$H_0 = -\log\left(a_{H^+}\frac{f_B}{f_{BH^+}}\right) \qquad (3\text{-}4)$$

$$H_0 = pK_{BH^+} + \log\frac{[B]}{[BH^+]} \qquad (3\text{-}5)$$

hence is an extension of the pH scale. By overlapping successively weaker amine bases and using Eq. (3-5) to determine pK_{BH^+} and H_0 alternately, the aniline scale can be extended to 99.4% sulfuric acid, in which 2,4,6-trinitroaniline is 90% protonated (*18*). A much smaller concentration of perchloric acid ($\sim 80\%$) will protonate this compound to the same extent (*19*).

Provided that certain conditions are met the pK_{BH^+} values of the indicator bases determined as this procedure is carried out refer to the aqueous state, even though the measurements are carried out in mixed media. The analogous treatment of strongly basic systems is described in Chapter 2, Section IV,C,4.

The central requirement for the H_0 scale to be valid (i.e., that pK_{BH^+} values refer to aqueous solution) is that the activity coefficient ratio f_B/f_{BH^+} for the indicator bases be a function only of the structure of the base. An obvious test of this procedure is to see if parallel lines are obtained in regions of overlap when $\log([B]/[BH^+])$ is plotted against some function of the medium's acidity. When the function is expressed as a percentage of sulfuric acid, parallel curved lines will be obtained; after the H_0 scale has been constructed, it will give straight lines of unit slope.

In the case of aniline indicators, reasonable parallelism between overlapping indicators was obtained, although it was subsequently found that there was some divergence between primary, secondary, and tertiary amines (20–23). Effectively, each class of amine generates its own scale, and the symbol H_0 is now generally reserved for that derived using primary amines. If even such closely related compounds as these produce their own scales, it would be surprising if compounds containing other functional groups followed the H_0 function and, indeed, it turns out that few of them do.

The problem of measuring pK_{BH^+} values of other weakly basic compounds such as aldehydes and carboxylic acids will be discussed later, but attention should be drawn here to the two principal reasons for the strengths of very weak aniline bases being more accurately known than those of such other compounds. First, protonation of an arylamine causes a very large blue shift in the ultraviolet or visible absorption spectrum of the compound and, indeed, it causes absorption at λ_{max} of the amine almost to disappear, unlike the situation with other functional groups, where protonation often produces modest changes in both λ_{max} and ϵ_{max}. Thus, the quantity $[BH^+]/[B]$ for arylamines can usually be determined with good precision, although medium effects on spectra must still be accounted for.

Second, the basicity of the amino group in an aniline base is strongly dependent on the presence of ring substituents so that one has available an extensive series of compounds whose more basic members are protonated in the aqueous pH range and whose very weakly basic members require high concentrations of sulfuric or other acid for protonation to occur. 2,4,6-Trinitroaniline, for example, is ~ 14 pK units less basic than aniline itself. Thus, the H_0 scale and the pK values derived therefrom are firmly anchored in the standard state of water.

Indicator parallelism can be accurately observed only over a limited region of acidities (usually from 10 to 90% ionization for any indicator), and it is always conceivable that the lines will diverge when the degree of protonation is very low, although there is no reason to suppose that this occurs. A more likely source of error for even well-behaved series such as aromatic amines is that parallelism is not quite perfect and that small errors accumulate over the considerable number of overlaps that must be made in order to link an extremely weak base such as a polynitroaniline to the aqueous state. It is unlikely that this effect will give rise to errors of more than a few tenths of a pK unit even in the latter case, and one can take with some confidence the values listed in Table 3-4 as being valid for dilute aqueous solution. A curiosity near the bottom of the listings should be noted, however; a bromo or chloro substituent has a base-*strengthening* effect on 2,4,6-trinitroaniline. It may be that the steric effect of the halogen atoms prevents the nitro group from fully conjugating with the amino group, although a more recent determination gives pK_{BH^+} values for the 3-bromo and parent compounds in the expected order (2,4,6-trinitroaniline, -10.3; 3-bromo-2,4,6-trinitroaniline, -10.8) (30a).

The term *acidity function failure* is often used in the literature. It should be remembered that what has failed is the expectation that a single acidity function would govern the ionization of all classes of compound, that is, that the activity quotient f_{BH^+}/f_B would be independent of the structure of the base and would depend only on the properties of the medium. If one is interested primarily in determining pK_{BH^+} values of a set of organic bases, it should be possible to determine them with considerable precision by using the acidity function–overlap technique, as has been done with aromatic amines. For most organic series, unfortunately, anchoring the scale in water turns out to be very difficult since few include a member whose degree of protonation can be measured in the standard state. Furthermore, there is no guarantee that indicator parallelism will be found in any particular series. Alternative approaches that have been applied to such compounds are considered in Section V.

The use of mixtures of water and polar aprotic solvents such as dimethyl sulfoxide (DMSO) containing a strong base to produce strongly basic solutions is described in Chapter 2 (Section IV,C,4). A test of the validity of the H_- function and the pK_{HA} values of amines and other weak acids that were obtained thereby could be made by measuring the absolute values of the dissociation constant in pure DMSO. The pK *difference* between any pair of weak acids must be the same in DMSO as that obtained using the acidity function in mixed media and which refers ostensibly to the standard state water. We saw that for aromatic amines this condition is very nearly met, but for many other weak acids this is anything but so. There is no analogous test that can be made of the validity of H_0 for sulfuric acid mixtures and the pK_{BH^+} values of very weak bases derived therefrom. This is because sulfuric acid is not simply a second solvent component in the way that DMSO is; it is the acid source as well.

It was pointed out earlier that protonation of secondary and tertiary aromatic amines does not follow the H_0 function generated by primary arylamines, a condition that has been attributed to the difference in the number of acidic protons and consequent hydrogen-bonding capacity in the corresponding conjugate acids, $ArNH_3^+$, $ArNH_2R^+$, and $ArNHR_2^+$ (20). Table 3-5 lists the pK_{BH^+} values of a series of anilines and their N,N-dimethyl derivatives. It can be seen that dialkylation increases the base strength of aniline but that the difference diminishes with decreasing base strength and, indeed, becomes negative in the case of the 4-nitro compound. There is little

Table 3-5

EFFECT OF N,N-DIMETHYLATION ON THE BASICITY OF ANILINES IN
WATER AT 25°C[a]

Substituent	pK_{BH^+}		ΔpK_{BH^+}
	Aniline	N,N-Dimethylaniline	
4-Methyl	5.08	5.63	0.57
Unsubstituted	4.60	5.07	0.47
4-Chloro	4.00	4.34[b]	0.34
3-Chloro	3.52	3.80[b]	0.28
3-Nitro	2.50	2.63	0.13
4-Nitro	0.99	0.66	−0.33
2,6-Dinitro-4-methyl	−4.36	−1.66	2.70
2,6-Dinitro	−5.48	−2.38	3.10
2,6-Dinitro-4-chloro	−6.17	−3.12	3.05
2,4,6-Trinitro	−10.04	−6.55	3.49

[a] Values taken from refs. 22, 26, and 27.
[b] Corrected to 25°C using Eq. (3-6).

doubt about the validity of the listed pK_{BH^+} values for these compounds since they were determined in essentially aqueous solution. As the base strength is further decreased the effect is reversed and the N,N-dimethyl compounds become increasingly more basic than their unalkylated counterparts. This means that the difference in the number of hydrogen-bonding sites in the cation is critical only for very weak bases and is overshadowed by other effect(s) (23a) in those arylamines that are somewhat stronger. It is somewhat unsettling that the reversal in ΔpK for anilines and their N,N-dimethyl derivatives coincides with the transition from the known (aqueous) region to the unknown, or at least less well known (mixed-acid), region. It should be remembered that the pK_{BH^+} values of the weaker bases in Table 3-5 were determined with their own acidity functions and hence refer ostensibly to the dilute aqueous state. Is it possible that the distinction between the ionization and dissociation behavior of weak and moderately strong neutral acids discussed in Chapter 2 (Section II,A) is pertinent here as well and gives rise to the curious results shown in Table 3-5? Such a phenomenon would have serious implications for the fundamental acidity function postulate that has the pK values refer to dissociation in the dilute aqueous state. Fortunately, there is extensive thermochemical and related evidence to suggest that at least for primary amines there is no discontinuity as one proceeds from weak to very weak amine bases (ref. 24 and Section III,A). In this connection it is odd that most investigators in this general area have tended to focus their attention on the medium, that is, on acidity functions, rather than on the solutes, which, indeed, are often called "indicators." In actual fact an acidity function has no meaning without an associated set of pK values, whereas the latter always have relevance for purely aqueous solution, even though such quantities may be difficult to determine experimentally.

A. Temperature Effects on Amine Base Strength

The extent of dissociation of most amines is fairly sensitive to changes in temperature, unlike the situation with carboxylic acids (Chapter 2, Section II,E,2). In general, base strength declines with increasing temperature, and the stronger the base the larger is the effect. Perrin has expressed this relationship in the form of Eq. (3-6) (25).

$$\frac{-d(pK_{BH^+})}{dT} = \frac{pK_{BH^+} - 0.9}{T} \tag{3-6}$$

For amines that are protonated in the aqueous region Eq. (3-6) is a useful approximation over a wide pK range although, as will be seen, it tends to obscure differences that exist among various classes of amine. In practical terms it can be noted that at room temperature a relatively strong base such as piperidine (pK_{BH^+} = 11.1; Table 3-6) has its pK_{BH^+} lowered by 0.34 pK unit for a 10°C rise in temperature, whereas a weaker base such as aniline (pK_{BH^+} = 4.60) will have its pK_{BH^+} lowered by only 0.12 pK unit for the same rise in temperature. The pK_{BH^+} of amines whose basicity is just detectable in water is almost temperature independent.

How is the ionization of extremely weak bases affected by a change in temperature? Equation (3-6) predicts that the effect will be the reverse of that found for weak bases; that is, an increase in temperature will cause a substantial increase in base strength. The careful work of Johnson et al. (30) shows that this expectation is realized. The pK_{BH^+} of 2,4,6-trinitroaniline, for example, changes from -10.0 at 25°C to -9.29 at 60°C and to -8.67 at 90°C. This change is of the correct sign but is just over half that predicted by Eq. (3-6).

There are two common ways of determining ΔH of reaction: by direct calorimetric measurement or by determining the temperature coefficient of reaction and then using the van't Hoff equation [Eq. (3-7)].

$$\frac{d \ln K}{dT} = \frac{\Delta H}{RT^2} \tag{3-7}$$

It was noted earlier (Chapter 2, Section II,E,2) that these methods do not always give good agreement in practice. In the case of amine protonation there has been extensive work done using both methods over wide ranges of base strength and, although there are a number of discrepancies, there are enough

Table 3-6

pK_{BH^+} VALUES OF HETEROCYCLIC AMINES IN WATER AT 25°C[a]

Compound	Structure	pK_{BH^+}
Cryptand [1.1.1]	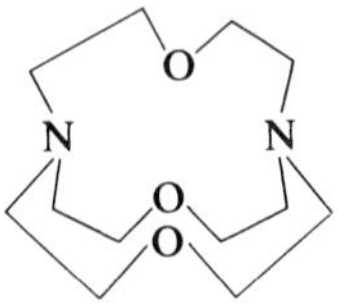	$\geq 17.8^b$
Azetidine		11.29
Pyrrolidine		11.30
Piperidine	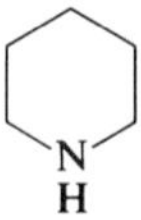	11.12
Hexahydroazepine	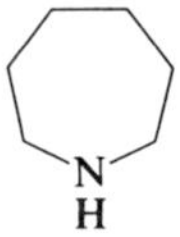	11.07
Quinuclidine	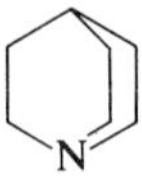	10.95
Piperazine	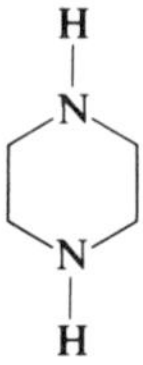	9.73
1,4-Thiazine	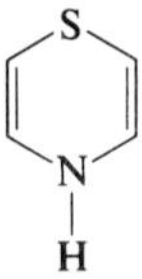	8.4
1,4-Diazabicyclo[2.2.2]octane (DABCO)	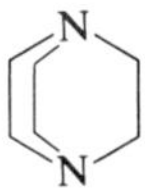	~8.4

(cont.)

Table 3-6 (*cont.*)

Compound	Structure	pK_{BH^+}
Morpholine		8.4
Aziridine		8.00
Imidazole[c]		6.99[d]
1,6-Diazabicyclo[4.4.4]tetradecane		6.6[e]
Thiazolidine		6.2
Benzimidazole		5.53
Acridine		5.5
Isoquinoline		5.3
Pyridine		5.23[f]
Quinoline		4.8
Pteridine		4.0

Table 3-6 (*cont.*)

Compound	Structure	pK_{BH^+}
N-Phenylazetidine		3.6^g
1,2,4-Triazole	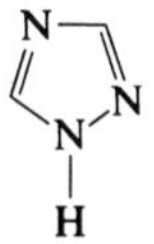	2.3
Cinnoline	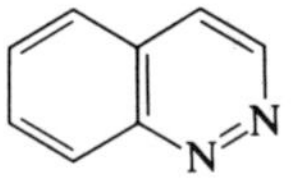	2.2
Pyridazine	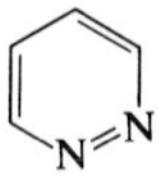	2.2
Isoxazole	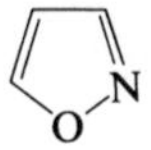	1.3
Pyrimidine	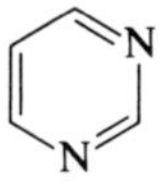	1.2
Phenazine	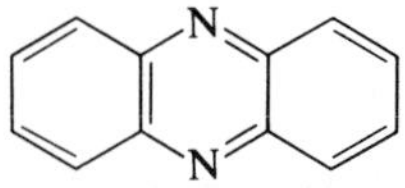	1.2
1,2,3-Triazole	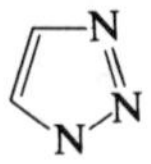	1.1
Pyrazine	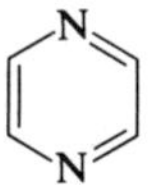	0.6
8-Carboxypurine	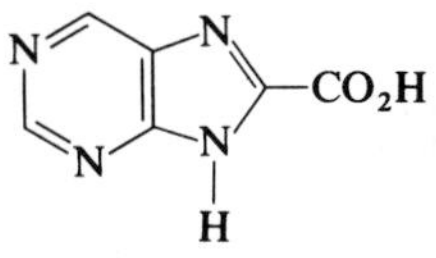	~ 0
Benzoxazole	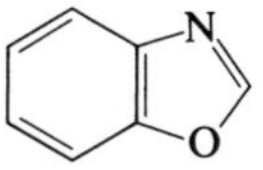	-0.1^h

(*cont.*)

Table 3-6 (cont.)

Compound	Structure	pK_{BH^+}
1,2,4,5-Tetrazine	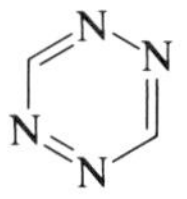	<0
1-Methylindole	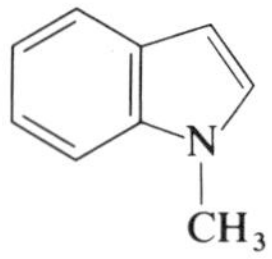	−2.3[i]
2,6-Dichloropyridine	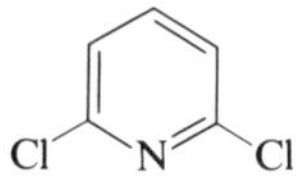	−3.4[j]
Pyrrole		−3.8[i]
Pentachloropyridine	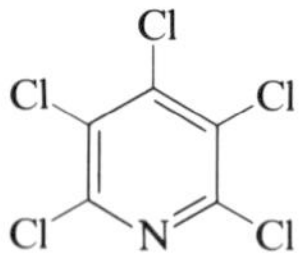	−5.1[j]

[a] Except where noted values are taken from Perrin (26). Precise measurements of the variation of pK_{BH^+} with temperature of a number of the compounds were made by Bates et al. (37–40).
[b] Reference 41. Value given is for internal protonation; external protonation, 7.1. See Section III,C.
[c] For imidazole derivatives, see ref. 43b and also J. Catalan and J. Elguero, J. Chem. Soc., Perkin Trans. 2 p. 1869 (1983).
[d] Temperature effects given in ref. 42.
[e] External protonation; pK_{BH^+} for internal protonation is very high (Section III,C).
[f] Temperature effects given in ref. 43.
[g] Reference 43a.
[h] Reference 64.
[i] C-Protonation; see Table 3-14 for pK_{BH^+} values of other indoles and pyrroles.
[j] Reference 30a.

data available to allow us to draw some very useful conclusions about the variation of ΔH with base strength and about the validity of pK_{BH^+} values determined for weakly basic amines using the H_0 function. Two approaches to this problem are considered here, and both support the idea that pK_{BH^+} values determined by acidity function techniques do, in fact, refer to the dilute aqueous standard state.

Arnett *et al.* determined the calorimetric heats of neutralization of 31 amines in fluorosulfuric acid and found an excellent linear relationship (correlation coefficient 0.992) between these values and the pK_{BH^+} values of the compounds measured either directly in water or indirectly by means of acidity functions [Eq. (3-8)] (*33*).

$$-\Delta H_{FSO_3H} = (1.78pK_{BH^+} + 28.1) \quad \text{kcal mol}^{-1} \qquad (3\text{-}8)$$

Hughes and Arnett measured the heats of neutralization of a number of amines over a very wide temperature range (as much as 170°C) using methanesulfonic acid in tetramethylene sulfone (sulfolane). They found ΔH to be remarkably insensitive to temperature change; that is, the heat capacity of ionization is negligible (*33a*).

A large number of heats of neutralization of amines have been measured in water, either calorimetrically or by measuring the variation of K_{BH^+} with temperature, and these values have been assembled by Christensen *et al.* (*31*). These results are treated here in the following way. First, all the primary amines were selected except for those that had amino groups on adjacent atoms or were α-amino acids. (It was found that these compounds tended to deviate from the relationship given later.) Second, an average value of the temperature coefficient and calorimetric values was taken when both were available. The values of ΔH were then combined with values that can be calculated from the temperature-dependence work of Johnson *et al.* (*30*) for extremely weak bases. A plot of all these heats of neutralization against pK_{BH^+} is given in Fig. 3-1 for

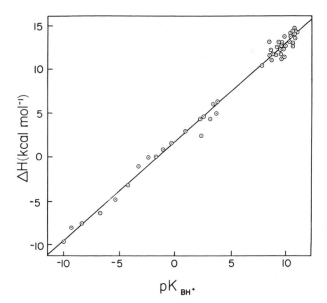

Fig. 3-1 Heats of neutralization of primary amine bases as a function of their pK_{BH^+}.

Table 3-7

THERMODYNAMIC PARAMETERS FOR DISSOCIATION OF THE
CONJUGATE ACIDS OF AMINES IN WATER[a]

Compound	pK_{BH^+}, 25°C	ΔH (kcal mol^{-1})	ΔS (cal deg^{-1} mol^{-1})
Primary amines			
CH_3NH_2	10.64	13.3	−4.1
$C_2H_5NH_2$	10.63	13.7	−2.7
$n\text{-}C_3H_7NH_2$	10.57	13.8	−2.0
Secondary amines			
$(CH_3)_2NH$	10.73	12.0	−8.7
$(C_2H_5)_2NH$	10.93	12.7	−7.3
$(n\text{-}C_3H_7)_2NH$	11.00	13.2	−6.2
Tertiary amines			
$(CH_3)_3N$	9.75	8.8	−15.1
$(C_2H_5)_3N$	10.72	10.3	−14.4
$(n\text{-}C_3H_7)_3N$	10.66	10.5	−14.7

[a] For the sake of internal consistency these values are taken directly from ref. *31*.
They differ in some instances from pK_{BH^+} values quoted elsewhere herein and from
ΔH values used in the construction of Fig. 3-1. See also Larson and Hepler (*32*).

56 primary amine bases. It can be seen that a remarkably linear relationship
exists (correlation coefficient 0.995) over a range of 21 pK_{BH^+} units. The
important aspect of these results and those of Arnett is that there is no
discontinuity or change in slope as one proceeds from the aqueous region to
that of mixed acid. It is interesting that the two most deviant points in Fig. 3-1
are those for the species $2\text{-}H_2NC_6H_4SO_3{}^-$ and $H_2NCH_2CH_2CH_2NH_3{}^+$,
each of which has a charge fairly near the group in question.

Values of ΔH and ΔS for the dissociation of the conjugate acids of a number
of aliphatic amines of similar base strength are given in Table 3-7. It can be
seen that primary amines tend to have the largest values of ΔH (largest
temperature dependence) and tertiary amines the smallest (smallest temperature dependence). The trend to more negative entropies of dissociation that
accompanies the change from primary to tertiary amines has been attributed
to the smaller degree of solvation of the cations that results from their
nitrogen-bound protons being replaced by alkyl groups (*20, 32, 35, 36*). As a
consequence, dissociation to give the neutral amine liberates fewer water
molecules to help solvate the concurrently liberated proton, and this results in
a more negative (i.e., less favorable) entropy of reaction.

It is interesting to note once again the tendency for changes in ΔH and ΔS to
cancel one another out. Ethylamine has almost the same pK_{BH^+} as tri-*n*-

propylamine at room temperature, although its ΔH is 3.2 kcal less favorable. The latter difference is almost exactly canceled at room temperature by the substantially more favorable entropy of dissociation of the primary ammonium ion that is associated with the higher degree of solvation of the latter ion. As a consequence of having the smaller enthalpy of dissociation the tertiary amine becomes the stronger of the two at higher temperatures (e.g., $\Delta pK_{BH^+} = 0.3$ at $75°C$).

B. Medium Effects on Amine Base Strength

Turning to the effect of medium on the base strengths of amines we find a situation that is comparable to, though in some ways simpler than, that of the carboxylic acids discussed in Chapter 2. If amine ionization is expressed in the form of Eq. (3-9), it is generally found that the degree of ionization declines

$$RNH_2 + H_2O \rightleftharpoons RNH_3^+ + HO^- \qquad (3-9)$$

rather sharply as the dielectric constant of the medium falls, since ions are being produced from neutral molecules. We normally express base strength, however, in terms of the acidity of the conjugate acid [Eq. (3-10)], for which

$$RNH_3^+ \rightleftharpoons RNH_2 + H^+ \qquad (3-10)$$

there is no change in ionic charge, and this means that the simple dielectric effect of the solvent is eliminated.

There remains to be considered the effect of differing degrees of solvation of the ammonium ion and the proton and, to a smaller extent, the neutral amine. Bell has shown, in fact, that for a series of primary amines a change of solvent from water to either methanol or ethanol has a fairly small effect on the equilibrium of Eq. (3-10) (44). An increase in pK of roughly one unit is observed, regardless of the strength of the amine. Carboxylic acids, on the other hand, are much weaker in alcohol and other solvents having dielectric constants lower than that of water (Chapter 2, Section II,E,1), with the result that ordinary amines frequently cannot neutralize carboxylic acids in such solvents.

In rationalizing the effect of substituents on the base strengths of amines in solution we rely heavily on the familiar electronic factors, resonance and induction, though, as we have seen, specific solvation effects are also important. In the gas phase, only the electronic effects are present but they are overlaid with a large polarizability effect, since the substituent group is now the only matter with which the positive charge can interact. If the latter factor is not considered, orders of base strength are obtained that seem incongruous (e.g., aniline being more basic than ammonia). However, if the polarizability of the groups attached to the basic site is made approximately the same by a choice of groups of comparable size, then the familiar electronic effects can

Table 3-8

ORDER OF GAS PHASE BASICITIES[a]

tBuNH$_2$ > iPrNH$_2$ > EtNH$_2$ > MeNH$_2$ > NH$_3$

CH$_3$CH$_2$NH$_2$ > FCH$_2$CH$_2$NH$_2$ > F$_2$CHCH$_2$NH$_2$ > NH$_3$ > F$_3$CCH$_2$NH$_2$

4-CH$_3$OC$_6$H$_4$NH$_2$ > C$_6$H$_5$NH$_2$ > 4-ClC$_6$H$_4$NH$_2$ > 3-ClC$_6$H$_4$NH$_2$

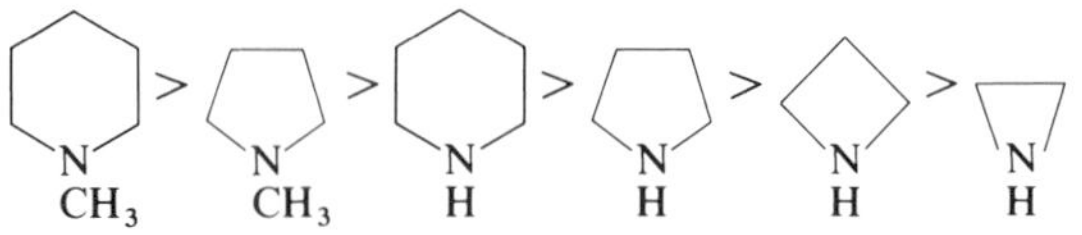

(CH$_3$CH$_2$CH$_2$)$_3$N > (CH$_3$CH$_2$)$_3$N > (CH$_3$)$_3$N > (CH$_3$)$_2$NH > CH$_3$CH$_2$CH$_2$CH$_2$NH$_2$ >

CH$_3$CH$_2$CH$_2$NH$_2$ > CH$_3$CH$_2$NH$_2$ > CH$_3$NH$_2$ > NH$_3$

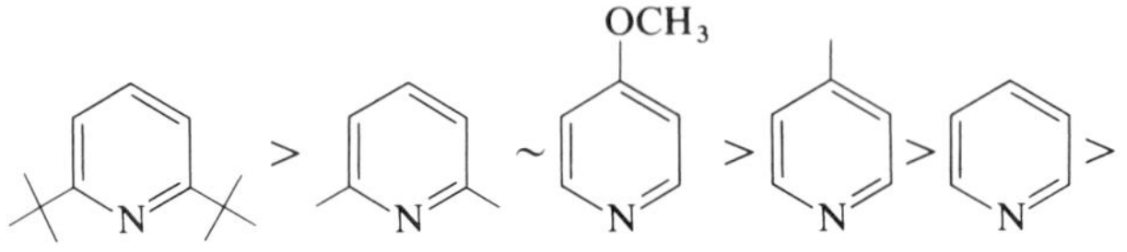

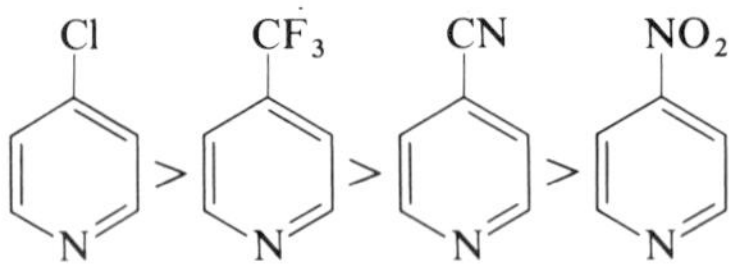

[a] References 46–48.

usually be seen. Thus, as Kebarle has pointed out, aniline may be a much stronger base in the gas phase than ammonia but it is much weaker than cyclohexylamine (45). Phosphine, PH$_3$, resembles ammonia in that both methyl and phenyl substitutions increase gas phase base strength, although the effect of a phenyl group is particularly large in the phosphine case (45a).

The order of gas phase basicity of amines shown in Table 3-8 reflects a mix of inductive, resonance, and polarizability effects.

Although aniline and most of its derivatives are protonated on nitrogen in the gas phase, m-toluidine (8) and m-anisidine (9) are preferentially protonated on carbon (49).

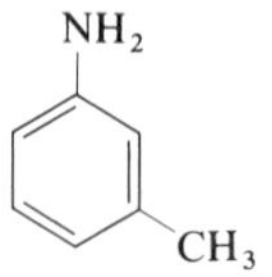

8 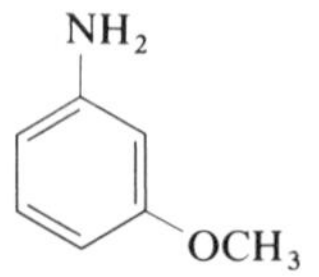   9

The extra conjugative stabilization that a methoxyl (or methyl) group can provide is presumably enough to tip the equilibrium from the N-protonated form to a C-protonated form, as illustrated in Eq. (3-11).

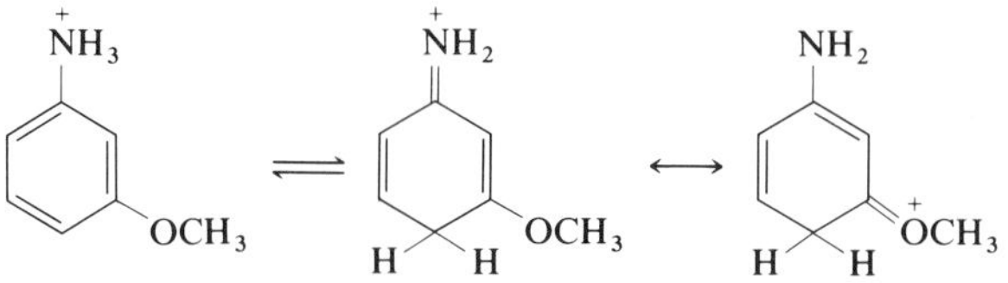

$$\tag{3-11}$$

It is interesting that the most basic of the pyridines listed in Table 3-8 is the 2,6-di-*tert*-butyl compound, in contrast to the situation in solution. This compound has an anomalously *low* basicity in aqueous alcohol, the order of base strength in that medium being 2,6-dimethylpyridine > 2,6-diisopropylpyridine > 2-methylpyridine > 2-ethylpyridine > 2-isopropylpyridine > 2-*tert*-butylpyridine > pyridine > 2,6-di-*tert*-butylpyridine (*50*). This effect has been attributed to the presence of a pair of *tert*-butyl groups in the latter compound preventing the pyridinium proton from forming a hydrogen bond to the molecules of solvent. Indeed, it appears that a single water molecule would be sufficient to raise the base strength of this compound substantially (*51*).

C. Strong Amine Bases

There are two groups of amino compounds that can be considered strong bases in water; that is, their pK_{BH^+} values are greater than ~ 14. These are (*a*) the hindered 1,8-diaminonaphthalenes and (*b*) those bicyclic diamines that can undergo "internal" protonation.

The unusually high basicity of 1,8-bis(dimethylamino)naphthalene (**14**, Table 3-9) was first noted by Alder *et al.* in 1968 (*52*). Its pK_{BH^+} of 12.1 (*2*) makes it more than seven units more basic than the monodimethylamino compound **11** (Table 3-9). Its high basicity [and relatively low heat of neutralization (*33a*)] have been attributed to the large degree of steric strain that is relieved by formation of the protonated form **17** in which the added

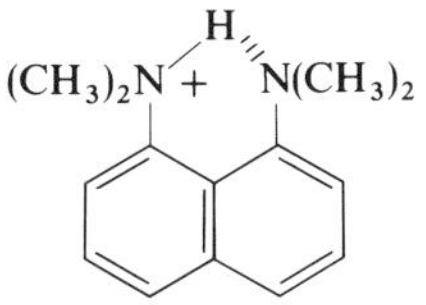

17

proton links the two nitrogen atoms by hydrogen bonding. Since the 1,8-

Table 3-9

BASICITY OF AMINONAPHTHALENES IN WATER AT 25°C

Compound	Number	pK_{BH^+}	Ref.
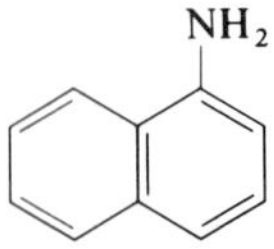	**10**	3.9	*54*
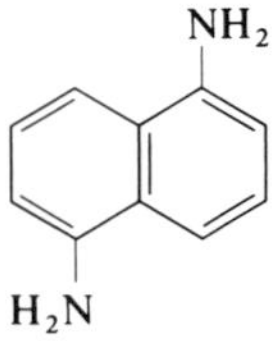	—	4.4	*55*
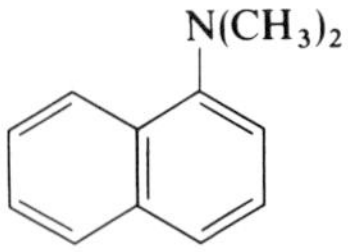	**11**	4.9	*56*
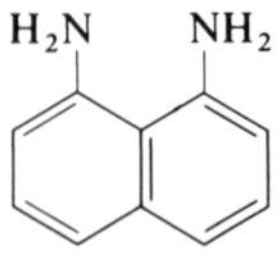	**12**	4.9	*55*
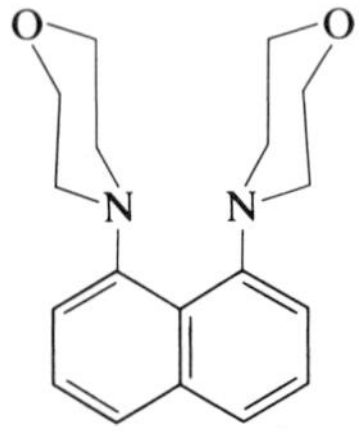	**13**	7.5	*2*
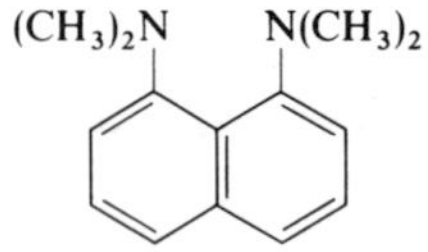	**14**	12.1	*2, 52*
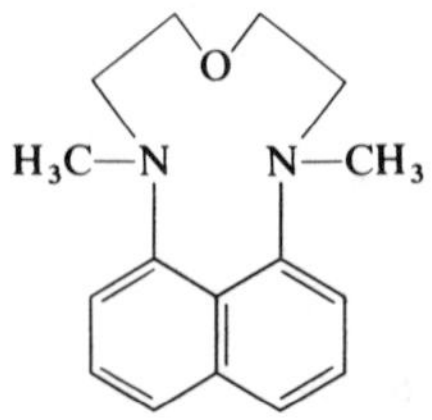	—	12.9	*2*

Table 3-9 (*cont.*)

Compound	Number	pK_{BH^+}	Ref.
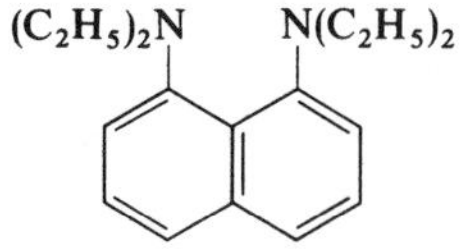 (C₂H₅)₂N N(C₂H₅)₂ [structure]	—	~13.3	2
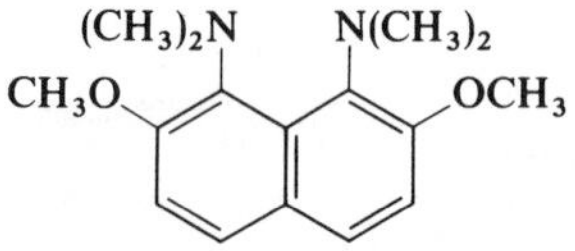 (CH₃)₂N N(CH₃)₂ CH₃O OCH₃ [structure]	15	~16.1	2, 53, 57
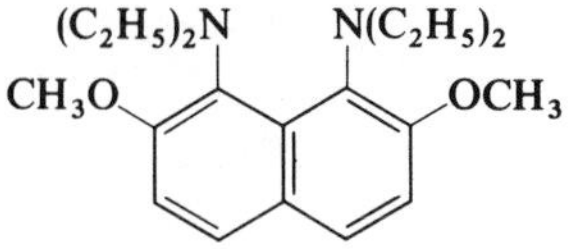 (C₂H₅)₂N N(C₂H₅)₂ CH₃O OCH₃ [structure]	16	~16.3	2, 53, 57

diamino compound is only slightly more basic than the monoamino compound it is clear that the presence of alkyl groups at nitrogen is required.

Although **14** is unusually basic for an aromatic amine, it does not quite qualify as a strong base in water. Increasing the size of the alkyl groups generally causes basicity to increase, although compound **13** is an exception in this respect. The introduction of groups such as alkoxyl at the 2 and 7 positions causes further increases in basicity, and the pK_{BH^+} of compounds **15** and **16** clearly shows them to be strong bases in water. Their pK_{BH^+} values were estimated by comparing their degrees of deprotonation with that of **14** in an aqueous alkaline DMSO mixture (*53*). The effect of 2,7-disubstitution is to force the nitrogen lone pairs of electrons into closer proximity, thus increasing strain in the neutral compound without adversely affecting the intramolecular hydrogen bond in the cation.

Compound **16a** is also stronger than **14**. In this case the geometrical constraints are such as to shorten the N–N distance to direct the nitrogen lone pairs toward one another to give an approximately linear hydrogen bond in the cation (*53a*).

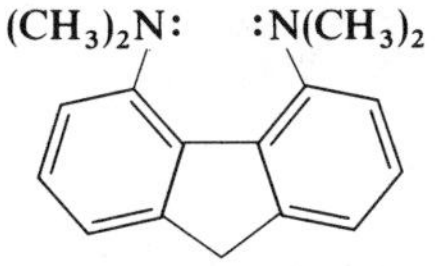

$(CH_3)_2N:$ $:N(CH_3)_2$

[structure]

16a

The high stabilities of the conjugate acids of compounds **15** and **16** are reflected not only in their high pK_{BH^+} values but also in their reduced rate of proton exchange with other bases such as hydroxide ion $(57-59)$. In the case of **16** the half-life of the latter reaction is in the range of minutes (57).

In 1968 Simmons and Park prepared a series of macrobicyclic diamines and showed that their diprotonated forms could exist in three stereoisomeric forms (60). The equilibrium between the three forms, designated *out–out* or o^+o^+, *out–in* or o^+i^+, and *in–in* or i^+i^+, is shown in Eq. (3-12). The o^+i^+ form (**19**) is

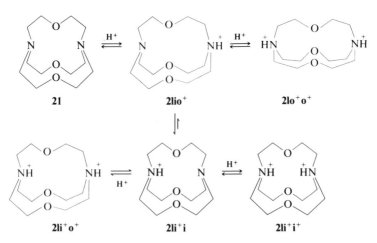

$$\text{H—N} \qquad \text{N—H} \rightleftharpoons \text{H—N} \quad \text{H—N} \rightleftharpoons \text{N—H} \quad \text{H—N} \qquad (3\text{-}12)$$

18   19   20

the least stable isomer, with preference for o^+o^+ (**18**) or i^+i^+ (**20**) being determined by the length of the methylene chains that link the two basic centers.

Soon thereafter Cheney and Lehn prepared trioxa derivatives of such diamines, which they called *cryptands* (61). In the case of the [1.1.1] cryptand **21**, the pertinent equilibria are shown in Scheme 3-1, although it should be noted that transfer of protons in and out of the cavity is extremely slow, and indeed, the internally monoprotonated species **2li$^+$i** cannot in practice be deprotonated without destroying the cryptand. However, measurements of protonation rates allow a pK_{BH^+} estimate of 17.8 or greater to be made for the internal protonation of **21** (3).

Scheme 3-1 Equilibrium between the [1.1.1] cryptand **21** and its internal (i$^+$) and external (o$^+$) protonated forms.

The external protonation of **21** to give **2lio⁺** is a fairly rapid process, with a pK_{BH^+} of 7.1, considerably lower than that of most tertiary aliphatic amines. This indicates that **21** indeed exists in the in–in form shown in Scheme 3-1. If the neutral molecule had an in–out or out–out conformation, a more normal value of K_{BH^+} would be expected for external protonation. The much greater stability of **2li⁺i** than **2lio⁺** means that the latter isomerizes slowly to the former on standing.

An even tighter inside-protonated structure (**22**) is formed from 1,6-diazabicyclo[4.4.4]tetradecane, a waxy solid compound with an ultraviolet absorption that is unusually strong for a saturated amine. Alder and Sessions have shown that the proton in this case can neither be inserted nor removed by simple proton transfer (*4, 5*). Heating the neutral compound to 200°C with *p*-toluenesulfonic acid is ineffective, but allowing it to stand in aqueous sulfuric acid, particularly in the presence of 1-equivalent oxidants, produces **22**. It thus appears that radical-cation intermediates are involved in the conversion. In this connection the radical-cation **23** is indefinitely stable in the solid state (*4*). Again the outside-protonated isomer of **22** is much less stable (pK_{BH^+} = 6.5) though it is, of course, much more easily formed.

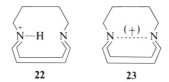

D. Polyaminoalkanes

The strength of polycarboxylic acids is considered in Section II,E,3 of Chapter 2, where it is seen that nearby carboxyl groups influence one another's acidities to a considerable degree. As a result of polar effects, symmetry, and hydrogen bonding, which in some cases is sterically enhanced, the extent of the first dissociation of dicarboxylic acids can be markedly greater and the second dissociation markedly less than that of an isolated carboxyl group.

What effects might be expected to be present in diaminoalkanes? First, the symmetry effect (Chapter 2, Section II,E,3) should increase the first pK and decrease the second pK by 0.3 unit in both cases. Second, the inductive effect of one amino group on the other should cause a fairly modest decrease in the first pK that drops off with distance. A more dramatic effect should be seen on the second dissociation, since the polar effect of the ammonio group, $-NH_3^+$, is quite substantial. Finally, there may be the possibility of internal hydrogen bonding between amino and ammonio groups in the monocation. From the results shown in Table 3-10 it can be seen that these expectations

Table 3-10

BASE STRENGTHS OF POLYAMINOALKANES IN WATER AT 25°C[a]

Compound	$pK_1{}^b$	$pK_2{}^c$	$pK_3{}^d$
$[CH_3(CH_2)_nNH_2]$	10.6–10.7	—	—
$H_2NCH_2CH_2NH_2$	9.93	6.85	—
$H_2N(CH_2)_3NH_2$	10.4	8.5	—
$H_2N(CH_2)_4NH_2$	10.7	9.2	—
$H_2N(CH_2)_6NH_2$	10.9	9.8	—
$H_2N(CH_2)_8NH_2$	11.0	10.1	—
$(CH_3)_2C(CH_2NH_2)_2$	10.4	8.3	—
$(\pm)$-$CH_3CHNH_2CHNH_2CH_3$	10.0	6.9	—
meso-$CH_3CHNH_2CHNH_2CH_3$	10.0	6.9	—
$(CH_3)_2CNH_2CNH_2(CH_3)_2$	10.1	6.6	—
$HC(CH_2NH_2)_3$	10.5	8.9	6.9
$NH_2CH_2CHNH_2CH_2NH_2$	9.5	7.9	3.6
$C(CH_2NH_2)_4{}^e$	9.9	8.2	5.7
$N(CH_2CH_2NH_2)_3{}^f$	10.17	9.47	8.43
$(CH_2NHCH_2CH_2NH_2)_2{}^{f,g}$	9.68	9.09	6.58
$HN(CH_2CH_2NHCH_2CH_2NH_2)_2{}^{f,h}$	9.85	9.27	8.19

[a] Taken from Perrin (26), except where noted.
[b] pK_{HA} of monocation (pK_{BH^+} of amine).
[c] pK_{HA} of dication.
[d] pK_{HA} of trication.
[e] $pK_4 = 3.0$.
[f] Reference 65.
[g] $pK_4 = 3.28$.
[h] $pK_4 = 5.08$, $pK_5 = 3.43$.

are realized, except that hydrogen bonding in the monocations appears to be slight or absent.

Internal hydrogen bonding in the monocation would both increase the first and decrease the second pK. Although some quite large effects on the second dissociation constant are observed (and on the third in triamino compounds), it seems likely that these are due to the polar effect of ammonio since the first dissociation constants appear to be normal. Moreover, there is no evidence for the existence of the special effect of alkyl substitution, which causes anomalies in the case of polycarboxylic acids.

IV. Amides

We have seen that the range of base strengths spanned by amines is very great. A few amines are strong, most are weak, and some are very weak indeed, which provides a continuum that allows us to assign pK_{BH^+} values to the latter group with some confidence. The only other functional group that approaches the amino group in basicity is the carboxamido group.

Two problems arise with regard to the protonation of amides. First is the question of the structure of the conjugate acid, the consensus favoring the O-protonated form **24a** rather than the N-protonated form **24b**. (This subject is considered in more detail in Chapter 5.) Second, except for unusual bases such as the quinuclidone **25**, in which resonance is severely curtailed and which has

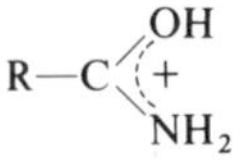

24a	**24b**	**25**

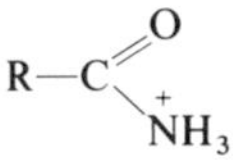

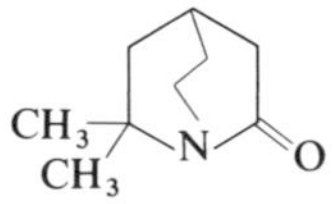

a pK_{HA^+} of 5.33 (*66*) the most basic amides are not quite basic enough for their pK_{BH^+} values to be accurately measured in water, thus presenting a problem with respect to anchoring an amide acidity function in the standard state of dilute aqueous solution. The necessity for such a scale became apparent soon after the first quantitative measurements of amide basicity were made (*67*), when it was found that plots of $\log([BH^+]/[B])$ against H_0 give slopes that are much less than the theoretical value of unity (*68*). That is, amides do not follow the H_0 function.

In 1964 Yates *et al.* (*68*) established an amide acidity scale, designated H_A, but were frustrated in their efforts to anchor it in water because their most basic amide, 2-pyrrolecarboxamide (**26**) requires 24 wt % sulfuric acid to attain half-protonation and so it is not detectably protonated in the aqueous region. They used 2- and 4-nitroaniline to link this compound to the aqueous standard state, since there were indications that the activity coefficient ratios of these species [Eq. (3-4)] were similarly affected in the dilute aqueous region, in effect making H_0 and H_A coincident there. The various acidity scales that have come into existence, however, all show signs of divergence when the concentration of sulfuric acid rises above about 1%, and it was suggested by Bunnett and Olsen that H_0 and H_A were not in fact coincident in the dilute acid region, with the consequence that the amide basicities produced by Yates *et al.* were ~0.3 pK unit too negative (*69*). That is, amides are slightly more basic than use of the original H_A scale would suggest.

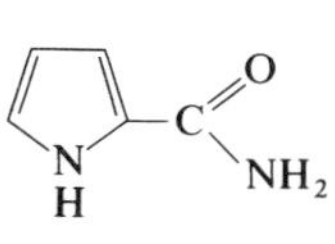

26	**27**

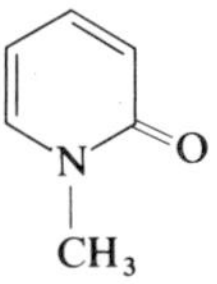

Table 3-11

pK_{BH^+} VALUES OF AMIDES IN WATER AT 25°C[a]

Compound	pK_{BH^+}
1-Methyl-2-pyridone	0.38
5-Chloro-2-pyridone	0.03
2-Pyridone-5-carboxylic acid	−0.64
Pyrrole-2-carboxamide	−0.90
3-Hydroxyphenylurea	−0.98
N-Phenylacetamide	−0.99[b]
4-Methoxybenzamide	−1.11
Benzamide	−1.35[c]
3,4,5-Trimethoxybenzamide	−1.49
Furan-2-carboxamide	−1.76[c]
Thiophen-2-carboxamide	−1.88[d]
3-Nitrobenzamide	−2.09
3,5-Dinitro-4-methylbenzamide	−2.36
3-Iodothiophene-2-carboxamide	−2.40[d]
Thioacetamide	−2.51[e]
2,3,6-Trichlorobenzamide	−2.97
Thiobenzamide	−3.20[e]
2,4-Dichloro-3,5-dinitrobenzamide	−3.40
2,4,6-Trinitrobenzamide	−3.75[f]

[a] Taken except where noted from refs. 68, 70, and 72 and corrected where necessary (see text).
[b] Recalculated from data of ref. 73.
[c] Recalculated from data of ref. 67.
[d] Reference 78.
[e] Reference 79.
[f] The extrapolative procedure of Cox and Yates (30a), discussed in Section V, gives a value of −3.68 for this compound.

In 1977 Edward and Wong found that 1-methyl-2-pyridone (27) could serve as an anchor since it is sufficiently basic that a plot of $\log([BH^+]/[B])$ − $\log[H^+]$ against $[H^+]$ can be accurately extrapolated to infinite dilution; the pK_{BH^+} thus obtained is 0.38 (70). Excellent overlap between this compound and 2-pyrrolecarboxamide was observed and, indeed, the H_A scale and the pK_{BH^+} of the amide indicators used in its construction were found to be 0.3 unit too negative. The basicity of a number of amides is given in Table 3-11, corrected where necessary to the modified H_A scale.

The basicity of aliphatic amides is not known with as much precision as those of most of the compounds in Table 3-11, but acetamide and propionamide are half-protonated in 13.0 and 16.6% sulfuric acid, respectively (71, 72), which if they follow H_A would place their pK_{BH^+} values in the range of -0.3 to -0.5. [Grant et al. have determined the pK_{BH^+} values of a number of aliphatic amides by a conductimetric method; their values are acetamide, -0.62; N-methylacetamide, -0.42; N,N-dimethylacetamide, -0.28; N,N-dimethylformamide, -1.2; urea, 0.05; and thiourea, -0.9 (83).] N-Phenylacetamide (acetanilide) is somewhat weaker, as might be expected, being half-protonated in 23.9% sulfuric acid (73). This compound has a near unit ionization slope using H_A and gives a value of approximately -1.0 for pK_{BH^+}.

Urea, which is the dicarboxamide of carbonic acid, might have been expected to be a moderately basic compound, given the additional resonance possibilities in the cation **28**. Although salts of urea and its derivatives can be readily isolated, the pK_{BH^+} values are little different from those of aliphatic amides, for which salts can also be usually prepared (74). Most ureas and aliphatic amides are more basic than benzamide ($pK_{BH^+} = -1.35$) by roughly one or so pK units, which means that they have no detectable basicity in water.

$$\underset{\textbf{28}}{\overset{\displaystyle\overset{OH}{\overset{\|\!\!\|}{C}}\;\;+}{H_2N^{\diagup}\;\;{}^{\diagdown}NH_2}}\qquad\qquad\underset{\textbf{29}}{\overset{\displaystyle\overset{OR}{\overset{|}{C}}}{H_2N^{\diagup}\;\;{}^{\diagdown}NH}}$$

Presumably the rather low basicity of urea is due to the high degree of resonance present in the neutral molecule. This would be in line with the high basicity observed for the O-alkylated ureas **29**. These compounds are $\sim 10^{10}$ times as basic as urea, even though their conjugate acids should have virtually the same amount of resonance stabilization as that of urea (75–77).

The H_A scale has been extended to fuming sulfuric acid using oxypyrimidine monocations as indicator bases (80). In 17% fuming sulfuric acid, H_A is -9.0, some four units less negative than H_0 (81).

V. Oxygen Bases

Amides, which were discussed in the previous section, differ from other compounds containing the carbonyl group in producing cations in which the positive charge is largely on nitrogen rather than oxygen. They also differ in that their most basic members are sufficiently basic (just) that their pK_{BH^+} values can be measured in what is essentially aqueous solution, thus giving rise to an acidity scale, designated H_A, that is firmly anchored in the standard state. None of the other common oxygen bases, aldehydes, ketones, carboxylic acids, esters, alcohols, ethers, and so on, are sufficiently basic, even when larded with electron-donating substituents, to be detectably protonated in solutions containing less than 20% sulfuric acid.

Can the H_0 function be used to measure the basicity of these compounds, or do they deviate in the same way as amides? It turns out that many oxygen bases deviate less than do amides in this respect but they, nonetheless, do not satisfy the first experimental test of adherence to an acidity function: plots of $\log([BH^+]/[B])$ for these compounds against H_0 do not give lines of unit slope, though they are, in fact, usually linear. For most carbonyl compounds the slopes of such lines are less than unity (H_0 as abscissa), which means that using H_0 at half-protonation leads to a pK_{BH^+} value that is too negative.

The first successful attempt to deal with this problem was made in 1966 by Bunnett and Olsen (69). They derived Eq. (3-13), which makes use of the linear relationship usually found when $\log([BH^+]/[B])$ is plotted against H_0.

$$\log([BH^+]/[B]) + H_0 = \phi(H_0 + \log[H^+]) + pK_{BH^+} \qquad (3-13)$$

The assumption underlying this derivation is that the activity coefficient term of the base B being studied and that of a primary aniline indicator An are proportional to one another [Eq. (3-14)].

$$\log\left(\frac{f_B f_{H^+}}{f_{BH^+}}\right) = (1 - \phi)\log\left(\frac{f_{An} f_{H^+}}{f_{AnH^+}}\right) \qquad (3-14)$$

The magnitude of the parameter ϕ is an indication of the extent of deviation from the acidity function H_0, since if ϕ is zero (as it must be for primary anilines) Eq. (3-13) reduces to the familiar equation for the Hammett acidity function [Eq. (3-5)]. For most carbonyl compounds ϕ is positive, as can be seen when their values of $\log([BH^+]/[B]) + H_0$ are plotted against the quantity $H_0 + \log[H^+]$. The intercept of such a plot is pK_{BH^+}.

There have been a number of other attempts to extrapolate ionization data from mixed acid media to the dilute aqueous state. Perhaps the simplest of these is due to Yates and McClelland (84), who used the empirical relationship shown in Eq. (3-15).

$$\log([BH^+]/[B]) = -mH_0 + pK_{BH^+} \qquad (3-15)$$

If a plot of experimental values of $\log([BH^+]/[B])$ for an oxygen base against H_0 is linear, as it very often is, and if this linearity extends right through the region of undetectable protonation, that is, to the aqueous region, then pK_{BH^+} can be readily obtained using Eq. (3-15). (The quantity m is analogous to ϕ in the Bunnett–Olsen equation in that it represents departure from adherence to H_0.) This method is very easy to apply and, provided that the ionization data are fairly precise, undoubtedly gives pK_{BH^+} values that are more reliable than those obtained by simply taking H_0 at half-protonation, a common practice in earlier days. There are indications that in certain cases extrapolations depart from linearity at low acid concentrations (87), however, and the more elaborate Bunnett–Olsen extrapolative procedures, or those due principally to Cox and Yates and to Marziano et al., are preferable if more precise estimates of pK_{BH^+} are desired (82, 88–92).

In 1973 Marziano's group introduced the idea of using a generalized activity coefficient term and subsequently developed the M_c function, which is derived using a large number of bases of differing chemical type. Function M_c, which is equal to $-\log(f_{B^*}f_{H^+}/f_{B^*H^+})$, where B^* is some standard base, effectively takes the place of the term $H_0 + \log[H^+]$ in the Bunnett–Olsen equation, but since it is not based on a single set of indicators it possibly provides a more general relationship [Eq. (3.16)]. A single quantity, M_c, can

$$\log([BH^+]/[B]) - \log[H^+] = -nM_c + pK_{BH^+} \qquad (3\text{-}16)$$

be used in this way for different kinds of compounds because their activity coefficient ratios are linearly related (93, 94). The quantity n in Eq. (3-16) varies, of course, from compound to compound.

Cox and Yates (82, 88, 89) used an approach similar to that of Marziano et al. but used the term *excess acidity*, symbol X, in their equation [Eq. (3.17)], the slope parameter m^* being clearly analogous to n in Eq. (3.15).

$$\log([BH^+]/[B]) - \log[H^+] = m^*X + pK_{BH^+} \qquad (3\text{-}17)$$

The idea of excess acidity, first put forward by Perrin (95), is attractive in that it represents the difference between the observed acidity of a solution and that which the system would have if it were ideal, that is, if all activity coefficients were unity. The X function is available for a considerable number of acid systems.

Which system should one use? Cox and Yates in their review of acidity functions (89) have addressed this question, as have Arnett and Scorrano (96) and Kresge et al. (87), who examined the accuracy of the extrapolation procedure in quite dilute acid solution. It has been shown that the Bunnett–Olsen and excess acidity methods give essentially the same pK_{BH^+} values in sulfuric acid and that m^* in the one is equal to $1 - \phi$ in the other (97). The method of choice for other acid systems, however, will depend on the availability of X data for the excess acidity method or $H_0 + \log[H^+]$ data for the Bunnett–Olsen method.

Cox and Yates stated that in less well studied systems, where only H_0 may be available, the Bunnett–Olsen method is still unsurpassed (89). It should be noted that all extrapolative methods depend on rather precise ionization data being available. If the ionization data are imprecise, as they will be for compounds that undergo small spectral or other changes on protonation, the extrapolative technique is likely to fail badly. In these cases if the aqueous

pK_{BH^+} is desired it may be better to determine the acidity at which the compound is approximately half-protonated and then choose an acidity function that on structural grounds seems the most appropriate, that is, the function whose indicator members most resemble the compound in question. If an astute choice is made, then the value of the acidity function at half-protonation should be close to the pK_{BH^+} of the compound in question.

A. Carbonyl and Carboxyl Bases

Early attempts to measure the basicity of aldehydes, ketones, and carboxylic acids were based on the assumption that these compounds were Hammett bases, that is, that their ionization is governed by H_0. It was realized that their apparent ionization slopes were not linear, but it was thought that this might be due to medium effects on the spectra of the neutral compounds and their conjugate acids. Such medium effects can be important, and attempts to compensate for them have been made using characteristic vector analysis by Reeves (98), Edward (99, 100), Wold (100), and others (100a).

It became clear in the mid-1960s that carbonyl and carboxyl compounds, like amides, are not governed by the H_0 function, although many appear to deviate less than do amides in this respect. Amides, however, have the advantage of having their most basic members protonated in what is essentially water, and hence a scale can be constructed using the overlap procedure that firmly anchors it in the standard state. The most basic carbonyl compounds that have been studied, apart from a few special cases such as cycloheptatrienone (tropone), $pK_{BH^+} \approx 0$ (26, 101), are the polyalkoxy-benzophenones. But even these compounds require 50% sulfuric acid to effect half-protonation. Bonner and Phillips, who constructed a scale based on the protonation of benzophenones using the overlap procedure, were thus forced to attach it to the H_0 scale at 60% sulfuric acid (102). They showed that their scale was congruent with H_0 down to 40% sulfuric acid, the limit of measurement in their systems, but deviated from H_0 at higher concentrations of acid. Since a real anchor in the standard state is lacking, all we can safely infer from this is that the benzophenone scale and H_0 are parallel in the region of 40 to 60% sulfuric acid.

Listed in Table 3-12 are a large number of aldehydes, ketones, carboxylic acids, and esters in order of decreasing base strength in aqueous sulfuric acid. If none of these compounds' ionization curves intersect on the long descent to the aqueous state, then their pK_{BH^+} values follow the same order.

Can the extrapolative techniques described in the previous section be used to evaluate pK_{BH^+} for the compounds in Table 3-12? This has been done for a number of benzophenones by Cox and Yates using both the Bunnett–Olsen and excess acidity methods (82). The pK_{BH^+} values that they list are close to

Table 3-12

ORDER OF BASE STRENGTHS OF CARBONYL AND CARBOXYL COMPOUNDS
IN AQUEOUS SULFURIC ACID

Compound	Weight % H_2SO_4 to half-protonate	Ref.
2,2′,4,4′-Tetramethoxybenzophenone	50.0	102,
2,4,4′-Trimethoxybenzophenone	51.4	102
Mesityl oxide	55.9	103
4,4′-Dihydroxybenzophenone	57.6	104
4,4′-Dimethoxybenzophenone	59.5	104
4-Hydroxy-3,3′,4′-trimethoxybenzophenone	60.4	102
2,4,6-Trimethylbenzaldehyde	62.0	106, 107
4-Hydroxyacetophenone	62.4	105
4-Methoxyacetophenone	63.0	105
4-Ethoxyacetophenone	63.8	105
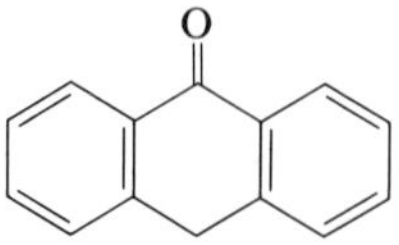	64.7	104
2,4-Dihydroxybenzophenone	64.9	102
4-Methoxybenzophenone	65.7	102
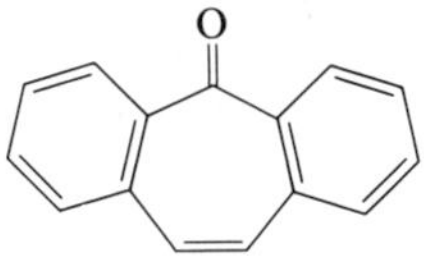	66.6	104
4-Hydroxybenzophenone	66.9	102
2-Cyclohexen-1-one	68.0	99
4-Methylacetophenone	68.7	105
4-Chloro-4′-methoxybenzophenone	68.9	102
4-Methoxybenzaldehyde	69.2	106
4-Ethylacetophenone	69.7	105
4,4′-Dimethylbenzophenone	70.0	104
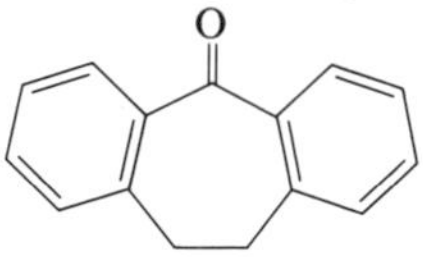	70.2	104
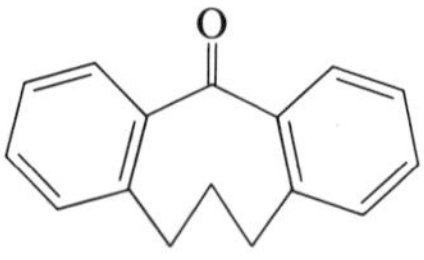	71.1	104
2-Acetylnaphthalene	71.1	105
3-Chloro-4-methoxybenzophenone	72.3	102
3-Methylacetophenone	73.0	105
4-Fluoroacetophenone	73.4	105

(cont.)

Table 3-12 (*cont.*)

Compound	Weight % H_2SO_4 to half-protonate	Ref.
Crotonic acid	73.7	108
Benzophenone	73.8	104
Acetophenone	74.0	105
4-Methylbenzaldehyde	75.2	106
Cycloheptanone	75.7	109
Cyclooctanone	75.7	109
Cyclohexanone	76.3	109
Crotonaldehyde	76.5	108
2-Methylbenzaldehyde	76.5	106
4-Chloroacetophenone	76.6	105
4-Bromoacetophenone	76.6	105
3-Hydroxyacetophenone	77.2	105
4-Ethoxybenzoic acid	77.3	110
Cyclopentanone	77.4	109
4-Chlorobenzophenone	77.6	104
Fluorenone	77.6	104
4-Hydroxybenzoic acid	77.7	110
4-Methoxybenzoic acid	77.8	110
3-Methoxyacetophenone	78.0	105
3-Ethoxyacetophenone	78.0	105
2-Butanone	78.0	109
Acetone	78.3	109
3-Chlorobenzophenone	78.6	102
3-Methyl-2-butanone	78.9	109
3,3-Dimethyl-2-butanone	79.3	109
3-Pentanone	79.3	109
3-Bromoacetophenone	79.4	105
4-Phenylbenzoic acid	79.5	110
4-*tert*-Butylbenzoic acid	79.5	110
4-Methylbenzoic acid	79.6	110
4-Ethylbenzoic acid	79.6	110
4,4'-Dichlorobenzophenone	80.0	102
3-Chloroacetophenone	80.3	105
3-Methylbenzaldehyde	80.6	106
β-Naphthoic acid	80.8	110
Benzaldehyde	80.9	106
3-Methylbenzoic acid	81.4	110
Methyl 4-methylbenzoate	82.0	111
Benzoic acid	82.1	110
4-Chlorobenzaldehyde	82.1	106
4-Fluorobenzoic acid	82.4	110
2-Methylbenzoic acid	83.0	110
3-Methoxybenzoic acid	83.4	110
3-Hydroxybenzoic acid	83.6	110

Table 3-12 (cont.)

Compound	Weight % H_2SO_4 to half-protonate	Ref.
4-Chlorobenzoic acid	83.6	110
4-Iodobenzoic acid	83.6	110
Methyl 2-methylbenzoate	84.0	111
3-Ethoxybenzoic acid	84.3	110
3-Fluorobenzoic acid	84.7	110
3-Iodobenzoic acid	84.9	110
3-Bromobenzoic acid	85.0	110
3-Chlorobenzaldehyde	85.2	106
Methyl benzoate	85.4	111
3-Chlorobenzoic acid	85.6	110
3-Nitroacetophenone	85.7	105
Methyl 2,6-dimethylbenzoate	86.0	111
4-Nitroacetophenone	87.3	105
3-Nitrobenzoic acid	87.6	110
4-Nitrobenzaldehyde	91.5	106
Cyclobutanone	>96	109
3,3'-Dinitrobenzophenone	99.3	104
4,4'-Dinitrobenzophenone	99.6	104

those given by Bonner and Phillips for the more basic alkoxybenzophenones ($pK_{BH^+} \approx -3.5$), which are directly attached to the H_0 scale (as described earlier), but they are more than one pK unit less negative for the less basic halo ketones. The extrapolative methods produce pK_{BH^+} values that have beguilingly small standard errors; they must nonetheless be treated with caution since a number of rather serious anomalies emerge when pK_{BH^+} values are obtained by extrapolating different sets of experimental data, even when the same extrapolative technique is used. Thus, the value for benzophenone using the excess acidity approach can be either -3.62 ± 0.22 or -4.77 ± 0.11, depending on which set of data is chosen (82). The discrepancy here arises largely because of difficulties in correcting for medium effects on spectra when converting absorbance data to concentration data.

The reasonableness of the pK_{BH^+} values derived by various means for the benzophenone series can be assessed in another way; those with meta or para substituents should give a linear Hammett plot (e.g., see refs. 27 and 112–115). The conjugation present in the conjugate acid of a compound such as 4-methoxybenzophenone [Eq. (3-18)] makes it uncertain whether σ, σ^+, or, more likely, a combination of the two should be used, but for the present purposes this distinction need not prevent us from applying this test. When the pK_{BH^+} values obtained by Bonner and Phillips (102) are plotted against σ and σ^+, reasonably linear relationships are obtained in each case with correlation coefficients of 0.976 and 0.974, respectively. This, of course, does not mean that the absolute pK values are correct but merely that they are in a reasonable relationship with one another. (The pK_{BH^+} values, anchored to the H_0 scale, range from -4.41 for the 4,4'-dimethoxy to -6.46 for the 4,4'-dichloro compound.)

$$(3\text{-}18)$$

When the pK_{BH^+} values derived by the Bunnett–Olsen and excess acidity methods are subjected to the same test, there is much more scatter in the Hammett plots, as reflected in the unsatisfactory correlation coefficients obtained: Bunnett–Olsen, 0.839 for σ and 0.843 for σ^+; excess acidity, 0.747 for σ and 0.716 for σ^+ (115a). [The pK_{BH^+} values range from -3.68 (Bunnett–Olsen) or -3.55 (excess acidity) for the 4,4'-dimethoxy compound to -5.55 (Bunnett–Olsen) or -4.82 (excess acidity) for the 3-chloro compound (82).] The poor correlation with the Hammett equation suggests that the experimental error in determining pK values by extrapolative techniques is much greater than the statistical treatment would suggest, though this is not to say that any individual value is necessarily farther from the true pK than those obtained by the overlap–acidity function technique described earlier.

It should be remembered that medium effects on pK are of no relevance with regard to these discrepancies, since the object of both the acidity function and extrapolative procedures is to obtain a standard state pK for each compound, that is, pK_{BH^+}. It is not infrequently found that different values of pK are obtained for the same compound using different aqueous acid systems (e.g., aqueous sulfuric and aqueous perchloric acids), and it is easy to forget that such disagreements cannot legitimately be ascribed to medium effects, although these are, of course, largely responsible for the errors in concentration measurements in the first place. In this connection it would be helpful if authors did not use such terms as "pK values in sulfuric acid" when what are actually being referred to are pK values in water determined by means of measurements in aqueous sulfuric acid mixtures.

Because of the uncertainties outlined, pK_{BH^+} values are not given for all the compounds assembled in Table 3-12. Instead, listed in Table 3-13 are the pK_{BH^+} values of a few typical carbonyl and carboxyl compounds determined by Edward and Wong (99) by the Bunnett–Olsen procedure and using characteristic vector analysis to correct for medium effects on spectra.

Table 3-13

pK_{BH^+} VALUES OF SELECTED
CARBONYL AND CARBOXYL
COMPOUNDS IN WATER AT
25°C[a]

Compound	pK_{BH^+}
2-Cyclohexen-1-one	−3.11
Benzaldehyde	−3.91
Acetophenone	−4.32
Benzophenone	−4.97
Fluorenone	−5.98
Benzoic acid	−4.73
Cinnamic acid	−4.31
Ethyl benzoate	−6.16
Ethyl cinnamate	−5.06

[a] Measurements of Edward and
Wong (99) in aqueous sulfuric
acid with medium effects corrected by means of characteristic
vector analysis and using the
Bunnett–Olsen procedure.

Dicinnamalacetone (1,9-diphenyl-1,3,6,8-nonatetraen-5-one), **30**, has roughly the same basicity as 2,2′,4,4′-tetramethoxybenzophenone, the most

$$(C_6H_5CH=CH—CH=CH—)_2C=O$$

30

basic ketone listed in Table 3-12. Because of its extended conjugation it
absorbs in the visible part of the spectrum and is used in nonhydroxylic
solvents as an indicator for the presence of hydrogen halides. Protonation
produces the conjugate acid and results in a color change from yellow to bright
red (115b).

There are indications that some carbonyl compounds form complexes with
the hydrated proton in aqueous sulfuric acid [e.g., B $\cdots$ H$^+$ (aq)] (116–118).
Attempts have been made to disentangle the small spectral changes that such a
step would produce from medium effects on the spectra of the discrete entities
B and BH$^+$. These uncertainties have given rise to widely varying estimates of
the base strength of acetone, for example (119), although a consensus seems to
be appearing among a group of investigators that this compound has a pK_{BH^+}
close to −2.9 (109, 120, 120a). (This view is not by any means unanimous and
the base strength of alkanones remains a vexing problem.)

Ultraviolet spectroscopy and, to a much lesser extent, nmr spectroscopy [including ^{17}O spectroscopy (121)] have been the most frequently used tools for investigating the strength of very weak carbonyl and carboxyl bases, although a number of other methods have also been used, with there being little agreement, unfortunately, in the results obtained (33).

A method that spans an enormous range of basicities is calorimetry. Arnett et al. measured the heat of protonation of 52 carbonyl bases in pure fluorosulfuric acid and found a roughly linear relationship between this quantity and free energy of protonation as expressed by (aqueous) pK_{BH^+} values from the literature (122). As we have seen, however, the latter quantities are often not known with certainty. The similar relationship that these workers found with amine bases is more firmly established since that series can be anchored in the aqueous state (Section III,A). Although different functional groups appear to generate different lines in such free-energy–enthalpy plots, there are indications that they are nearly parallel (124).

It can be seen in Table 3-13 that ethyl benzoate and ethyl cinnamate are weaker bases than their carboxylic acid counterparts, despite the very similar charge distributions that must be present in the cations of each species and despite the fact that alkyl is usually electron donating relative to hydrogen. The explanation likely lies in the capacity of the cationic protons to form strong hydrogen bonds to water (or to H_2SO_4, since in more concentrated acid essentially all the water molecules are bound to protons). It can be seen that the carboxyl system has one more hydrogen-bonding site than does the ester in both the neutral and cationic forms [Eq. (3-19)]. The strength of this

$$RCO_2H\text{\tiny IIIII}OH_2 \;\overset{H^+}{\rightleftharpoons}\; R-C\begin{smallmatrix}OH\text{\tiny IIIII}OH_2\\[2pt]+\\[2pt]OH\text{\tiny IIIII}OH_2\end{smallmatrix}$$

$$RCO_2Et \;\overset{H^+}{\rightleftharpoons}\; R-C\begin{smallmatrix}OH\text{\tiny IIIII}OH_2\\[2pt]+\\[2pt]OEt\end{smallmatrix}$$

(3-19)

additional hydrogen bond, however, will be much greater in the cation than in the neutral carboxylic acid, thus favoring the conjugate acid of the carboxylic acid relative to that of its ethyl ester.

It is interesting that an analogous effect is found when polyhydroxylic acids are esterified; the *acid* strength is invariably *increased* (Chapter 2, Section II,C). Differential solvation of some sort is probably also at the root of that effect.

B. Hydroxyl and Alkoxyl Bases

Water, the prototypical hydroxyl base, has a formal pK_{BH^+} of -1.74, which is simply the negative logarithm of the water molarity and which derives, in turn, from the cancellation in Eq. (3-20) of the terms $[H^+]$ and $[H_3O^+]$, both of which represent the solvated proton.

$$K_{BH^+} = \frac{[H_2O][H^+]}{[H_3O^+]} \qquad (3\text{-}20)$$

On the other hand, it requires $\sim 73\%$ sulfuric acid before the stoichiometric concentrations of H_2O and H_3O^+ become equal (125). At this acidity an amide of $pK_{BH^+} = -1.74$ would be more than 99% protonated and an arylamine of the same pK would be more than 99.99% protonated.

If water is a typical hydroxyl base we can expect alcohols, and also ethers, to be governed by an acidity function that changes only gradually with increasing concentration of sulfuric acid. Alternatively, water may be atypical in that its molecular state is continuously changing as it is transformed from solvent, where it is highly self-associated, to solute, where the molecules exist largely in the proton solvation sphere; in Arnett's phrase we may be shooting at a moving target (125). Probably both of these conclusions are valid.

Lee and Cameron have used nmr spectroscopy to study the protonation of ethanol and have concluded that its protonation is, indeed, governed by a very shallow acidity function (126). The pK_{BH^+} that they, and later Perdoncin and Scorrano (120), derived for ethanol is -1.94, a value that is only slightly more negative than the formal value of -1.74 for water. We know that when ethanol is the solvent water acts as a base since the conductivity of an HCl–ethanol solution plummets when a small amount of water is added. This result is attributed to the Grotthus proton-jump mechanism being suppressed by the capture of protons by the more basic solute molecules. The H_3O^+ ions so produced can carry the current only by drifting in the electric field or by means of an uphill proton transfer to a molecule of ethanol (127, 128).

We can conclude that aliphatic alcohols are comparable to water in basicity but tend to be slightly less basic. Aliphatic ethers appear to be slightly less basic still. Such alcohols and ethers are half-protonated near 80% sulfuric acid (120, 132). [Earlier estimates of alcohol and ether basicity based on distribution methods indicated that they were half-protonated at appreciably lower sulfuric acid concentrations (129–131).] The pK_{BH^+} values at 25°C given by Perdoncin and Scorrano for methanol, ethanol, dimethyl ether, and diethyl ether are -2.05, -1.94, -2.48, and -2.39, respectively, values that are close to being temperature independent (120). Note that the ethers are weaker bases than the alcohols; that is, an alkyl group is base weakening. [This effect is opposite to that found in the isoelectronic amines among which $(CH_3)_2NH$ is more basic than CH_3NH_2, although $(CH_3)_3N$ is the weakest of the lot.] The same base-weakening effect of O-alkylation is present in the carboxylic acid–ester system (previous section), and the same explanation probably applies.

Phenol, anisole, and their *p*-halo and *p*-methyl derivatives are half-protonated (on oxygen) in 75 to 82% aqueous sulfuric acid (*132a,b*). The pK_{BH^+} values of phenol and anisole have been estimated to be -6.4 and -6.0, respectively, using acidity function techniques.

The gas phase basicities of a number of polyethers have been determined by Meot-Ner and by Kebarle *et al.* (*132c*).

VI. Carbon Bases

There are very few neutral organic compounds that are appreciably protonated on carbon in dilute aqueous acid solution. One such compound is 2,3,4-trimethylpyrrole, whose pK_{BH^+} is 3.94 and which is protonated chiefly at position 2 [Eq. (3-21)] (*133*). Curiously, 2,3,4,5-tetramethylpyrrole is a slightly weaker base (Table 3-14).

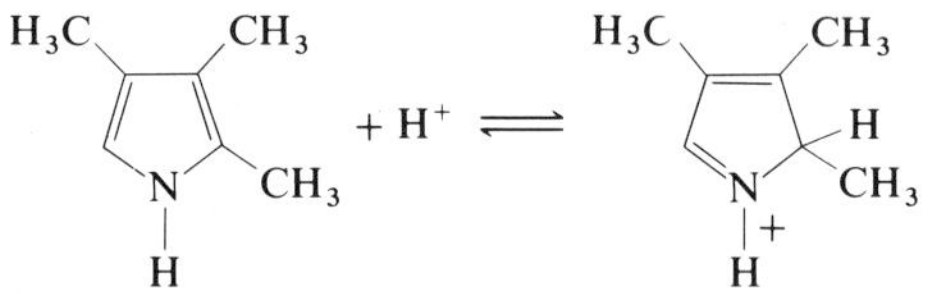

$$\text{(structures)} + H^+ \rightleftharpoons \text{(structures)} \qquad (3-21)$$

Pyrrole, itself, is very much weaker, requiring 5.34 *M* sulfuric acid (38% sulfuric acid) to effect half-protonation; here, again, the most stable cation has the added proton at position 2 (*134*). Enough alkyl derivatives of pyrrole are available that the overlap procedure can be used to link the parent compound, pyrrole, to the aqueous state; its pK_{BH^+} is -3.80. If the H_0 value of 5.34 *M* sulfuric acid had been used directly, a pK_{BH^+} of -2.43 would have been obtained. It is clear, therefore, that pyrroles do not follow the H_0 function but, unlike the situation with most carbonyl compounds, where taking the H_0 at half-protonation gives a pK_{BH^+} that is too negative, such a procedure would give a pK_{BH^+} that is not negative enough.

Indoles resemble pyrrole in undergoing preferential protonation on carbon, though with both kinds of substrate the reactions are not instantaneous; that is, the less stable N-protonated forms presumably dominate briefly upon mixing the heterocycles with acid. Indoles differ from pyrroles in preferring to be protonated at the 3 position [Eq. (3-22)] (*135*), an example of the operation of the Mills–Nixon effect (*136, 136a, 137*).

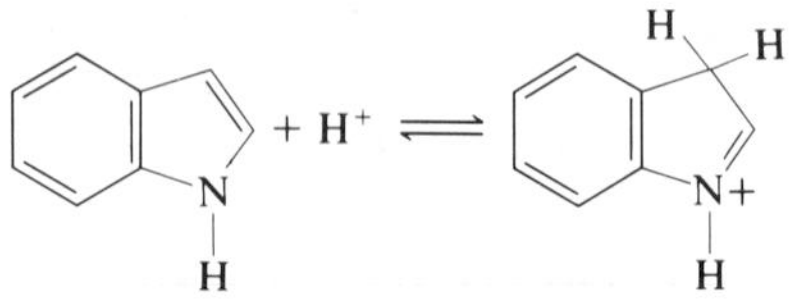

$$\text{(structure)} + H^+ \rightleftharpoons \text{(structure)} \qquad (3-22)$$

Table 3-14

pK_{BH^+} VALUES OF CARBON BASES IN WATER AT 25°C

Compound	pK_{BH^+}	Molarity of H_2SO_4 to half-protonate	Ref.
1,3,5-Triaminobenzene	5.5	—	138, 142
2,3,4-Trimethylpyrrole	3.94	—	133
2,3,4,5-Tetramethylpyrrole	3.77	—	133
2,4-Dimethylpyrrole	2.55	—	134
2,3,5-Trimethylpyrrole	2.00	—	134
3,4-Dimethylpyrrole	0.66	—	134
1,2-Dimethylindole	0.30	0.09	135
2,5-Dimethylindole	0.26	0.10	135
2-Methylpyrrole	−0.21	0.70	134
2-Methylindole	−0.28	0.76	135
2-Ethylindole	−0.41	0.92	135
1,2,3-Trimethylindole	−0.66	1.22	135
2,5-Dimethylpyrrole	−0.71	1.10	134
1-Methylazulene	−0.83	1.50	150, 152
3-Methylpyrrole	−1.00	1.50	134
2,5-Dimethyl-3-n-propylindole	−1.03	1.78	135
2,3-Dimethylindole	−1.49	2.46	135
Azulene	−1.70	2.60	150
1-Ethylindole	−2.30	3.61	135
1-Methylindole	−2.32	3.70	135
1-Methylpyrrole	−2.90	4.14	134
1,2-Dimethyl-5-nitroindole	−2.94	4.60	135
1-Chloroazulene	−3.25	4.70	150
2-Methyl-5-nitroindole	−3.58	5.54	135
Pyrrole	−3.80	5.34	134
3-$tert$-Butylindole	−3.84	5.91	135
3-Ethylindole	−4.25	6.53	135
3-n-Propylindole	−4.34	6.66	135
1,1-Di-p-anisylethylene	−4.45	6.02	150
3-Methylindole	−4.55	6.95	135
1,3,5-Trimethoxybenzene	−6.14[a]	7.80	150
1-p-Anisyl-1-phenylethylene	−6.30	8.00	150
1,1-Di-p-tolylethylene	−8.1	10.0	145
1-Cyanoazulene	−8.41	10.5	150
1,1-Diphenylethylene	−9.4	11.6	145, 150
2,6-Dimethoxytoluene	−10.3	12.6	150
1,1-Bis(4-chlorophenyl)ethylene	−10.3	12.6	145, 150
1,1-Diphenylpropene	−10.5[b]	13.7	145
Hexamethylbenzene	−12.3[b,c]	16.8[d]	145

[a] Reference 82 gives −5.71.
[b] Corrected values; see text.
[c] Reference 82 gives −10.7.
[d] 91% Sulfuric acid.

As with the pyrroles there are enough derivatives available to link the more basic indoles such as the 1,2-dimethyl compound ($pK_{BH^+} = 0.30$, and thus eligible to be an anchor compound) with less basic compounds such as the 3-methyl derivative ($pK_{BH^+} = -4.55$). Effectively, an acidity scale was constructed using a series of alkylindoles. The 3-methyl compound is half-protonated in 6.95 M sulfuric acid and, were the H_0 value of this solution to be taken as the pK_{BH^+}, a value of -3.30 would be obtained. As with the pyrroles the appropriate acidity function rises more steeply with increasing concentration of acid than does H_0.

Curiously, indole itself, which is half-protonated in 5.5 M sulfuric acid and hence appears to be more basic than 3-methylindole, overlaps very poorly with its neighboring indicators. One might suspect that the apparent anomaly associated with indole requiring a lower concentration of acid than the 3-methyl compound is somehow associated with its deviant ionization pattern. It is clear from the slopes of the ionization curves, however, that the anomaly will be *more* pronounced in water. Cox and Yates's extrapolative procedure gives a value of -2.38 for indole, making it almost as basic as 1-methylindole, and a value of -4.49 for 3-methylindole, in good agreement with the value given in Table 3-14 (*82*). Thus, a 3-methyl substituent in indole is, indeed, base weakening.

Protonation on carbon also occurs with 1,3,5-trihydroxybenzene (phloroglucinol) and its ethers (*140, 141*) and with 1,3,5-triaminobenzene (*138, 139*) [Eqs. (3-23) and (3-24)].

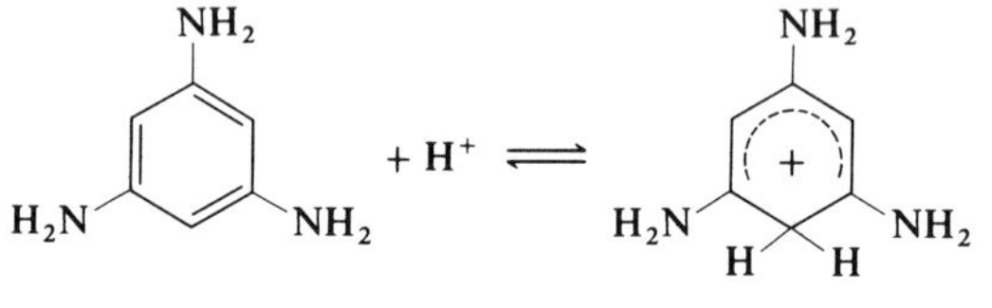

$$(3\text{-}23)$$

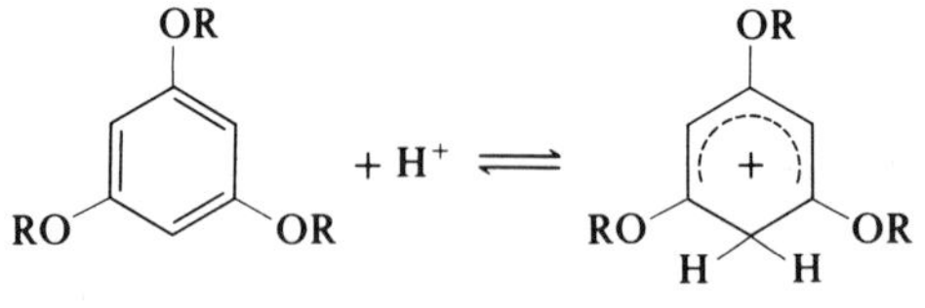

$$(3\text{-}24)$$

1,3,5-Triaminobenzene has a pK_{BH^+} of 5.44 in water at 25°C for protonation on carbon; the N-protonated monocation, which is a minor component at room temperature, becomes a major component at lower temperatures (*142*). 1,3-Diaminobenzenes are C-protonated to a minor extent at room temperature; monoaminobenzenes, of course, are N-protonated in the condensed phase. There is evidence to indicate that monoalkoxybenzenes undergo C-protonation in very strongly acidic solution (*143, 144*), unlike the situation in aqueous sulfuric acid (previous section). The factors governing the choice of protonation sites in organic molecules are discussed in Chapter 5.

The construction of an acidity scale based on carbon bases has been attempted with pyrroles and indoles (discussed earlier) and with azulenes and diarylethylenes. The latter compounds are more truly characterized as carbon bases since in most of them the positive charge in the conjugate acid resides on carbon, in contrast to pyrrole and indole cations, in which the charge resides principally on nitrogen.

The $H_{R'}$ acidity function, which refers to diarylalkene protonation [Eq. (3-25)], was derived indirectly by Deno et al. (145) in 1959 from measurements of the carbinol–carbonium ion equilibria [Eq. (3-26)], a well-

$$Ar_2C{=}CH_2 + H^+ \qquad Ar_2\overset{+}{C}CH_3 \qquad\qquad (3\text{-}25)$$

$$ROH + H^+ \qquad R^+ + H_2O \qquad\qquad (3\text{-}26)$$

established system that is governed by the H_R (sometimes called J_0) function (146, 147). (See refs. 146–149 for accounts of the relationship between $H_{R'}$ and H_R.)

Reagan (150) subsequently used a variety of carbon bases, including diarylalkenes, azulenes, and alkoxyarenes, to construct a scale designated H_C that was anchored in water with 1-methylazulene, which is half-protonated (at the 3 position) (151) in 1.5 M sulfuric acid [Eq. (3-27)].

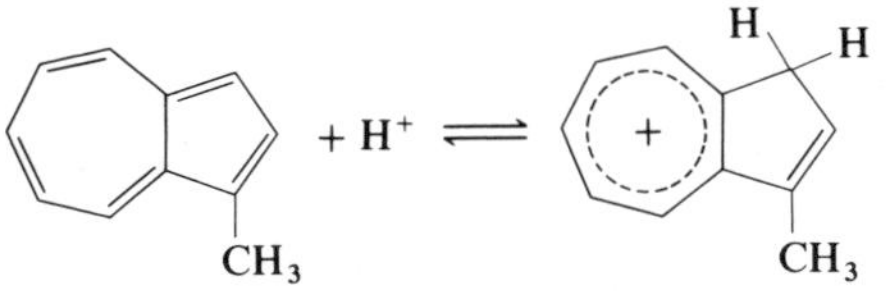

$$(3\text{-}27)$$

Although 1.5 M sulfuric acid hardly constitutes the dilute aqueous state, Long and Schulze had shown previously by means of extrapolation to dilute aqueous solution that a satisfactory pK_{BH^+} could be obtained (152). The value of -0.83 at 25°C that they obtained for 1-methylazulene is almost 0.3 unit more negative than that which would be obtained using the H_0 function. This gap widens as one goes to weaker and weaker carbon bases. That is, the $H_{R'}$ and H_C scales rise more sharply than does the H_0 scale, principally because of the absence of hydrogen bonding in the cations. In the case of the primary arylammonium ions that are used to define the H_0 scale, interactions of the type shown in **31** make the protonation process less favorable than it otherwise would be as the concentration of acid increases and that of water decreases (20).

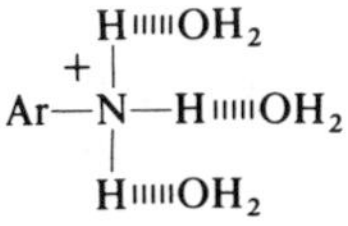

31

Tertiary arylamines, with only one hydrogen-bonding site in the cation, generate a function that is roughly midway between the H_0 and $H_{R'}$ or H_C scales. The sign and magnitude of the Bunnett–Olsen ϕ parameter (Section V) is a convenient way of expressing the direction and extent of departure of an acidity function from H_0 and the resemblance, or more correctly the lack thereof, between the bases used in the construction of the scales. The values of ϕ for carbon bases are in the range of -0.3 to -0.8; for primary anilines they are zero by definition; and for amides they are in the range of 0.4 to 0.5 (153).

Listed in Table 3-14 are the pK_{BH^+} values of carbon bases that are linked to the aqueous state by overlap procedures. These are mostly indoles, azulenes, and diarylethylenes, and they have been listed in order of decreasing base strength. It can be seen by scanning the columns giving the molarity of sulfuric acid at half-protonation that the pyrroles, indoles, and the more basic azulenes form a coherent set, although since each of these series has been determined by its own overlap procedure there is no requirement that they do so. The values are listed in the table to two decimal places, as is the custom, but for those pK_{BH^+} values that are more negative than about -2.0 the second decimal place (and in some cases the first) has no real significance.

The diarylethylenes, which are linked to the standard state via the azulenes, appear to have different activity coefficient terms than do indoles of comparable base strength. Thus, 1,1-di-p-anisylethylene has a more negative pK than 3-n-propylindole but is protonated to a greater degree in 6 to 7 M sulfuric acid. The values that Deno, Groves, and Saines derived indirectly for diarylethylenes using the H_R function are about one pK unit more negative than the values obtained by Reagan by overlap with the azulenes, although there is satisfactory agreement with respect to the concentration of sulfuric acid required to achieve half-protonation. Accordingly, the pK_{BH^+} values of the two weakest bases in Table 3-14 have been corrected by adding one pK unit to the values given by Deno et al.

An estimate of the pK_{BH^+} of 4-methylstyrene has been made by Richard and Jencks (154) on the basis of measurements of its acid-catalyzed hydration rate (155). Their value of -11.2 would make this base some three pK units weaker than the analogous diaryl compound, 1,1-di-p-tolylethylene, $pK_{BH^+} = -8.1$ (Table 3-14). As is usual with carbon bases the addition and removal of the proton is a much slower process than for oxygen or nitrogen bases, where, in most cases, the reaction is instantaneous for all practical purposes.

Super-acid media, as designated by Gillespie, are those systems in which the hydrogen ion activity is greater than in pure sulfuric acid (165). The most powerful and useful super acids are made by combining strong Brønsted acids such as H_2SO_4, HF, FSO_3H, or CF_3SO_3H with strong Lewis acids such as BF_3 or SbF_5 (166, 167). Super acids readily protonate aromatic rings to give benzenonium ions [Eq. (3-28)] (156–158), but it is difficult to assign pK_{BH^+}

values to such compounds. Diazomethane is protonated by super acids to give, depending on the conditions, the C- or N-protonated species [Eq. (3-29)] (*159*). Hydrazoic acid, HN_3, which is isoelectronic with diazomethane, gives the aminodiazonium ion with super acids [Eq. (3-29a)], and alkyl azides react similarly (*160*).

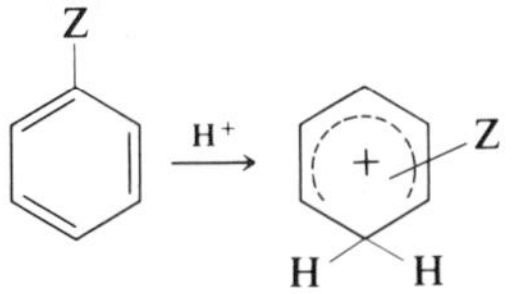

$$\text{(3-28)}$$

$$CH_2N_2 \xrightarrow{H^+} \begin{cases} H_3C-\overset{+}{N}\equiv N \\ H_2C=\overset{+}{N}=NH \end{cases} \qquad \text{(3-29)}$$

$$HN_3 \xrightarrow{H^+} H_2N-\overset{+}{N}\equiv N \qquad \text{(3-29a)}$$

Super acids may also function as oxidants with aromatic systems, generating radical-cations or dications (*161*). For example, naphthalene and its 1,4,5,8-tetramethyl derivative in SbF_5/SO_2ClF at $-78°C$ generate the radical-monocation and the radical-dication **32** and **33**, respectively (*162–164*).

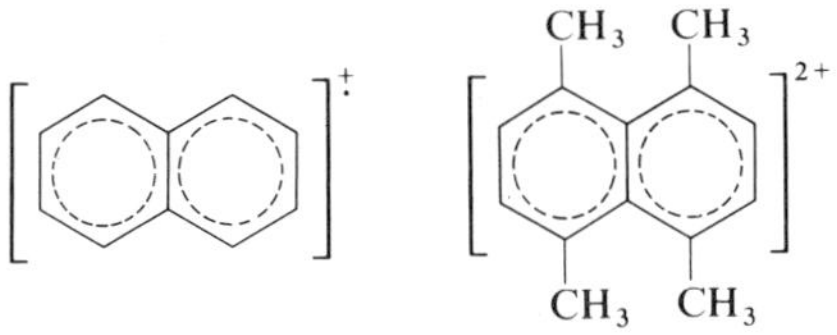

32   **33**

The weakest hydrocarbon bases are the alkanes. There is no doubt that protonated alkanes (e.g., CH_5^+) exist in the gas phase, but the extent of their lifetimes in the condensed state, even in super-acid systems, is still a matter of debate.

Gold *et al.* have estimated the H_0 value of mixtures of HF and SbF_5 to be as low as -26, which would make it $\sim 14 \, pK$ units more acidic than pure sulfuric acid (*168*). Regardless of the numerical value of the acidity function for such super acids the question that must be asked is the following. Does the proton prefer to protonate, even to a slight extent, a molecule of alkane to produce a nonclassical ion in which five hydrogen atoms are held to the carbon atom by eight electrons, as in Eq. (3-30), or would it prefer to form exclusively normal electron-pair bonds, such as those shown in Eq. (3-31), even if these are of rather high energy?

$$HF + SbF_5 + CH_4 \rightleftharpoons CH_5^+ + SbF_6^- \qquad \text{(3-30)}$$

$$2\,HF + SbF_5 \rightleftharpoons H_2F^+ + SbF_6^- \qquad \text{(3-31)}$$

Fabre, Devynck, and Tremillon in their review of the behavior of alkanes in super-acid media concluded that methane and ethane are not protonated under such conditions (167, 169) but that the higher alkanes probably do form the species RH_2^+, though they conceded that there is no spectral evidence to support this view.

Olah was the first to suggest that protonated alkanes had a discrete existence in solution (170), after it had been recognized that alkanes can be cleaved by super acids to give carbonium ions and either molecular hydrogen or alkanes, as in Eqs. (3-32) and (3-33) (171–173).

$$R{-}H \xrightarrow[\text{(FSO}_3\text{H/SbF}_5)]{H^+} R^+ + H_2 \qquad\qquad (3\text{-}32)$$

$$R{-}R \xrightarrow[\text{(FSO}_3\text{H/SbF}_5)]{H^+} R^+ + RH \qquad\qquad (3\text{-}33)$$

The transition states for these processes doubtless have the composition $[RH_2]^+$ and $[RRH]^+$, respectively, but whether such ions have a finite lifetime or, if they do, whether it is sufficiently long for them to be detected remains to be seen.

VII. Sulfur Compounds

The most basic sulfur bases are the thioamides, the strongest of which are nevertheless only slightly protonated in dilute aqueous acid. Edward et al. (175) have shown that 1,3-di-tert-butylthiourea (34), has a pK_{BH^+} of -1.32, just barely strong enough to serve as an anchor for the construction of an acidity scale for thioamides and thion esters (176). These compounds are protonated on sulfur (178, 179) and hence are, by definition, sulfur acids, even though the positive charge is largely dispersed by resonance to other sites in the ion [Eqs. (3-34) and (3-35)].

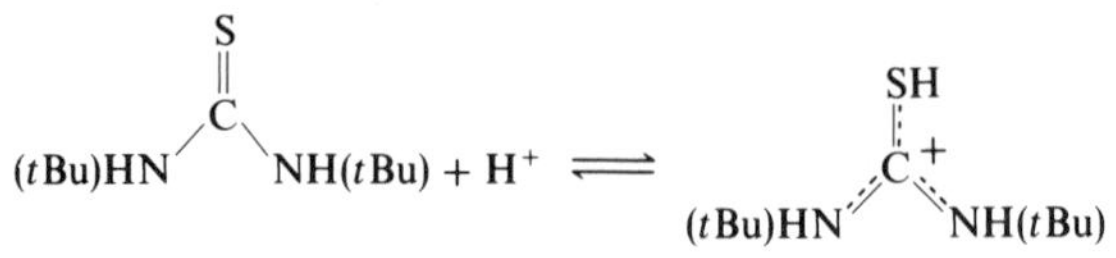

$$\underset{\textbf{34}}{(t\text{Bu})\text{HN}-\overset{\overset{\textstyle S}{\|}}{\underset{}{C}}-\text{NH}(t\text{Bu})} + H^+ \;\rightleftharpoons\; (t\text{Bu})\text{HN}-\overset{\overset{\textstyle SH}{|}}{\underset{}{\overset{+}{C}}}-\text{NH}(t\text{Bu}) \qquad (3\text{-}34)$$

$$\underset{\textbf{35}}{4\text{-O}_2\text{NC}_6\text{H}_4\text{C}\!\!\overset{\displaystyle S}{\underset{\displaystyle OC_2H_5}{}}} + H^+ \;\rightleftharpoons\; 4\text{-O}_2\text{NC}_6\text{H}_4\text{C}\!\!\overset{\displaystyle SH}{\underset{\displaystyle OC_2H_5}{}}\!\!+ \qquad (3\text{-}35)$$

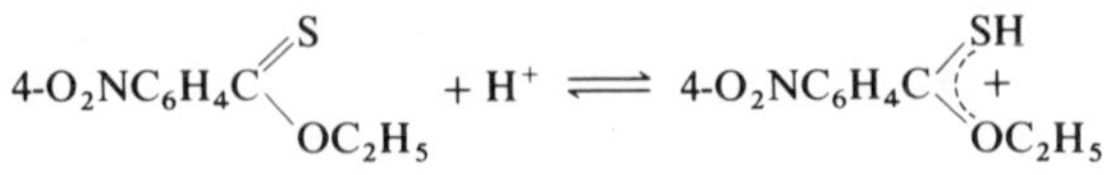

Satisfactory overlap was obtained by Edward for the series of 16 thioureas, thioamides, and thion esters, with the weakest base being ethyl 4-nitro-thionbenzoate (35), which is half-protonated in 83% sulfuric acid and has a

Table 3-15

pK_{BH^+} VALUES OF SULFUR BASES IN WATER AT 25°C

Compound	pK_{BH^+}	Ref.
1,3-Di-*tert*-butylthiourea[a]	−1.32	*175*
1,3-Diethylthiourea[a]	−1.70	*175*
Thiocaprolactam	−2.24	*175*
Thioacetamide	−2.51	*79*
4-Methoxythiobenzamide	−2.96	*175*
Thiobenzamide	−3.20	*79*
4-Chlorothiobenzamide	−3.67	*175*
4-Nitrothiobenzamide	−4.02	*175*
Dimethyl sulfide	−7.0	*120, 180*
Ethyl 4-methoxythionbenzoate	−7.63	*175*
Ethyl thionbenzoate	−8.55	*175*
Ethyl 3-bromothionbenzoate	−9.63	*175*
Ethyl 4-nitrothionbenzoate	−10.55	*175*

[a] Reference *177* gives a value of −1.19 for thiourea.

pK_{BH^+} of −10.6. Interestingly enough, when the extrapolative procedures of Bunnett and Olsen and Marziano *et al.* (Section V) were applied to the data, it was found that in most cases the pK values bracketed the values given by the Hammett overlap technique.

The Bunnett–Olsen parameter ϕ is between −0.3 and −0.4 for most thioamides, as it is for most of the other sulfur bases listed in Table 3-15. As we have seen, the sign and magnitude of ϕ are a measure of the difference in extent of hydration of the hydronium ion and the conjugate acid of the substrate [Eq. (3-36)].

$$B + H_3O^+(aq) \rightleftharpoons BH^+(aq) + H_2O \tag{3-36}$$

The negative ϕ for sulfur bases indicates that their conjugate acids are less hydrated than those of primary anilines on which the H_0 scale is based, not an unexpected result since hydrogen atoms that are attached to sulfur are not particularly effective hydrogen-bonding sites.

Both thioamides and thion esters are weaker bases than their oxygen analogues. The same condition obtains with ethers and thio ethers and with alcohols and thiols. Thiols, of course, are several units more *acidic* than alcohols (Chapter 2, Section III).

Scorrano *et al.* (*120, 180*), using the Bunnett–Olsen method, have estimated the base strength of dimethyl sulfide to be more than four pK units less than that of dimethyl ether: pK_{BH^+} values of −7.0 and −2.5, respectively. However, it takes a *higher* concentration of sulfuric acid to half-protonate dimethyl ether, 86% compared with 70% for dimethyl sulfide. This is because ethers have a very shallow, and sulfides a very steep, ionization curve. That is, the acidity function governing ether (and alcohol) protonation rises only gradually with increasing sulfuric acid concentration, as reflected in the substantial positive value for ϕ, the Bunnett–Olsen parameter, which is nearly 0.8 for these compounds. For thio ethers, on the other hand, ϕ is negative (approximately -0.3), as it is for the other sulfur acids listed in Table 3-15.

Thiol esters (37), in which sulfur replaces the alkoxyl oxygen atom, appear to be considerably more basic than thion esters (36), at least in aqueous solution. (As with ethers and thio ethers, reversal of basicity order is seen under strongly acidic conditions, where the protonated sulfur species suffer less than do their oxygen analogues from being deprived of water of hydration.) The pK_{BH^+} value of ethyl thiolbenzoate is estimated to be near -5.0, compared with -8.55 for the thion analogue (Table 3-15) and -6.16 for ethyl benzoate (Table 3-13). These results lead to a curious paradox, outlined in the following paragraphs.

It is generally agreed that thion esters (36) are less stable than their thiol isomers (37), the hypothetical equilibrium being shown in Eq. (3-37) (181).

$$
\begin{array}{cccc}
 & \overset{\displaystyle S}{\underset{\displaystyle OR}{R-C}} & \rightleftharpoons & \overset{\displaystyle O}{\underset{\displaystyle SR}{R-C}} & \hspace{2cm}(3\text{-}37) \\
 & \mathbf{36} & & \mathbf{37} &
\end{array}
$$

$$
\begin{array}{cc}
\overset{\displaystyle SH}{\underset{\displaystyle OR}{R-C}}{\scriptstyle +} & \overset{\displaystyle OH}{\underset{\displaystyle SR}{R-C}}{\scriptstyle +} \\
\mathbf{38} & \mathbf{39}
\end{array}
$$

If both classes of compound are protonated at the unsaturated atom, as is generally believed to be the case and as seems reasonable from the sign and magnitude of the Bunnett–Olsen ϕ or related parameters (182, 183), then the conjugate acids have the structures 38 and 39 with very similar charge distributions and, one would have thought, very similar energies. If this were in fact so, 39 would have to be the stronger acid and its conjugate base, the thiol ester, the weaker base, the reverse of what is found. (This is a variation of the general system discussed in Chapter 5 in which the minor component of a tautomeric pair that produce a common ion is always the stronger acid or stronger base, depending on the sort of equilibrium being considered.)

If we assume that the derived pK_{BH^+} values of the isomeric thiol and thion esters are not grossly in error and that protonation indeed occurs at the unsaturated atom in both cases, then we must conclude that there is a large difference in energy between the isomeric ions 38 and 39, with 38 being very much less stable than 39.

The equilibrium constant for the hypothetical equilibrium $38 \rightleftharpoons 39$ can be easily shown to be equal to $(K_{taut}K_{BH^+}^{thion})/K_{BH^+}^{thiol}$, where $K_{taut} = [37]/[36]$, and $K_{BH^+}^{thiol}$ and $K_{BH^+}^{thion}$ are the acidity constants of the thiol and thion forms respectively. In the case of the thioethyl benzoates, $K_{BH^+}^{thion}$ is more than three pK units greater than $K_{BH^+}^{thiol}$, which means that whatever factor by which the neutral thiol form is favored over its thion isomer will be magnified more than a thousandfold in the conjugate acids. The greater capacity of **39** to hydrogen-bond to water (*174*) is undoubtedly a factor in this regard, but whether it is sufficient to account for these results is uncertain.

Thiolcarboxylic acids undergo rapid hydrolysis in aqueous solution [Eq. (3-38)], but estimates of their base strengths indicate that they are not greatly different from those of the corresponding thiol esters (*184*). Both classes of compound are, of course, oxygen, not sulfur, bases.

$$R-C{\overset{\displaystyle O}{\underset{\displaystyle SH}{}}} + H_2O \;\rightleftharpoons\; R-C{\overset{\displaystyle O}{\underset{\displaystyle OH}{}}} + H_2S \qquad (3\text{-}38)$$

Dimethyl sulfoxide is an important sulfur-containing oxygen base. It is half-protonated in 42% sulfuric acid and has a pK_{BH^+} of -1.54, making it much more basic than acetone, its carbon analogue, which is half-protonated in $\sim 80\%$ sulfuric acid (*120*). Like all other oxygen acids it has a positive Bunnett–Olsen parameter ϕ, meaning that its degree of protonation increases less rapidly with increasing acid concentration than do those of the aniline bases with which the H_0 scale is built. (An aniline base with the same pK_{BH^+} would be half-protonated in 26% sulfuric acid.)

Sulfonamides are much less basic than carboxamides, as might be expected. N-Methylmethanesulfonamide is half-protonated in 71% sulfuric acid, whereas the analogous carboxamide, N-methylacetamide, requires only 15% sulfuric acid to be protonated to the same extent (*185*). There seems to be general agreement that sulfonamides are protonated on nitrogen rather than oxygen [Eq. (3-39)] (*186–188*).

$$R-\overset{\displaystyle O}{\underset{\displaystyle O}{\overset{\|}{\underset{\|}{S}}}}-NH_2 + H^+ \;\rightleftharpoons\; R-\overset{\displaystyle O}{\underset{\displaystyle O}{\overset{\|}{\underset{\|}{S}}}}-\overset{+}{N}H_3 \qquad (3\text{-}39)$$

Probably the most basic functional unit containing sulfur is the thioimidate group. Methyl benzthioimidate (**40**), has a pK_{BH^+} of 5.84, making it slightly more basic than its oxygen analogue, methyl benzimidate (**41**), which has a pK_{BH^+} of 5.60 (*189*).

$$H_5C_6-C{\overset{\displaystyle NH}{\underset{\displaystyle SCH_3}{}}} \qquad\qquad H_5C_6-C{\overset{\displaystyle NH}{\underset{\displaystyle OCH_3}{}}}$$

$$\textbf{40} \qquad\qquad\qquad \textbf{41}$$

VIII. Effect of Oxidation State on Basicity

It is well known that the strength of both inorganic and organic oxy acids increases with the oxidation state of the central atom (e.g., nitrous–nitric acid, benzenesulfinic–benzenesulfonic acid). The situation with organic bases is more far reaching since there are a large number of ways in which the formal oxidation state of these compounds can be increased, although dehydrogenation or oxygen atom addition is the usual result. We shall see that oxidation almost always produces a compound of lower basicity, provided that the site of protonation is not altered. The connection between the concepts of acidity and oxidation is, of course, a very old one, going back to the time of Lavoisier, who regarded oxygen as the "acidifying principle."

A. Carbon and Oxygen Bases

Alkanes are the first members of the oxidation sequences alkane–alcohol–aldehyde–carboxylic acid, and alkane–alkene–alkyne. They are unique, however, in having all their valence electrons tied up in single bonds, and their basicity, as a consequence, is extremely low, much lower than that of any of their oxidation products.

With regard to alcohols, aldehydes and ketones, and carboxylic acids it is curious how close their basicities appear to be if one's criterion is the amount of acid required for half-protonation. Thus, methanol, ethanol, acetone, benzaldehyde, and benzoic acid are all half-protonated in the region 78–82% sulfuric acid. (Acetophenone and benzophenone appear to be somewhat stronger in being half-protonated in $\sim 74\%$ sulfuric acid.) Since these compounds respond differently to medium changes, the half-protonation point can serve only as a rough indication of the pK_{BH^+} value of a compound in water. The estimated pK_{BH^+} values for benzyl alcohol, benzaldehyde, and benzoic acid, -3.15, -3.91, and -4.73, respectively, show a steady, though fairly modest, decrease in basicity with increasing degree of oxidation. [The latter two values come from Table 3-13; the value for benzyl alcohol is obtained by correcting the pK_{BH^+} of methanol, -2.05, (120) for phenyl substitution on the assumption that the difference present in the $CH_3NH_3^+/C_6H_5CH_2NH_3^+$ pair, 1.1 unit, is also present here.]

It should be noted that the protolytic equilibria of aryl alcohols, aldehydes, and carboxylic acids have different Hammett ρ values, and it is likely that with certain substituents the order of base strength alcohol > aldehyde > carboxylic acid will not be found to hold. In any case the different classes of oxygen base are not particularly different in strength, unlike the situation that obtains with nitrogen bases (Section VIII,B).

With respect to the series alkane–alkene–alkyne there is no doubt that alkanes are the weakest bases. The available evidence with regard to alkenes

Table 3-16

COMPARISON OF RATES OF ACID-CATALYZED HYDRATION OF ALKENES AND ALKYNES IN AQUEOUS SULFURIC ACID[a]

Compound	$\log k_{H^+}$[b]	k_{alkene}/k_{alkyne}
$CH_3CH=CH_2$	-8.62	360
$CH_3C\equiv CH$	-11.17	
$C_6H_5CH=CH_2$	-6.49	1.1
$C_6H_5C\equiv CH$	-6.52	
$CH_3OCH=CH_2$	-0.12	0.01
$CH_3OC\equiv CH$	1.89	

[a] Data from ref. 193.
[b] Rate constants in units of M^{-1} sec^{-1}; 25°C.

and alkynes is somewhat ambiguous since there are no alkyne analogues of the more basic alkenes (e.g., 1,1-diarylethylenes or azulenes) and since equilibrium protonation in solution cannot be achieved for other alkene–alkyne pairs that are available. Protonation produces ordinary carbonium ions from alkenes and vinyl cations from alkynes [Eqs. (3-40) and (3-41)].

$$RHC=CH_2 + H^+ \rightleftharpoons R\overset{+}{C}HCH_3 \tag{3-40}$$

$$RC\equiv CH + H^+ \rightleftharpoons R\overset{+}{C}=CH_2 \tag{3-41}$$

Richey and Richey (190) pointed out that vinyl cations have a greater stability than is often realized and, indeed, they are often formed faster than carbonium ions are formed from the analogous alkenes, particularly if strongly electron-donating groups are present. Protonation is the rate-controlling step in hydration of both alkenes and alkynes (191, 192), and we can therefore use hydration rates to compare their kinetic basicities. Allen, Chiang, Kresge, and Tidwell measured the rates of hydration of a large number of alkenes and alkynes in aqueous sulfuric acid (193). As would be expected they found that both series are strongly activated by the presence of electron-donating groups, as reflected in the extremely large negative slopes of plots of the logarithms of the proton catalytic constant ($\log k_{H^+}$) against the Hammett–Brown substituent constant σ^+ for substituted styrenes and phenylacetylenes. The slopes ρ^+ are -10.5 for the styrenes and -14.1 for the phenylacetylenes.

Table 3-16 gives a comparison of the rates of hydration of several alkenes and alkynes, and it is clear that the order of basicity of a pair of alkenes or alkynes depends on the groups attached to the unsaturated centers. It is uncertain to what extent kinetic results can be taken to indicate the order of equilibrium basicities of alkenes and alkynes. If, indeed, the more reactive alkynes have greater equilibrium basicities than their alkene counterparts, it would be in marked contrast to the analogous imine–nitrile system, where a drastic drop in basicity appears always to accompany sp^2- to sp-bonding change (next section).

B. Nitrogen Compounds

An amine RCH_2NH_2 has five immediate oxidation products: hydroxyl-amine (RCH_2NHOH), imine ($RHC{=}NH$), carbinolamine ($RCHOHNH_2$), oxyamine (RCH_2ONH_2), and amine oxide ($RCH_2\overset{+}{N}H_2{-}O^-$. These com-pounds, some of which have very limited stability, can each be considered products of a two-electron (2-equivalent) (204) oxidation of an amine. Of these compounds the only one that is not protonated on nitrogen is the amine oxide; that is, it is not a nitrogen base (stable amine oxides are discussed later).

In Scheme 3-2 nitrogen bases are set out according to their oxidation levels, together with estimates of their pK_{BH^+} values. The connections shown between compounds in different oxidation levels indicate potential pathways involving addition or removal of either one oxygen atom or two hydrogen atoms. In many cases, of course, no practical way has yet been found of bringing about the direct conversions.

The second oxidation level contains six classes of compound: nitrosoalkane (RCH_2NO), oxime ($RHC{=}NOH$), nitrile (RCN), amide enol [$RC(OH){=}NH$], amide hydrate [$RC(OH)_2NH_2$], and amide ($RCONH_2$). The only one of these classes of compound that is believed to be an oxygen base is the amide. It has been included in Scheme 3-2 because its oxygen basicity places an upper limit on its nitrogen basicity.

A fourth oxidation level ($-6\ e^-$ on the scale shown in Scheme 3-2) would include nitrile oxides ($R{-}C{\equiv}\overset{+}{N}{-}O^-$) and nitro compounds ($RCH_2NO_2$). Little is known about the basicity of nitrile oxides but they, like nitro compounds, which are discussed later, lack a pair of electrons on the nitrogen atom and must be protonated on oxygen.

It can be seen in Scheme 3-2 that oxidation invariably decreases basicity, with the most drastic decline by far, $\sim 18\ pK$ units, coming with the formation of the cyano group. Fuming sulfuric acid is required in many cases to effect protonation (7) of nitriles, and it is clear that the change from sp^2 bonding in imines to sp bonding in nitriles has an enormous effect on the availability of the electron pair on nitrogen. The corresponding effect in going from sp^3 to sp^2 bonding is much smaller, being less than three orders of magnitude for the three cases shown in the scheme (amine to imine, hydroxylamine to oxime, and carbinolamine to amide enol).

Oxidation
level[a]

0

−2 e⁻

−4 e⁻

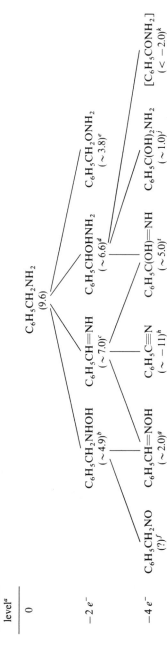

Scheme 3-2 Basicity as a function of oxidation level of nitrogen bases. The pK_{BH^+} values are in parentheses; connecting lines refer to dehydrogenation or oxygen atom addition. The lowercase letters indicate the following: [a] Oxidation levels correspond to two-electron and four-electron (or 2- or 4-equivalent) changes, with the amino level being arbitrarily taken as zero. [b] Based on pK_{BH^+} of CH_3NHOH, 5.96 (196) and a ΔpK of −1.1 (assumed to be the same as the $CH_3NH_2/C_6H_5CH_2NH_2$ difference). [c] Based on pK_{BH^+} of imine of benzophenone, 7.2 (197); the 4-chloro-N-phenyl derivative has a pK_{BH^+} of 2.80 (205); see also ref. 205a. [d] Calculated using the Taft equation (198). [e] Based on pK_{BH^+} of methoxyamine, 4.60 (196) and a ΔpK_{BH^+} of −0.8 (assumed to be the same as the $CH_3CH_2NH_2/C_6H_5CH_2CH_2NH_2$ difference). [f] Nitrosoarenes are not protonated in the aqueous region (203). [g] Based on pK_{BH^+} of acetone oxime, 1.92 (199). [h] Half-protonated in 99.8% sulfuric acid (7, 201). [i] Based on pK_{BH^+} of O-methyl derivative, 5.60 (189). [j] Based on pK_{BH^+} of $CH_3COH(OC_2H_5)NMe_2$, 1.1 (202). [k] Value for O-protonation is −1.35 (Table 3-11).

Table 3-17

pK_{BH^+} VALUES OF NITROARENES AND
PYRIDINE N-OXIDES

Compound	pK_{BH^+}
Nitroarenes[a]	
4-Nitrotoluene	−11.4
3-Nitrotoluene	−12.0
Nitrobenzene	−12.1
4-Nitrofluorobenzene	−12.4
4-Nitrochlorobenzene	−12.7
3-Nitrochlorobenzene	−13.2
2,4-Dinitrotoluene	−13.7
2,4-Dinitrofluorobenzene	−14.0
Pyridine N-oxides[b]	
4-Aminopyridine 1-oxide	3.65
4-Hydroxypyridine 1-oxide	2.36
3-Aminopyridine 1-oxide	1.47
4-Methylpyridine 1-oxide	1.29
3-Methylpyridine 1-oxide	1.08
Isoquinoline N-oxide	1.01
Pyridine 1-oxide	0.79
4-Chloropyridine 1-oxide	0.33
Nicotinic acid N-oxide	0.09
3-Chloropyridine 1-oxide	−0.13
2-Acetylpyridine 1-oxide	−0.45
Isonicotinic acid N-oxide	−0.48
2-Chloropyridine 1-oxide	−0.77
3,5-Dibromopyridine 1-oxide	−0.85
3-Nitropyridine 1-oxide	−1.07
4-Nitropyridine 1-oxide	−1.73
2-Cyanopyridine 1-oxide	−2.08
2-Nitropyridine 1-oxide	−2.71

[a] Data of Gillespie et al. (206), based on overlap
with the H_0 scale through picramide.
[b] Data of Jaffé and Doak (207) and Johnson
et al. (208).

The most basic compound in the third oxidation level is the enol of
benzamide, which appears to be about one pK unit more basic than
benzyloxyamine, even though the latter is in a more reduced state. The
proximity of the electronegative oxygen atom to the basic center in the latter
case is clearly important here.

Listed in Table 3-17 are the pK_{BH^+} values of a number of nitroarenes (206)
and amine oxides (207, 208), both series being determined by the overlap

procedure (Section III). Only the amine oxides are firmly linked to the aqueous state, however. Because of their extremely low basicity (nitrobenzene is half-protonated in 100.1% sulfuric acid) it was necessary to link the nitro compounds to picramide, $pK_{BH^+} = -10.0$ (Table 3-4), and although the overlap at the junction seems to be satisfactory there is no guaranty that the two classes of compound have parallel ionization curves all the way down to the standard state of water.

Nitroalkanes are weaker than their aryl counterparts and, moreover, tend to decompose in concentrated sulfuric acid, although they can be fully protonated in super-acid media (209). Nitromethane appears to be half-protonated in 23% fuming sulfuric acid.

N-Oxidation of pyridines decreases the pK_{BH^+} by between four and five pK units (207). The site of protonation changes, of course, from nitrogen to oxygen in these cases. A comparable difference is found when azobenzene (42) is converted to azoxybenzene (43) (210, 211), although since neither of these

$$C_6H_5N{=}NC_6H_5 \qquad C_6H_5\underset{+}{N}{=}NC_6H_5$$
$$\overset{O^-}{\underset{\mid}{\;}}$$

42   43

compounds is protonated in the aqueous region the pK difference is not known with as much precision. The protonation site changes from nitrogen in 42 to oxygen in 43. Organic phosphine and arsine oxides are also considerably weaker than their unoxidized precursors (212–214).

References

1. R. W. Alder, P. S. Bowman, W. R. S. Steele, and D. R. Winterman, *Chem. Commun.* p. 723 (1968).
2. F. Hibbert and K. P. P. Hunte, *J. Chem. Soc., Perkin Trans. 2* p. 1562 (1981).
3. P. B. Smith, J. L. Dye, J. Cheney, and J.-M. Lehn, *J. Am. Chem. Soc.* **103**, 6044 (1981).
4. R. W. Alder, and R. B. Sessions, *J. Am. Chem. Soc.* **101**, 3651 (1979).
5. R. W. Alder, A. Casson, and R. B. Sessions, *J. Am. Chem. Soc.* **101**, 3652 (1979).
6. N. C. Deno, R. W. Gaugler, and T. Schulze, *J. Org. Chem.* **31**, 1968 (1966).
7. N. C. Deno, R. W. Gaugler, and M. J. Wisotsky, *J. Org. Chem.* **31**, 1967 (1966).
8. G. A. Olah, "Halonium Ions," Chap. 2. Wiley (Interscience), New York, 1975.
9. E. M. Arnett, "Progress in Physical Organic Chemistry" (S. G. Cohen, A. Streitwieser, and R. W. Taft, eds.), Vol. I, p. 223. Wiley (Interscience), New York, 1963.
10. D. H. R. Barton, J. D. Elliott, and S. D. Gero, *J. Chem. Soc., Perkin Trans. 1* p. 2085 (1982).
10a. D. Sternbach, M. Shibuya, F. Jaisli, M. Bonetti, and A. Eschenmoser, *Angew. Chem., Int. Ed. Engl.* **18**, 634 (1979).
11. S. J. Angyal and W. K. Warburton, *J. Chem. Soc.* p. 2492 (1951).
12. G. Schwarzenbach and K. Lutz, *Helv. Chim. Acta* **23**, 1162 (1940).
13. A. Albert, J. A. Mills, and R. Royer, *J. Chem. Soc.* p. 1452 (1947).
14. L. P. Hammett and A. J. Deyrup, *J. Am. Chem. Soc.* **54**, 2721 (1932); **55**, 1901 (1933).

15. L. P. Hammett, "Physical Organic Chemistry." McGraw Hill, New York, 1st ed., 1940; 2nd ed., 1970.
16. C. H. Rochester, "Acidity Functions." Academic Press, London, 1970.
17. M. A. Paul and F. A. Long, *Chem. Rev.* **57**, 1 (1957).
17a. See also ref. *87*.
18. M. J. Jorgenson and D. R. Hartter, *J. Am. Chem. Soc.* **85**, 878 (1963).
19. K. Yates and H. Wai, *J. Am. Chem. Soc.* **86**, 5408 (1964).
20. R. W. Taft, *J. Am. Chem. Soc.* **82**, 2965 (1960).
21. R. H. Boyd, *J. Am. Chem. Soc.* **85**, 1555 (1963).
22. E. M. Arnett and G. W. Mach, *J. Am. Chem. Soc.* **86**, 2671 (1964); **88**, 1177 (1966).
23. J. T. Edward, *Trans. R. Soc. Can.* **2**, 313 (1964).
23a. J. W. Eastes, M. H. Aldridge, and M. J. Kamlet, *J. Chem. Soc. B* p. 922 (1969).
24. E. M. Arnett and G. Scorrano, *Adv. Phys. Org. Chem.* **13**, 83 (1976).
25. D. D. Perrin, *Aust. J. Chem.* **17**, 484 (1964).
26. D. D. Perrin, "Dissociation Constants of Organic Bases in Aqueous Solution" Butterworths, London, 1965 (1st ed.); 1972 (Suppl.).
27. L. L. Kuznetsov and B. V. Gidaspov, *J. Org. Chem. USSR (Engl. Transl.)* **18**, 595 (1982).
28. R. S. Ryabova, I. M. Medvetskaya, and M. I. Vinnick, *Russ. J. Phys. Chem. (Engl. Transl.)* **40**, 182 (1966).
29. K. Yates and H. Wai, *Can. J. Chem.* **43**, 2131 (1965).
30. C. D. Johnson, A. R. Katritzky, and S. A. Shapiro, *J. Am. Chem. Soc.* **91**, 6654 (1969).
30a. R. A. Cox and K. Yates, *Can. J. Chem.* **62**, 2155 (1984).
31. J. J. Christensen, R. M. Izatt, D. P. Wrathall, and L. D. Hansen, *J. Chem. Soc. A* p. 1212 (1969); see also ref. *34*.
32. J. W. Larson and L. G. Hepler, *in* "Solute–Solvent Interactions" (J. F. Coetzee and C. D. Ritchie, eds.), Chap. 1. Dekker, New York, 1969.
33. E. M. Arnett, P. P. Quirk, and J. J. Burke, *J. Am. Chem. Soc.* **92**, 1260 (1970).
33a. D. L. Hughes and E. M. Arnett, *J. Am. Chem. Soc.* **105**, 4157 (1983).
34. J. J. Christensen, L. D. Hansen, and R. M. Izatt, "Handbook of Proton Ionization Heats and Related Thermodynamic Quantities." Wiley, New York, 1976.
35. A. G. Evans and S. D. Hamann, *Trans. Faraday Soc.* **47**, 34 (1951).
36. E. M. Arnett, *in* "Proton Transfer Reactions" (E. Caldin and V. Gold, ed.), p. 89. Chapman and Hall, London, 1975.
37. H. B. Hetzner, R. A. Robinson and R. G. Bates, *J. Phys. Chem.* **66**, 2696 (1962).
38. H. B. Hetzner, R. G. Bates, and R. A. Robinson, *J. Phys. Chem.* **67**, 1124 (1963).
39. H. B. Hetzner, R. G. Bates, and R. A. Robinson, *J. Phys. Chem.* **70**, 2869 (1966).
40. H. B. Hetzner, R. A. Robinson, and R. G. Bates, *J. Phys. Chem.* **72**, 2081 (1968).
41. P. B. Smith, J. L. Dye, J. Cheney, and J.-M. Lehn, *J. Am. Chem. Soc.* **103**, 6044 (1981).
42. S. P. Datta and A. K. Gryzbowski, *J. Chem. Soc. B* p. 136, (1966).
43. R. W. Green, *Aust. J. Chem.* **22**, 721 (1969).
43a. M. D. Rozeboom, K. N. Houk, S. Searles, and S. E. Seyedrezai, *J. Am. Chem. Soc.* **104**, 3448 (1982).
43b. M. R. Grimmett *Adv. Heterocycl. Chem.* **27**, 241 (1980).
44. R. P. Bell, "The Proton in Chemistry," 2nd ed., p. 51. Cornell Univ. Press, Ithaca, New York, 1973.
45. J. P. Briggs, R. Yamdagni, and P. Kebarle, *J. Am. Chem. Soc.* **94**, 5128 (1972).
45a. S. Ikuta and P. Kebarle, *Can. J. Chem.* **61**, 97 (1983).
46. R. W. Taft, ref. *36*, Chapter 2.
47. E. M. Arnett, F. M. Jones, M. Taagepera, W. G. Henderon, J. L. Beauchamp, D. Holtz, and R. W. Taft, *J. Am. Chem. Soc.* **94**, 4724 (1972).

48. M. T. Bowers, D. H. Aue, H. M. Webb, and R. T. McIver, *J. Am. Chem. Soc.* **93**, 4314 (1971).

49. K. D. Summerhays, S. K. Pollack, R. W. Taft, and W. J. Hehre, *J. Am. Chem. Soc.* **99**, 4585 (1977).

50. H. C. Brown and B. Kanner, *J. Am. Chem. Soc.* **88**, 986 (1966); see also E. M. Arnett and B. Chawla, *J. Am. Chem. Soc.* **101**, 7141 (1979); **103**, 7036 (1981); and E. M. Arnett, B. Chawla, L. Bell, M. Taagepera, W. J. Hehre, and R. W. Taft, *J. Am. Chem. Soc.* **99**, 5729 (1977).

51. M. Meot-Ner and L. W. Sieck, *J. Am. Chem. Soc.* **105**, 2956 (1983); see also H. P. Hopkins, D. V. Jahagirdar, P. S. Moulik, D. H. Aue, H. M. Webb, W. R. Davidson, and M. D. Pedley, *J. Am. Chem. Soc.* **106**, 4341 (1984).

52. R. W. Alder, P. S. Bowman, W. R. S. Steele, and D. R. Winterman, *Chem. Commun.* p. 723 (1968).

53. R. W. Alder, N. C. Goode, N. Miller, F. Hibbert, K. P. P. Hunte, and H. J. Robbins, *Chem. Commun.* p. 89 (1978).

53a. H. A. Staub, T. Saupe, and C. Krieger, *Angew. Chem., Int. Ed. Engl.* **22**, 731 (1981).

54. A. Bryson, *J. Am. Chem. Soc.* **82**, 4862 (1960).

55. E. M. Arnett, K. G. Venkatasubramaniam, R. T. McIver, Jr., E. K. Fukuda, F. G. Bordwell, and R. D. Press, *J. Am. Chem. Soc.* **104**, 325 (1982).

56. N. F. Hall and M. R. Sprinkle, *J. Am. Chem. Soc.* **54**, 3469 (1932).

57. F. Hibbert and K. P. P. Hunte, *J. Chem. Soc., Perkin Trans. 2* p. 1895 (1983).

58. F. Hibbert, *J. Chem. Soc., Perkin Trans. 2* p. 1862 (1974).

59. A. Awwal and F. Hibbert, *J. Chem. Soc., Perkin Trans. 2* p. 1589 (1977).

60. H. E. Simmons and C. H. Park, *J. Am. Chem. Soc.* **90**, 2428 (1968).

61. J. Cheney and J.-M. Lehn, *Chem. Commun.* p. 487 (1972); see also refs. *41*, *62*, and *63*.

62. J. Cheney, J. P. Kintzinger, and J.-M. Lehn, *Nouv. J. Chim.* p. 411 (1978).

63. B. G. Cox and H. Schneider, *J. Chem. Soc., Perkin Trans. 2* p. 1293 (1979).

64. P. F. Jackson, K. J. Morgan and A. M. Turner, *J. Chem. Soc., Perkin Trans. 2* p. 1582 (1972).

65. R. J. Motekaitis and A. E. Martell, *Can. J. Chem.* **60**, 168 (1982).

66. H. Pracejus, M. Kehlen, H. Kehlen, and H. Matschiner, *Tetrahedron* **21**, 2257 (1965).

67. J. T. Edward, H. S. Chang, K. Yates, and R. Stewart, *Can. J. Chem.* **38**, 1518 (1960).

68. K. Yates, J. B. Stevens, and A. R. Katritzky, *Can. J. Chem.* **42**, 1957 (1964).

69. J. F. Bunnett and F. P. Olsen, *Can. J. Chem.* **44**, 1899 (1966).

70. J. T. Edward and S. C. Wong, *Can. J. Chem.* **55**, 2492 (1977).

71. J. T. Edward and I. C. Wang, *Can. J. Chem.* **40**, 966 (1962).

72. K. Yates and J. B. Stevens, *Can. J. Chem.* **43**, 529 (1965).

73. J. W. Barnett and C. J. O'Connor, *J. Chem. Soc., Perkin Trans. 2* p. 220 (1973).

74. R. Stewart, L. J. Muenster, and J. T. Edward, *Chem. Ind.* p. 1906, (1961).

75. R. Stewart and L. J. Muenster, *Can. J. Chem.* **39**, 401 (1961).

76. M. Zief and J. T. Edsall, *J. Am. Chem. Soc.* **59**, 2245 (1937).

77. S. Basterfield and J. W. Tomecko, *Can. J. Res.* **8**, 458 (1933).

78. G. Alberghina, S. Fisichella, S. Occhipinti, G. Consiglio, D. Spinelli, and R. Noto, *J. Chem. Soc., Perkin Trans. 2* p. 1223 (1982).

79. P. Lemetais and J.-M. Carpentier, *J. Chem. Res. (S)* p. 282 (1981).

80. G. D. Frederick and C. D. Poulter, *J. Am. Chem. Soc.* **97**, 1797 (1975).

81. R. J. Gillespie, T. E. Peel, and E. A. Robinson, *J. Am. Chem. Soc.* **93**, 5083 (1971).

82. R. A. Cox and K. Yates, *J. Am. Chem. Soc.* **100**, 3861 (1978).

83. H. M. Grant, P. McTigue, and D. G. Ward, *Aust. J. Chem.* **36**, 2211 (1983).

84. K. Yates and R. A. McClelland, *J. Am. Chem. Soc.* **89**, 2686 (1967); see also refs. *85* and *86*.

85. R. F. Cookson, *Chem. Rev.* **74**, 5 (1974).

86. U. L. Haldna, *Russ. Chem. Rev. (Engl. Trans.)* **49**, 623 (1980).

87. A. J. Kresge, H. J. Chen, G. L. Capen, and M. F. Powell, *Can. J. Chem.* **61**, 249 (1983).

88. R. A. Cox and K. Yates, *Can. J. Chem.* **59**, 2116 (1981).
89. R. A. Cox and K. Yates, *Can. J. Chem.* **61**, 2225 (1983).
90. N. C. Marziano, G. M. Cimino, and R. C. Passerini, *J. Chem. Soc., Perkin Trans. 2* p. 1915, (1973).
91. N. C. Marziano, P. G. Traverso, A. Tomasin, and R. C. Passerini, *J. Chem. Soc., Perkin Trans. 2* p. 309 (1977).
92. N. C. Marziano, A. Tomasin, and P. G. Traverso, *J. Chem. Soc., Perkin Trans. 2* p. 1070 (1981).
93. R. Passerini, N. C. Marziano, and P. G. Traverso, *Gazz. Chim. Ital.* **105**, 901 (1975).
94. N. C. Marziano, P. G. Traverso, and R. C. Passerini, *J. Chem. Soc., Perkin Trans. 2* p. 306 (1977).
95. C. Perrin, *J. Am. Chem. Soc.* **86**, 256 (1964).
96. Ref. *24*, p. 94.
97. V. Lucchini, G. Modena, G. Scorrano, R. A. Cox, and K. Yates, *J. Am. Chem. Soc.* **104**, 1958 (1982).
98. R. L. Reeves, *J. Am. Chem. Soc.* **88**, 2240 (1966).
99. J. T. Edward and S. C. Wong, *J. Am. Chem. Soc.* **99**, 4229 (1977).
100. J. T. Edward, M. Sjöstrom, and S. Wold, *Can. J. Chem.* **59**, 2350 (1981).
100a. R. I. Zalewski, A. Y. Sarkice and Z. Geltz, *J. Chem. Soc., Perkin Trans. 2* p. 1059 (1983).
101. K. M. Harmon and T. T. Coburn, *J. Am. Chem. Soc.* **87**, 2499 (1965).
102. T. G. Bonner and J. Phillips, *J. Chem. Soc. B* p. 650 (1966).
103. C. C. Greig and C. D. Johnson, *J. Am. Chem. Soc.* **90**, 6453 (1968).
104. R. Stewart, M. R. Granger, R. B. Moodie, and L. J. Muenster, *Can. J. Chem.* **41**, 1065 (1963).
105. R. Stewart and K. Yates, *J. Am. Chem. Soc.* **80**, 6355 (1958).
106. K. Yates and R. Stewart, *Can. J. Chem.* **37**, 664 (1959).
107. W. M. Schubert and R. E. Zahler, *J. Am. Chem. Soc.* **76**, 1 (1954).
108. R. I. Zalewski and G. E. Dunn, *Can. J. Chem.* **46**, 2469 (1968).
109. R. A. McClelland and W. F. Reynolds, *Can. J. Chem.* **54**, 718 (1976).
110. R. Stewart and K. Yates, *J. Am. Chem. Soc.* **82**, 4059 (1960).
111. R. A. Cox, M. F. Goldman, and K. Yates, *Can. J. Chem.* **57**, 2960 (1979).
112. G. Alberghina, M. E. Amato, S. Finichella, and S. Occhipinti, *Gazz. Chim. Ital.* **112**, 281 (1982).
113. D. Dolman and R. Stewart, *Can. J. Chem.* **45**, 903 (1967).
114. M. Azzaro, J. F. Gad, and S. Geribaldi, *J. Org. Chem.* **47**, 4981 (1982).
115. M. R. Sharif and R. I. Zalewski, *Bull. Acad. Pol. Sci. Ser. Sci., Chim.* **29**, 385 (1981).
115a. R. Stewart, unpublished results.
115b. E. Paspaleev, A. Kozhukarova, *Monatsh. Chem.* **100**, 1213 (1969).
116. V. A. Palm, Ü. L. Haldna, and A. J. Talvik, *in* "The Chemistry of the Carbonyl Group" (S. Patai, ed.), p. 439. Wiley, New York, 1966.
117. R. A. Cox, C. R. Smith, and K. Yates, *Can. J. Chem.* **57**, 2952 (1979).
118. H. J. Campbell and J. T. Edward, *Can. J. Chem.* **38**, 2109 (1960).
119. P. Lemetais and J.-M. Carpentier, *J. Chem. Res. (S)* p. 204 (1982).
120. G. Perdoncin and G. Scorrano, *J. Am. Chem. Soc.* **99**, 6983 (1977).
120a. Ref. *24*, p. 105.
121. G. F. Olah, A. L. Berrier, and G. K. S. Prakash, *J. Am. Chem. Soc.* **104**, 2373 (1982).
122. E. M. Arnett, R. P. Quirk, and J. W. Larsen, *J. Am. Chem. Soc.* **92**, 3977 (1970); see also ref. *123*.
123. R. F. Childs, D. L. Mulholland, A. Varadarajan, and S. Yeroushalmi, *J. Org. Chem.* **48**, 1431 (1983).
124. Ref. *24*, p. 131.

125. Ref. *9*, p. 284.

126. D. G. Lee and R. Cameron, *J. Am. Chem. Soc.* **93**, 4724 (1971).

127. R. W. Gurney, "Ionic Processes in Solution," p. 75. McGraw-Hill, New York, 1953.

128. L. Thomas and E. Marum, *Z. Phys. Chem.* **143**, 213 (1929).

129. N. C. Deno and J. O. Turner, *J. Org. Chem.* **31**, 1969 (1966).

130. N. C. Deno and M. J. Wisotsky, *J. Am. Chem. Soc.* **85**, 1735 (1963).

131. E. M. Arnett and C. Y. Wu, *J. Am. Chem. Soc.* **84**, 1680 (1962); see also ref. *132.*

132. J. T. Edward, J. B. Leane, and I. C. Wang, *Can. J. Chem.* **40**, 1521 (1962).

132a. A. Koeberg-Telder, H. J. A. Lambrechts, and H. Cerfontain, *J. R. Neth. Chem. Soc.* **102**, 293 (1983).

132b. H. J. A. Lambrechts and H. Cerfontain, *J. R. Neth. Chem. Soc.* **102**, 299 (1983).

132c. M. Meot-Ner, *J. Am. Chem. Soc.* **105**, 4906 (1983); R. B. Sharma, A. T. Blades, and P. Kebarle, *J. Am. Chem. Soc.* **106**, 510 (1984).

133. R. J. Abraham, E. Bullock, and S. S. Mitra, *Can. J. Chem.* **37**, 1859 (1959).

134. Y. Chiang and E. B. Whipple, *J. Am. Chem. Soc.* **85**, 2763 (1963).

135. R. L. Hinman and J. Lang, *J. Am. Chem. Soc.* **86**, 3796 (1964).

136. W. H. Mills and I. G. Nixon, *J. Chem. Soc.* p. 2510 (1930).

136a. R. H. Mitchell, P. D. Slowey, T. Kamada, R. V. Williams, and P. J. Garratt, *J. Am. Chem. Soc.* **106**, 2431 (1984) and refs. therein.

137. For a theoretical discussion of the protonation of indoles and pyrroles, see J. Catalan and M. Yanez, *J. Am. Chem. Soc.* **106**, 421 (1984).

138. H. Kohler and H. Scheibe, *Z. Anorg. Allg. Chem.* **285**, 221 (1956).

139. R. J. Highet and F. E. Chou, *J. Am. Chem. Soc.* **99**, 3538 (1977).

140. A. J. Kresge, G. W. Barry, K. R. Charles, and Y. Chiang, *J. Am. Chem. Soc.* **84**, 4343 (1962).

141. W. M. Schubert and R. H. Quacchia, *J. Am. Chem. Soc.* **85**, 1278 (1963).

142. T. Yamaoka, H. Hosoya, and S. Nagakura, *Tetrahedron* **26**, 4125 (1970).

143. T. Birchall, A. N. Bourns, R. J. Gillespie, and P. J. Smith, *Can. J. Chem.* **42**, 1433 (1964).

144. D. M. Brouwer, E. L. Mackor, and C. MacLean, *Rec. Trav. Chim. Pay-Bas* **85**, 109 (1966).

145. N. C. Deno, P. T. Groves, and G. Saines, *J. Am. Chem. Soc.* **81**, 5790 (1959).

146. N. C. Deno, H. E. Berkheimer, W. L. Evans, and H. J. Peterson, *J. Am. Chem. Soc.* **81**, 2344 (1959).

147. N. C. Deno, J. J. Jaruzelski, and A. Schriesheim, *J. Am. Chem. Soc.* **77**, 3044 (1955).

148. Ref. *16*, p. 88.

149. M. Liler, "Reaction Mechanisms in Sulphuric Acid," p. 30. Academic Press, London, 1971.

150. M. T. Reagan, *J. Am. Chem. Soc.* **91**, 5506 (1969).

151. J. Schulze and F. A. Long, *J. Am. Chem. Soc.* **86**, 322 (1964).

152. F. A. Long and J. Schulze, *J. Am. Chem. Soc.* **86**, 327 (1964).

153. Ref. *24*, p. 104.

154. J. P. Richard and W. P. Jencks, *J. Am. Chem. Soc.* **106**, 1373 (1984).

155. See also G. W. Ellis and C. D. Johnson, *J. Chem. Soc. Perkin Trans. 2* p. 1025 (1982) and refs. therein.

156. R. J. Gillespie, *Acc. Chem. Res.* **1**, 202 (1968).

157. D. Farcasiu, *Acc. Chem. Res.* **15**, 46 (1982).

158. G. A. Olah, J. S. Staral, G. Asencio, G. Liang, D. A. Forsyth, and G. D. Mateescu, *J. Am. Chem. Soc.* **100**, 6299 (1978).

159. J. M. McGarrity and D. P. Cox, *J. Am. Chem. Soc.* **105**, 3961 (1983).

160. A. Mertens, K. Lammertsma, M. Arvanaghi, and G. A. Olah, *J. Am. Chem. Soc.* **105**, 5657 (1983).

161. G. K. S. Prakash, T. N. Rawdah, and G. A. Olah, *Angew. Chem., Int. Ed. Engl.* **22**, 390 (1983).

162. G. A. Olah and D. A. Forsyth, *J. Am. Chem. Soc.* **98**, 4086 (1976).

163. K. Lammertsma, G. A. Olah, C. M. Berke, and A. Streitwieser, Jr., *J. Am. Chem. Soc.* **101,** 6658 (1979).
164. See also K. Lammertsma and H. Cerfontain, *J. Am. Chem. Soc.* **102,** 3257 (1980).
165. R. J. Gillespie, *Endeavour* **32,** 3 (1973).
166. R. J. Gillespie and T. E. Peel, *J. Am. Chem. Soc.* **95,** 5173 (1973).
167. P.-L. Fabre, J. Devynck, and B. Trémillon, *Chem. Rev.* **82,** 591 (1982).
168. V. Gold, K. Laali, K. P. Morris, and L. Z. Zdunek, *Chem. Commun.* p. 769 (1981).
169. See also D. E. Smith and B. Munson, *J. Am. Chem. Soc.* **100,** 497 (1978).
170. G. A. Olah and R. H. Schlosberg, *J. Am. Chem. Soc.* **90,** 2726 (1968).
171. G. A. Olah and J. Lukas, *J. Am. Chem. Soc.* **89,** 2227 (1967).
172. G. A. Olah and J. Lukas, *J. Am. Chem. Soc.* **89,** 4743 (1967).
173. H. Hogeveen and A. F. Bickel, *Chem. Commun.* p. 635 (1967).
174. See E. M. Arnett and J. F. Wolf, *J. Am. Chem. Soc.* **95,** 978 (1973).
175. J. T. Edward, I. Lantos, G. D. Derdall, and S. C. Wong, *Can. J. Chem.* **55,** 812 (1977).
176. See also C. Tissier and M. Tissier, *Bull. Soc. Chim. Fr.* p. 2109 (1972) and also ref. *177.*
177. M. J. Janssen, *Rec. Trav. Chim. Pay-Bas* **81,** 650 (1962).
178. T. Birchall and R. J. Gillespie, *Can. J. Chem.* **41,** 2642 (1963).
179. G. A. Olah and A. T. Ku, *J. Org. Chem.* **35,** 331 (1970).
180. C. Capobianco, F. Magno, and G. Scorrano, *J. Org. Chem.* **44,** 1654 (1979).
181. J. Voss, *in* "The Chemistry of Acid Derivatives" (S Patai, ed.), Suppl. B, Part 2, p. 1040. Wiley, Chichester, 1979.
182. J. T. Edward, S. C. Wong, and G. Welch, *Can. J. Chem.* **56,** 931 (1978).
183. R. A. Cox and K. Yates, *Can. J. Chem.* **60,** 3061 (1982).
184. J. T. Edward, G. Welch, and S. C. Wong, *Can. J. Chem.* **56,** 935 (1978).
185. M. Liler, *J. Chem. Soc. B* p. 385 (1969).
186. T. Birchall and R. J. Gillespie, *Can. J. Chem.* **41,** 2642 (1963).
187. R. G. Laughlin, *J. Am. Chem. Soc.* **89,** 4268 (1967).
188. Ref. *149,* p. 110.
189. R. H. Hartigan and J. B. Cloke, *J. Am. Chem. Soc.* **67,** 709 (1945).
190. H. G. Richey and J. M. Richey, *in* "Carbonium Ions" (G. A. Olah and P. R. Schleyer, eds.), Vol. II, p. 900. Wiley (Interscience), New York, 1970.
191. D. S. Noyce, M. A. Matesich, M. D. Schiavelli, and P. E. Peterson, *J. Am. Chem. Soc.* **87,** 2295 (1965).
192. G. W. L. Ellis and C. D. Johnson, *J. Chem. Soc., Perkin Trans. 2* p. 1025 (1982).
193. A. D. Allen, Y. Chiang, A. J. Kresge, and T. T. Tidwell, *J. Org. Chem.* **47,** 775 (1982); see also ref. *194–195.*
194. G. H. Schmid, *in* "The Chemistry of the Carbon–Carbon Triple Bond" (S. Patai, ed.), Chap. 8. Wiley, New York, 1978.
194a. P. J. Stang, Z. Rapaport, M. Hanack, L. R. Subramanian, "Vinyl Cations," Chap. 4. Academic Press, New York, 1979.
194b. G. H. Schmid, A. Modro, and K. Yates, *J. Org. Chem.* **45,** 665 (1970).
195. F. Marcuzzi, G. Melloni and, G. Modena, *J. Org. Chem.* **44,** 3022 (1979).
196. T. C. Bissot, R. W. Parry, and D. H. Campbell, *J. Am. Chem. Soc.* **79,** 796 (1957).
197. J. B. Culbertson, *J. Am. Chem. Soc.* **73,** 4818 (1951).
198. D. D. Perrin, B. Dempsey, and E. P. Sergeant, "pK_a Prediction for Organic Acids and Bases." Chapman and Hall, London, 1981.
199. J. B. Conant and P. D. Bartlett, *J. Am. Chem. Soc.* **54,** 2881 (1932); see also ref. *200.*
200. P. R. Ellefsen and L. Gordon, *Talanta* **14,** 409 (1967).
201. M. Liler and D. Kosanovic, *J. Chem. Soc.* p. 1084, (1958).
202. Y. N. Lee and G. L. Schmir, *J. Am. Chem. Soc.* **100,** 6700 (1978).

203. E. Y. Belyaev, L. M. Gornostaev, M. S. Tovbis, and L. E. Borina, *Zh. Obshch. Khim.* **44,** 856 (1974); *Chem. Abstr.* **81,** 12921 (1974).

204. R. Stewart, "Oxidation Mechanisms: Applications to Organic Chemistry," Chap. 2. Benjamin, New York, 1964.

205. E. H. Cordes and W. P. Jencks, *J. Am. Chem. Soc.* **84,** 832 (1962).

205a. R. L. Reeves and W. F. Smith, *J. Am. Chem. Soc.* **85,** 724 (1963).

206. R. J. Gillespie, T. E. Peel, and E. A. Robinson, *J. Am. Chem. Soc.* **93,** 5083 (1971).

207. H. H. Jaffé and G. O. Doak, *J. Am. Chem. Soc.* **77,** 4441 (1955).

208. C. D. Johnson, A. R. Katritzky, B. J. Ridgewell, N. Shakir, and A. M. White, *Tetrahedron* **21,** 1055 (1965).

209. G. A. Olah and T. E. Kiovsky, *J. Am. Chem. Soc.* **90,** 6461 (1968).

210. C.-S. Hahn and H. H. Jaffé, *J. Am. Chem. Soc.* **84,** 949 (1962).

211. H. H. Jaffé and R. W. Gardner, *J. Am. Chem. Soc.* **80,** 319 (1958).

212. Ref. *149*, p. 112 and p. 114.

213. P. Haake, R. D. Cook, and G. H. Hurst, *J. Am. Chem. Soc.* **89,** 2650 (1967).

214. B. Silver and Z. Luz, *J. Am. Chem. Soc.* **83,** 786 (1961).

4

Proton Transfer, Hydrogen Atom Transfer, and Hydride Transfer

I. Introduction

The transfer of a hydrogen nucleus in a chemical reaction may take place by means other than proton transfer, that is, as the bare nucleus. Transfers of hydrogen atoms and hydride ions are also well-known processes, and these entities share with the proton the property of lacking an independent existence in solution. Their importance in solution is as transferred entities, and drawing distinctions between reactions involving these three sorts of process amounts to examining the structures of starting materials and products (or intermediates) and deducing the identity of the transferred entity by this means, that is, by difference.

Most proton transfer reactions between electronegative atoms can be characterized as very rapid processes in which equilibrium is established essentially instantaneously. [Reactions of cryptands are the most striking, but by no means the only, exceptions to this rule of thumb (1–7).] Hydrogen atom transfers and hydride transfers are sometimes very fast but, unlike proton transfers, they are seldom part of a mobile equilibrium, although a number of intramolecular hydride exchange reactions qualify in this respect and can be followed by dynamic nmr techniques (8–10); furthermore, biological systems have, in the form of coenzyme Q, for example, essentially hydride buffers (11).

There is no difficulty in identifying the reaction of an amine with acetic acid, for example, as proton transfer, and one can be reasonably confident, on the evidence, that the conversion of carbonium ions to alkanes by formate ion and other reagents is simple hydride transfer (12–17). However, there are many instances where the distinction is not as easily drawn. Such reactions are invariably oxidation–reduction processes. Whereas acid–base chemistry is just that—the transfer of a proton (adopting the Brønsted–Lowry definition of acids)—oxidation–reduction reactions may take place by a variety of processes, including electron transfer, hydrogen atom transfer, hydride transfer, and, provided that other conditions are met, even proton transfer. These conditions are that some other process, often one of those just referred to, take place elsewhere in the molecule.

146

By a commonly accepted though seldom enunciated definition, oxidation processes are those that involve the transfer either of electrons or of ions or molecules in what is not their "normal valence state," that is, not as protons, chloride ions, water molecules, and so on (18). Thus, proton transfer can only be a part of an oxidation process; it must be coupled to electron reorganization elsewhere that produces a net transfer of species such as a hydride ion, a chlorine atom, or a hydroxyl radical. That is not to say that proton transfers, where they occur in an oxidation reaction, are incidental to the oxidation process. They may occur in a preequilibrium step, as in Eq. (4-1),

$$
\begin{array}{c}
\text{OH} \\
| \\
C_6H_5\text{CHCF}_3 + \text{OH}^-
\end{array}
\rightleftharpoons
\begin{array}{c}
\text{O}^- \\
| \\
C_6H_5\text{CHCF}_3 + H_2O
\end{array}
$$

$$
\begin{array}{c}
\text{O}^- \\
| \\
C_6H_5\text{CHCF}_3
\end{array}
\xrightarrow{\text{Mn(VII)}}
\begin{array}{c}
\text{O}^- \\
| \\
H_5C_6-\overset{\cdot}{C}-CF_3
\end{array}
\longrightarrow
\begin{array}{c}
\text{O} \\
\| \\
H_5C_6-C-CF_3
\end{array}
$$

(4-1)

where activation is accomplished by deprotonation of the alcohol, or in the rate-controlling step, as in Eq. (4-2). (The degree of activation brought about

$$ C_6H_5CH_2OH + HNO_3 \rightleftharpoons C_6H_5CH_2ONO_2 + H_2O $$

$$
C_6H_5CH_2ONO_2 + NaOC_2H_5 \longrightarrow H_5C_6-C\overset{\diagup O}{\diagdown H} + NaNO_2 + C_2H_5OH
$$

(4-2)

by protonation or deprotonation is considered in some detail in Chapter 7 for both oxidative and nonoxidative processes.)

Equation (4-2), which represents the oxidation of an alcohol by the sequential use of nitric acid and sodium ethoxide, is analogous to the classical mechanism for chromic acid oxidation of alcohols (19, 20) in that a base removes a proton from the α-carbon atom of the ester while, simultaneously, an inorganic species whose oxidation state is lower than that originally used is expelled from the molecule. [Although the classical chromate ester mechanism in which a proton is transferred to a water molecule has wide currency, particularly in undergraduate textbooks, its validity is arguable (21–25). There is general agreement that chromate esters are intermediates in these reactions; it is the mode of C—H bond breaking that has been questioned (18, 24–27).]

Might the oxidant be organic rather than inorganic in proton-triggered reactions? The well-known Moffat–Pfitzner oxidation (28) using dimethyl sulfoxide as the oxidant is an example of such a process [Eq. (4-3)].

$$
\begin{array}{c}
\text{H} \\
| \\
R-C-O \\
| \quad \overset{+}{S}-CH_3 \\
H \quad CH_2
\end{array}
\longrightarrow RCHO + (CH_3)_2S
$$

(4-3)

Reactions triggered by proton transfer have also been suggested as likely routes for the enzymatic oxidation of various organic functional groups, particularly by Hamilton (29); these and alternative mechanisms are discussed in Section IV,A.

II. The Geometry of Hydrogen Transfer

It is generally agreed that the favored geometry for proton transfer is linear; that is, in the transfer of the proton from AH to B the transition state has linear geometry [Eq. (4-4)], although we shall see that deviations from linearity are not uncommon and need not be prohibitively expensive in energy terms.

$$A—H + :B \longrightarrow A \cdots H \cdots B \longrightarrow A: + H—B \qquad (4-4)$$

The situation with regard to hydrogen atom transfer is less conclusive, although these reactions also appear to favor linear geometry. On the other hand, there is not general agreement regarding the preferred geometry for hydride transfer. These three kinds of hydrogen transfer process are considered in some detail in the following sections.

A. The Geometry of Proton Transfer

Some of the most telling evidence regarding the geometry of proton transfer processes comes from studies of the hydrogen bond (30–32). The latter can be regarded as containing a partly transferred proton. If structural modifications are made to the proton donor or acceptor, their roles can be reversed. Thus, the more stable form can be either A—H $\cdots$ B or A $\cdots$ H—B, depending on the relative energies of the acids A—H and B—H and their conjugate bases A and B. Indeed, if enough pairs of related compounds A—H and B—H are available, the transition from reactants to products becomes continuous, in the same way that a series of snapshots becomes a film strip.

By such means Johnson and Rumon (33) examined the infrared spectra of 18 adducts of substituted benzoic acids and substituted pyridines and were able to relate the observed spectral changes to the difference in pK between the pyridine (pK_{BH^+}) and the carboxylic acid (pK_{HA}). In the pyridine–benzoic acid adduct there is a strong hydrogen bond between the carboxylic acid group and the nitrogen atom of the pyridine, but the infrared spectrum shows that the components of the adduct are essentially un-ionized, as in **1**. As the basicity of

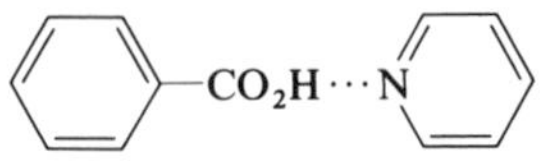

1

the pyridine and the acidity of the benzoic acid are increased, proton transfer takes place to produce salts, which are also capable of hydrogen bonding. If the difference in pK between the carboxylic acid and the pyridinium ion is sufficiently great, as, for example, with 2,6-dimethylpyridine and 2,4-dinitrobenzoic acid, the proton in question can be identified as being covalently attached to the nitrogen atom and hydrogen-bonded to the oxygen atoms of the carboxylate anion, as in **3**. There is a narrow region of pK difference, however, where the proton appears to be in a single potential energy well, bonded to both components, and this is the situation with, for example, the adduct of 3,5-dinitrobenzoic acid and 3,5-dimethylpyridine (**2**), for which

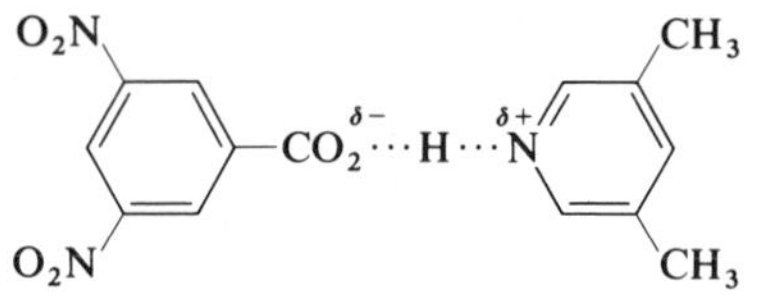

the designations "donor" and "acceptor" cannot be made. The shape of the potential energy well in strong hydrogen bonds has received a great deal of attention. The potential well may be asymmetrical and single (*as*), asymmetrical and double (*ad*), symmetrical and single (*ss*), or symmetrical and double (*sd*) (*34*).

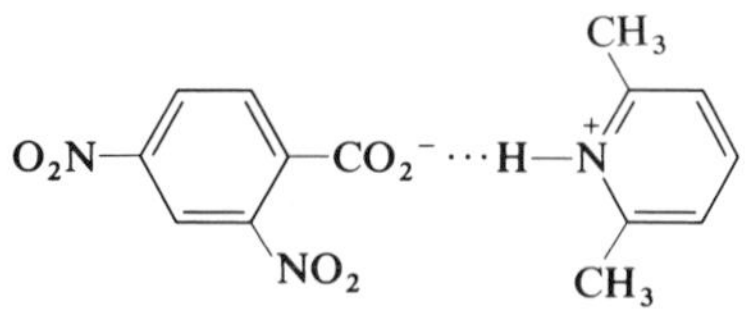

The basis for assigning the structures **1–3** to the hydrogen-bonded complexes is the appearance below 1700 cm^{-1} in the infrared spectra of **1** and **3** of two N–H and O–H bending frequencies, which can be attributed to a double-minimum potential function (*as*) for the proton stretching motion (*31, 35–40*). Although the benzoic acid–pyridine adducts were studied principally in the solid state, dissolution of the adducts in acetonitrile did not appear to change the results significantly.

Although the hydrogen bond seems to be a reasonable analogue of the partly transferred proton in an acid–base reaction, there are molecules in which the sites of hydrogen bonding and protonation differ. Thus, in carbon tetrachloride, 3-methyl-4-pyrimidone forms hydrogen bonds at the carbonyl oxygen atom with phenols of pK between 10.3 and 4.5; with stronger acids such as picric acid or hydrogen bromide, however, protonation occurs at N-1 of the ring. With phenols of intermediate pK both hydrogen bonding at oxygen and protonation at nitrogen take place [Eq. (4-5)] (*41*).

Presumably, a similar situation occurs with compounds such as phloroglucinol (**4**) that undergo ring protonation in strong acids but which would be expected to hydrogen-bond at oxygen [Eq. (4-6)] (*42, 43*).

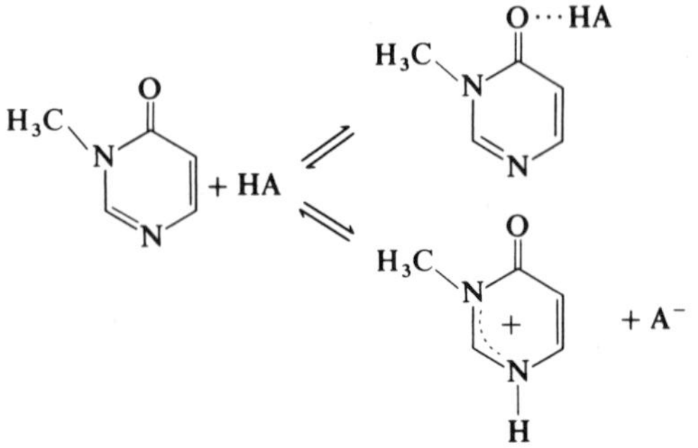

$$(4\text{-}5)$$

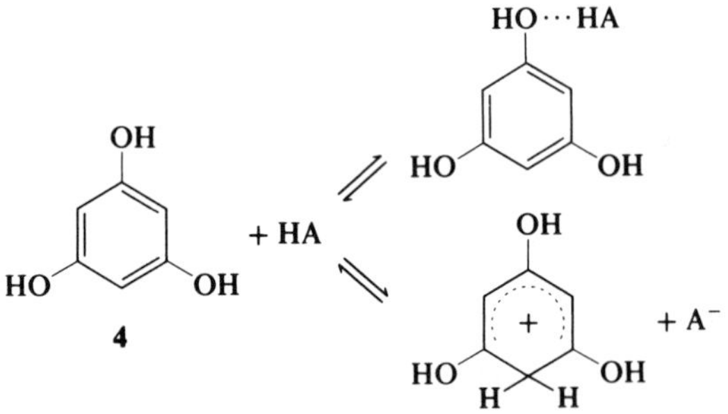

$$(4\text{-}6)$$

Taking these results as an indication that, on the whole, the hydrogen bond is a good model for proton transfer we can now look for evidence regarding the preferred geometry of the hydrogen bond in organic complexes.

1. Geometry of the Hydrogen Bond

The use of neutron diffraction and X-ray crystallography has provided an enormous amount of information regarding the geometry of the hydrogen bond in crystal lattices. Because of the importance of lattice energy effects the preferred geometry of a hydrogen bond may not be that which it would assume in isolation. Nonetheless, a linear or nearly linear arrangement is the one found most frequently in hydrogen bonds in crystalline compounds (44–46).

Though the frequency with which nonlinear hydrogen bonds are found shows that deviations from linearity are not excessively costly in terms of energy, the angular dependence of hydrogen bond interactions has been considered a factor in cooperative processes such as sharp melting of crystals and helical interactions in nucleic acids and polypeptides (47).

Systems with an X···H···Y bond angle of less than 150° have been designated by Brown et al. (48) as "highly nonlinear"; the quadrupole resonance spectra of a number of such systems have been examined. The relationship between bond distance and strength of hydrogen bonds has been reviewed by Novak (*49*).

The situation with regard to hydrogen bonds in solution is similar to that in crystals; a linear arrangement is in general preferred but not required. Luck has examined a number of cyclic lactams in carbon tetrachloride solution and shown them to form dimers only, in which the hydrogen bonds can attain a linear arrangement, as in **5** (*50*).

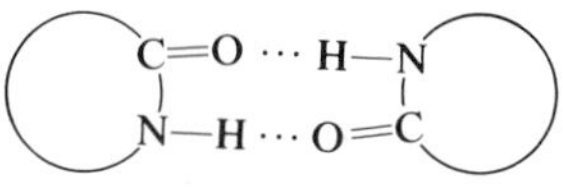

5

Carboxylic acids also prefer to exist as dimers with linear hydrogen bonds in low-polarity solvents; oximes form dimers in such solvents at low concentration but at higher concentrations form trimers, which can more easily form linear hydrogen bonds (*50–52*).

The preference for linear hydrogen bonds extends to the gas phase. Yamdagni and Kebarle have shown that interactions between protonated and unprotonated amino groups are greatest when a linear arrangement is possible (*53*). And it is not uncommon for those making theoretical studies of the charge-relay systems of certain enzymes simply to assume that the hydrogen bonds present in such systems are linear (*56*).

It was noted earlier that lattice effects can cause deviations from linearity for hydrogen bonds in crystals. Two other situations that can give rise to nonlinearity are the following: bifurcated bonds, as found in amino acids, and intramolecular bonds for which linearity is not possible, as found in *o*-halophenols (*35*) and *o*-hydroxythiophenols (*40*).

Bifurcated bonds are those in which a proton donor interacts with two acceptor atoms. Most such cases involve ammonio–carboxylate complexes (**6**)

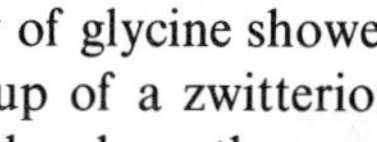

$$-\overset{|}{\underset{|}{N^+}}-H\cdot{}^-O_2C-$$

6

and appear to consist of one hydrogen bond that is almost normal in strength, though usually deviating from linearity, and one somewhat weaker interaction.

A neutron diffraction study of glycine showed that interactions between a proton on the ammonio group of a zwitterionic glycine molecule and the carboxylate anion of its neighbor have the geometry shown in **7** (*57, 58*).

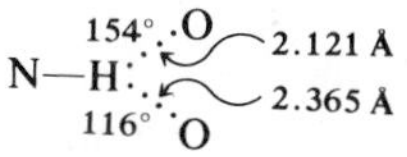

7

The hydrogen bonds of a number of other compounds, including tropolone, are also bifurcated in the solid state (59,60). (For more recent developments in the study of proton transfer in the solid state, see refs. 61 and 62 and refs. therein.)

The term *bifurcated* has also been used to describe hydrogen bonding of the type shown in **7a**, and the resulting ambiguity has led to suggestions that the latter interaction be called a *four-center* (or *trifurcated*) *hydrogen bond* and the type of bifurcation present in amino acids be called a *three-center hydrogen bond* (63, 64).

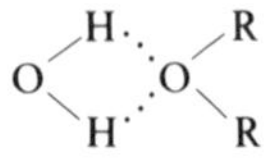

7a

In addition to lattice effects and bifurcation phenomena as causes of significant departure from linearity there exist a number of molecules whose geometry requires any interaction that is to take place to do so through sharply bent hydrogen bonds. Half a century ago Pauling showed that *o*-chlorophenol exhibits hydrogen bonding and that the difference in energy between the bonded and nonbonded forms is ~ 1.5 kcal [Eq. (4-7)] (35).

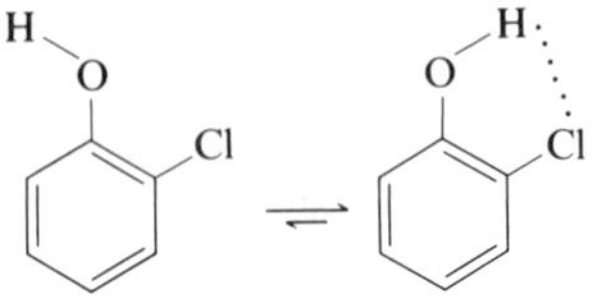

$$(4\text{-}7)$$

Clearly, linearity is not possible here. Nor is it possible in the case of the 1,2-diol monoacetates studied by Wright and Marchessault, where, as in the *o*-chlorophenol case, a hydrogen-bonded proton is a component of a five-membered ring (65).

Even though intramolecular hydrogen bonds are often found to be nonlinear there are indications that the strength of the bond increases if linearity can be attained. Thus, Fishman and Chen discovered that the bond strength of the intramolecular hydrogen bond found in the gas phase increases from ~ 1 kcal to ~ 3 kcal for 1,4-diols, where a nearly linear arrangement is possible because of the size of the ring (seven atoms including the binding hydrogen atom) (66, 67).

In the case of *N*-methyl-2-nitro-*p*-toluidine a shift from an internal hydrogen bond (undoubtedly nonlinear) to an external hydrogen bond (presumably linear) occurs as the hydrogen bond acceptor strength of the solvent increases [Eq. (4-8)]. Thus, Kamlet and Taft showed that the external bond is formed in hexamethylphosphoramide (HMPA) and the internal bond

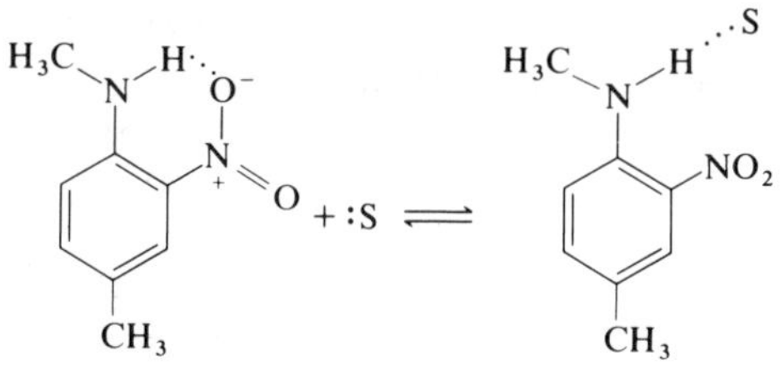

$$\text{(structures)} \quad \text{(4-8)}$$

is formed in anisole (68). Even in the strongly basic solvent N-methylpyrrolidone, which is a very efficient hydrogen bond acceptor (though not as efficient as HMPA), the hydrogen bond prefers to form internally, that is, to be bent.

In addition to the question of linearity of the three-atom triad in a hydrogen bond, there is the matter of orientation of the orbitals on the atom (usually oxygen) that is the proton acceptor. There are strong indications that hydrogen bonds formed between NH and carbonyl groups tend to occur in the direction of the conventionally viewed oxygen sp^2 pairs, as in **7b** (69).

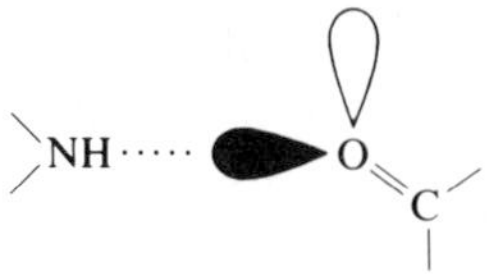

$$\text{NH} \cdots \text{O} = \text{C}$$

7b

2. Geometry of the Transition State for Proton Transfer

The conclusions that can be drawn from the evidence cited in the previous section are that the preferred geometry of the hydrogen bond is linear, though such an arrangement is by no means obligatory. In most cases where nonlinearity is observed a somewhat weaker bond results, though this does not apply to bifurcated bonds, where nonlinearity allows a secondary interaction to contribute to the overall strength of the bond. A hydrogen bond represents a partial transfer of a proton; is a linear pathway also preferred when the proton is completely transferred? Much of the evidence in this regard comes from studies of bifunctional catalysis, or tautomeric catalysis (70), and this work indicates a strong preference for a collinear arrangement of donor atom, proton, and acceptor atom.

Cox and Jencks studied the reaction of methoxyamine with phenylacetate catalyzed by general acids [Eq. (4-9)] (71, 72). A Brønsted plot for a wide

$$CH_3C\overset{O}{\underset{OC_6H_5}{\diagdown}} + CH_3ONH_2 \overset{HA}{\longrightarrow} CH_3C\overset{O}{\underset{NHOCH_3}{\diagdown}} + C_6H_5OH \quad (4\text{-}9)$$

selection of general acids shows a linear and a curved region. The linear section, found with the stronger acid catalysts ($pK_{HA} = 2$ to -2) has a slope of 0.16, indicating a rather modest degree of stabilization of the intermediate **8** shown in Eq. (4-10). (The rate-limiting step here is attack by the methoxyamine

$$
\begin{array}{c}
\underset{C_6H_5O}{\overset{H_3C}{>}}C\underset{\overset{+}{N}H_2}{\overset{O^-}{<}} + HA \longrightarrow \underset{C_6H_5O}{\overset{H_3C}{>}}C\underset{\overset{+}{N}H_2}{\overset{O^-\cdots HA}{<}} \longrightarrow \text{products} \qquad (4\text{-}10) \\
\underset{\textbf{8}}{\overset{|}{OCH_3}} \qquad\qquad \overset{|}{OCH_3}
\end{array}
$$

on the carbonyl group, assisted by a nearby molecule of general acid.) As the acid strength of the catalyst is decreased the reaction, of course, slows down and the rate-controlling step becomes the protonation of the oxygen anion by the catalyst. As the strength of the catalyzing acid is further decreased the slow step becomes the diffusion of the anion of the catalyst away from the complex. The net effect of these changes is a curved Brønsted plot. (See Chapter 7 for a fuller account of the Brønsted relation.)

There are a number of general acids that do not follow the curved Brønsted plot but, instead, fall on an extension of the linear portion of the curve. These catalysts, of course, are much more effective than their pK_{HA} values would suggest (up to 10^3 times more effective), and all of these catalysts have the potential of bifunctionality. They are cacodylic acid, bicarbonate ion, and the monoanions of phosphate, phosphonate, and methylarsonate. It is suggested that the protonation of the oxygen anion and the deprotonation of the nitrogen cation in **8** are concerted, as shown in **9** with phosphate monoanion

$$
\underset{C_6H_5O}{\overset{H_3C}{>}}C\underset{\underset{\overset{|}{OCH_3}}{\overset{|}{H}}}{\overset{O^-\cdots H-O}{\underset{N-H\cdots{}^-O}{<}}}\underset{OH}{\overset{O}{>}}P
$$

9

acting as the bifunctional catalyst. [Enolization of ketones, which requires proton removal from carbon, does not appear to make use of an analogous route, at least in aqueous solution (73).]

The common structural feature of such bifunctional catalysts is their capacity to form an eight-membered ring, which allows two proton transfers to occur in a linear fashion. Similar results have been observed in other systems where bifunctional catalysis appears to operate (74–82).

Lee and Schmir examined the effect of bifunctional catalysts on the hydrolysis of the imidate ester **9a** (83). In this reaction the "catalyst" changes the distribution of products rather than the rate of reaction [Eq. (4-11); asterisk indicates "catalyzed route"].

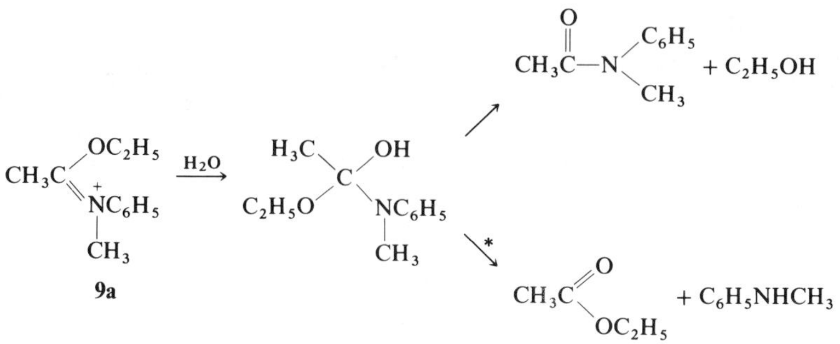

$$CH_3C \overset{O}{\underset{}{\parallel}} -N \overset{C_6H_5}{\underset{CH_3}{\diagup}} + C_2H_5OH$$

$$CH_3C \overset{OC_2H_5}{\underset{\overset{+}{NC_6H_5}}{\diagup}} \overset{H_2O}{\longrightarrow} \overset{H_3C}{\underset{C_2H_5O}{\diagup}} C \overset{OH}{\underset{NC_6H_5}{\diagdown}}$$

$$\overset{CH_3}{} \quad\quad \overset{CH_3}{}$$

9a

$$CH_3C \overset{O}{\underset{OC_2H_5}{\diagup}} + C_6H_5NHCH_3$$

(4-11)

In the presence of low concentrations of buffer catalyst the amide is the principal product, whereas at higher concentrations of buffer the ester is the principal product. Indeed, with arsenate buffer a concentration as low as 2×10^{-3} M arsenate is sufficient to produce a yield of ester that is greater than 95%.

As with the reaction studied by Cox and Jencks the most effective catalysts (those with rates much greater than the Brønsted relation predicts) are those that can add and remove protons simultaneously, that is, those that are bifunctional. They are arsenate, phosphate, and the hydrates of fluoro ketones, all of which bear acidic and basic groups in a 1,3 relationship that allows an eight-membered ring to form and a linear proton transfer to take place. The one exception is acetone oxime (10), which is a very effective bifunctional catalyst despite having the acidic and basic centers in a 1,2 relationship and hence being able to form only a seven-membered ring in the transition state.

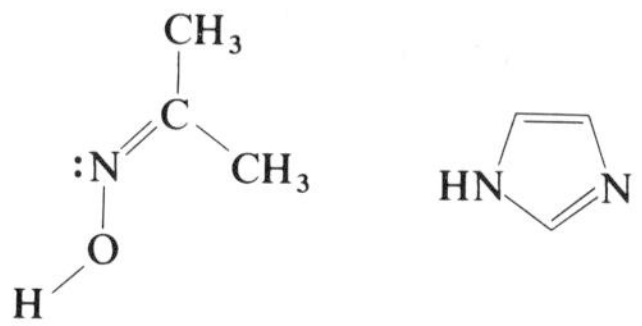

$$\overset{CH_3}{\underset{:N}{\overset{|}{C}}} \overset{}{\underset{\overset{O}{\underset{H}{\diagdown}}}{\diagdown}} CH_3 \quad\quad HN \overset{}{\underset{}{\diagdown}} N$$

10   11

It is clear, however, that a 1,3 arrangement of acidic and basic groups is not of itself sufficient to produce bifunctional catalysis if the geometry is such as to prevent a cyclic transition state from being formed. Thus, imidazole (11) does not show augmented catalytic effects in the aforementioned reaction since an eight-membered ring cannot accommodate linear proton transfers with this compound.

Though an 8-membered ring appears to be the most suitable for the double proton transfer that takes place in the aforementioned cases, a large number of reactions take place by means of single proton transfer in 6- and 7-membered rings. Indeed, in the intramolecular catalysis of enolization brought about by carboxylate and other groups, transition states containing 6- or 7-membered rings are common. Bell *et al.* (*84*) studied the rates of enolization of *o*-carboxyacetophenone (**12**) by measuring the rate of its iodination. They

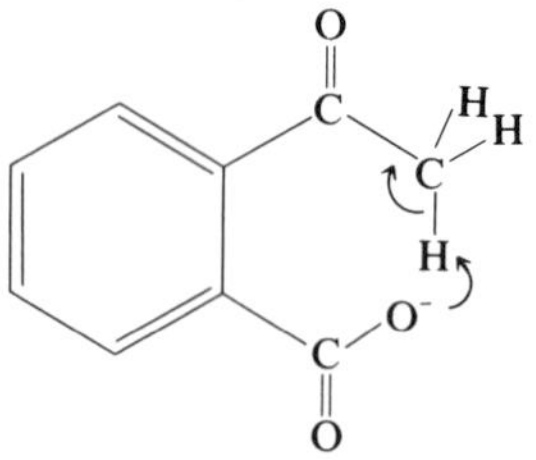

12

showed that intramolecular catalysis took place, involving in these cases proton transfer in a 7-membered ring [sometimes referred to as a $6\frac{1}{2}$-membered ring (*85*)]. Similar intramolecular effects have been observed with other ketones (*85–87*).

Though linearity is not as easily achieved with six- or seven-membered rings as with eight-membered rings, the deviations in bond angle appear in many cases to be only minor impediments to the reaction taking place by the intramolecular pathway, particularly when a single proton transfer is involved.

Intramolecular catalysis of enolization also takes place in aliphatic ketocarboxylic acids such as **13**, where both acid catalysis (involving the neutral carboxylic acid group) and base catalysis (involving the carboxylate ion) are observed. In this series the compound most susceptible to intramolecular attack by carboxylate has $n = 3$, in which the transition state for the base-catalyzed route (**14**) contains a six-membered ring (*88*).

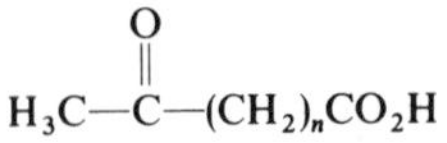

13 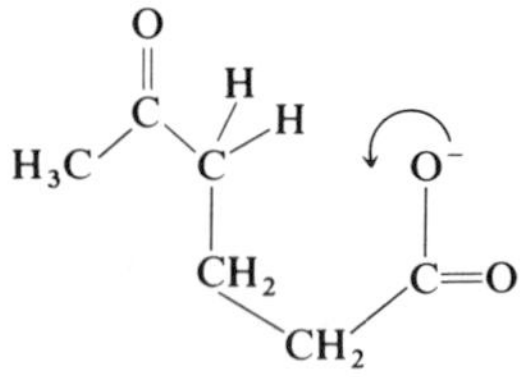   **14** 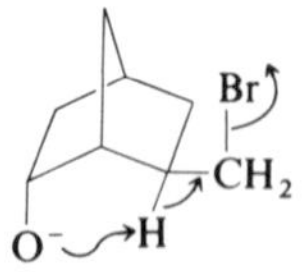   **15**

Menger considered the degree of variation in angle of attack that is possible in proton transfer and other reactions and concluded that the transition states frequently possess considerable flexibility. That is, modest deviations from preferred angles of attack need not involve significant penalties (*89*). Indeed, in the case of the proton transfer occurring in **15** a deviation from linearity of as much as 74° can be tolerated (*90, 91*).

3. Stereoelectronic Aspects of Proton Transfer Reactions

Although the term *stereoelectronic* (*92*) might be used in a broad sense to include the factors leading to linearity or nonlinearity in the triad $X \cdots H \cdots Y$, it seems preferable to restrict its use to questions of the preferred orientation of the orbitals on X or Y and their neighbors.

The necessity for proper orbital alignment in proton transfer reactions was pointed out by Corey and Sneen in 1956 in their studies of the enolization process (*93*). Furthermore, it is well known that the antiperiplanar arrangement **16** is favored in E2 eliminations (*94–97*). In the case of proton

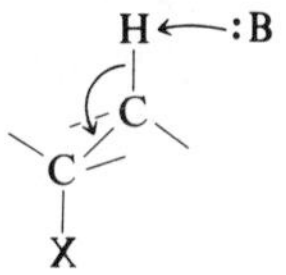

H ←—:B

16

abstraction from the α-carbon atom of carbonyl compounds there is now both experimental and theoretical evidence to support the conclusion that the most acidic α-hydrogen atom occupies a position orthogonal to the carbonyl plane, by which is meant that maximum overlap occurs between the σ C—H orbital and the *p* orbital of the carbonyl carbon atom (*98–100*). Such an arrangement is shown in **17**, in which the favored location is indicated by an asterisk. [There

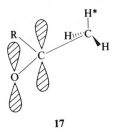

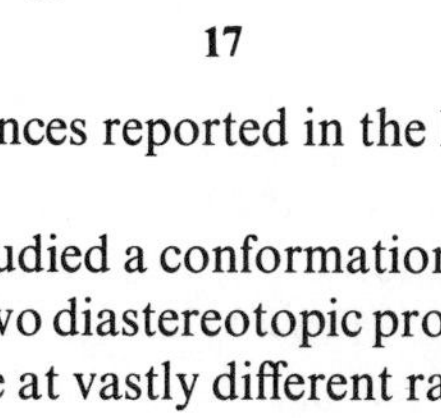

17

have been a number of instances reported in the literature where this effect is small or absent (*101*).]

Fraser and Champagne studied a conformationally biased ketone (twistan-4-one) and showed that the two diastereotopic protons α to the carbonyl group undergo deuterium exchange at vastly different rates (290:1), with the favored position corresponding to that singled out in **17** (*100*).

Can the assumed preference (perhaps slight) for linearity in the bonds of $C \cdots H \cdots O$ and the preference for orthogonality in the bonds of

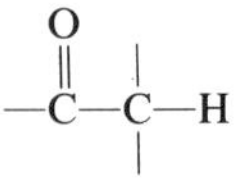

$$\begin{array}{c} O \\ \parallel \quad | \\ -C-C-H \\ | \end{array}$$

be reconciled with the size of ring favored in the intramolecular enolization processes described in the previous section? Molecular models indicate that the six-membered ring of **14** cannot easily attain a linear arrangement at the transferred proton, although the situation is no worse than with many rings of other size. With respect to the orthogonality preference there is no difficulty in **14** attaining such an arrangement. The seven-membered ring of **12** has somewhat different geometrical parameters. A linear arrangement for the transferred proton can easily be accommodated in this case through the boatlike conformation **18**, although the ketone carbonyl will be neither fully conjugated with the ring nor orthogonal to the transferred proton. Presumably, the transition state of **12** reaches some compromise with respect to these three factors.

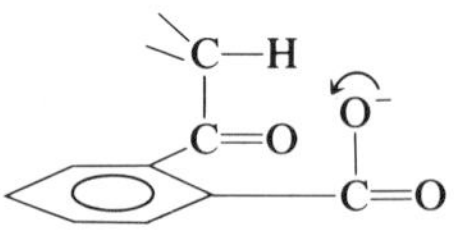

18 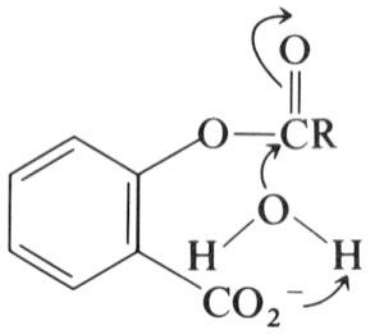   **19**

Many cases of intramolecular catalysis of nucleophilic attack at carbonyl carbon [e.g. aspirin hydrolysis (*102*)] involve the participation of a water molecule, as in **19**, and it is curious that this sort of process is so rare in intramolecular proton abstractions.

Effective molarity (EM) is a term used to express the degree of advantage that an internal process has over the analogous bimolecular process (*103, 104*), and Kirby pointed out that direct nucleophilic processes that do not involve an additional water molecule tend to have larger EM values than proton transfer reactions (*104*). The latter group includes, of course, simple proton transfers, as in enolization, and proton transfer from a water molecule that itself becomes a nucleophile, as in the case of aspirin hydrolysis.

When the acceptor and donor atoms are 1,3 with respect to one another, as in ethyl hydrogen malonate, a strained four-membered ring would be required for proton transfer, and in this case there is evidence that an external water molecule is involved in the reaction (*105, 106*).

4. Conclusions Regarding the Geometry of Proton Transfer

One can conclude that proton transfers prefer to be linear but that, in many cases, deviations from this condition are accomplished without undue strain, much as is the case for the hydrogen bond. The most stringent requirements for linearity are met in cases of bifunctional catalysis, in which two proton transfers take place.

B. Transition State Geometry of Hydrogen Atom and Hydride Transfer

There is general agreement that hydrogen atom transfer, like proton transfer, tends to be favored by linear geometry (*107*), though there is very much less evidence available to support this conclusion than in the case of proton transfer. A very simple reaction taking place by hydrogen atom transfer is that between molecular and atomic hydrogen, and this reaction has indeed been shown to have linear geometry [Eq. (4-12)] (*108, 109*).

$$H_2 + H\cdot \longrightarrow [H\cdots H\cdots H] \longrightarrow H\cdot + H_2 \qquad (4\text{-}12)$$

Kwart has pointed out that those hydrogen atom transfers that are believed to be linear on other grounds frequently show a pronounced dependence of the deuterium isotope effect on temperature. On the other hand, those reactions that require bent geometry for the hydrogen (or proton or deuterium) transfer usually show little or no dependence of k_H/k_D on temperature (*110*). For example, the [1,5]-sigmatropic rearrangement of **20** to **21** [Eq. (4-12a)],

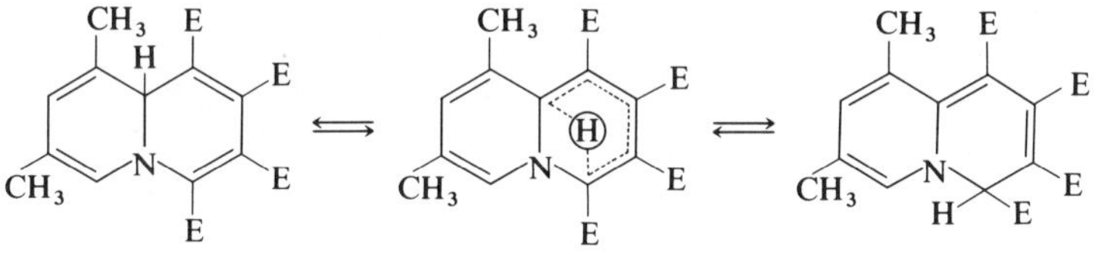

20 (E = —COOMe) **21**

which must have a bent transition state, proceeds with an isotope effect that does not stray from the range 4.9–5.0 over a temperature of 60°C (*111*).

There are a number of systems that undergo intramolecular hydrogen atom transfer upon photochemical excitation, for example, those investigated by Scheffer *et al.* (*112, 113*); their general form is shown in Eq. (4-13). Upon photochemical excitation the quinone **22** produces the triplet species **23**, whose carbonyl oxygen atom abstracts a hydrogen atom from the contiguous methylene group to give **24**, which, in turn, undergoes bond formation to give the cage structure **25**. The transition state for the reaction contains a

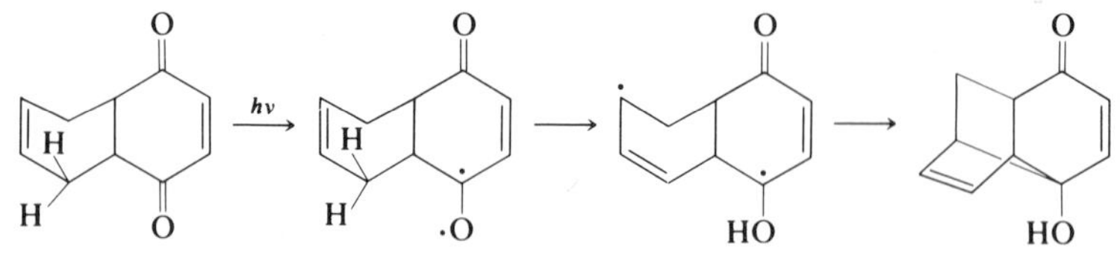

$$(4\text{-}13)$$

22   **23**   **24**   **25**

five-membered ring, and it is clear that a linear arrangement for the $O \cdots H \cdots C$ unit is quite impossible, indicating that departure from linearity is not a serious impediment to reaction taking place.

The most stringent angular requirement in these systems seems to be that of the orbitals on oxygen. The abstracted hydrogen atom in these, but apparently not in all, cases is essentially coplanar with the carbonyl group. This allows the new O—H bond to be formed using the half-occupied n orbital on oxygen that lies in the plane of the carbonyl group (114–117). Thus, this system, like some of the proton transfer reactions considered earlier, is governed by what can be designated as stereoelectronic effects rather than by the simple geometry of the three-atom system whose central unit is the transferred hydrogen.

The distances over which hydrogen atom transfer occurs in systems such as that shown in Eq. (4-13) range from 2.26 to 2.58 Å, which are somewhat greater distances than those found in other systems (118).

What is the situation with regard to hydride transfer? The matter was explored by a number of investigators, including Lewis, Symons, Hawthorne, and Olah, who concluded that hydride abstraction from carbon might be facilitated by a triangular rather than a linear transition state (119–121). The basis for this suggestion is the idea that the attacking reagent seeks the electrons in the bond rather than the nucleus and hence transition state **26** should be preferable to **27** in Eq. (4-14) (Z, the oxidant or electrophile, is arbitrarily shown as a cation in this equation).

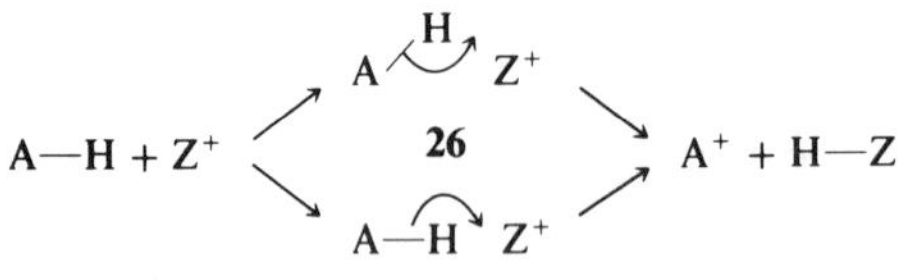

$$(4\text{-}14)$$

27

A consequence of the triangular arrangement in **26** would be the retention of stretching motion in the transition state at the expense of bending motion; that is, the translational mode of the transferred hydrogen would have its origin in a bend rather than in a stretch. There being less energy associated with bending than with stretching, there would be less energy to lose in going to the transition state and the difference introduced by deuterium substitution would be less, leading, in turn, to a lower isotope effect.

Using the magnitude of an isotope effect as a criterion of mechanism is explored more fully in Section III. Apart from the indirect and rather inconclusive support for nonlinear geometry that can be drawn from the results described there, what other evidence is available regarding the geometry of the three central atoms in a hydride transfer reaction?

The well-known Meerwein–Pondorff–Oppenauer oxidation–reduction system (122), which is believed to take place by means of hydride transfer from the alkoxide ion of an alcohol to the carbonyl group of an aldehyde or ketone, has an intramolecular analogue that allows the geometry of the transition state to be examined. Thus, under the influence of base, hydroxy ketones undergo hydride transfer that interchanges the carbonyl and alcohol groups [Eq. (4-15)].

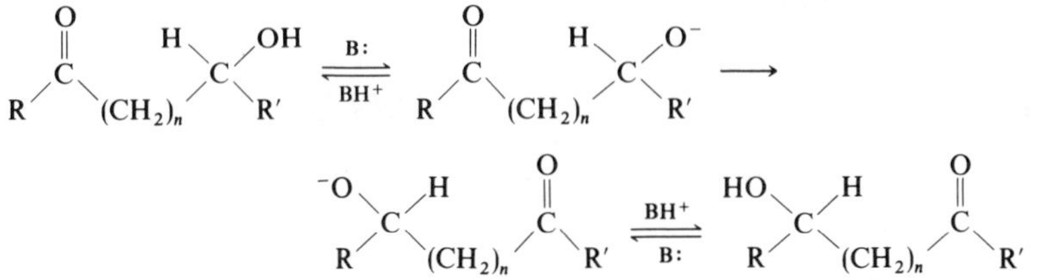

$$(4\text{-}15)$$

In the case of α-hydroxy ketones (acyloins), where n in Eq. (4-15) is zero, the reaction appears to take place via enediol intermediates rather than by hydride transfer (Section IV). When $n > 0$ there is no reasonable alternative to the hydride transfer route, and many examples of such reactions exist (123–128).

Craze and Watt studied the molecular geometry of such hydride transfer processes in a series of cage structures of general formula 28, where $n = 1, 2,$ or 3 (129). The rates of these processes were followed by dynamic ^{13}C-nmr spectroscopy, the reactant and product having identical structures.

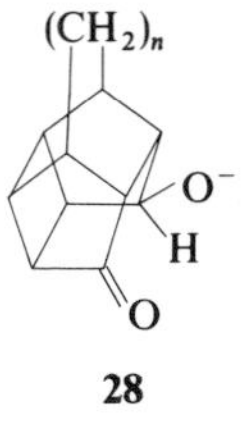

28 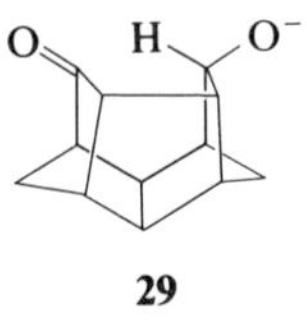   29

The size of n in 28 affects both the distance between the carbon atoms involved in the hydride transfer and the rate of reaction. When $n = 3$ the intercarbon distance is the least (2.52 Å, calculated) and the reaction rate is the greatest. Thus, as might be expected, proximity favors reaction. The angle made by the $C \cdots H \cdots C$ triad, however, does not appear to differ significantly in the three cases. It has been calculated to be close to 83°. Since the transition state for the reaction contains a five-membered ring (including the transferred hydrogen) there is no possibility of linearity in any case, and so it is difficult to draw conclusions regarding the preferred geometry for hydride transfer, other than the obvious one that lack of linearity does not provide a significant barrier to reaction taking place. An even faster intramolecular hydride shift (faster by a factor of 10^3) takes place in **29** (*130*). The $C \cdots H \cdots C$ bond angle is wider here (93°), but because of other geometrical differences between **28** and **29**, such as differences in angle of approach to the carbonyl group (*131*), one cannot infer that the wider angle is responsible for the greater rate of reaction.

The dynamic nmr technique has also been used by Verhoeven *et al.* to study the intramolecular hydride transfer between dihydropyridine and pyridinium units shown in Eq. (4-16) (*132*). The value of *n* was varied between 2

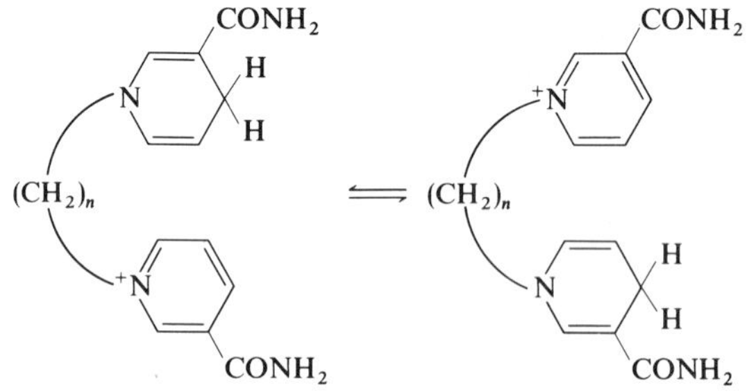

$$(4\text{-}16)$$

and 4 and the effect on the rate of hydride exchange determined. The maximum rate was found with three methylene groups in the bridge ($n = 3$), the length of bridge that allows a face-to-face orientation of the pyridine rings. [The EM of this reaction is 210 (Section II,A,3).] The transition state for this reaction is illustrated schematically by **30**, and it is clear that the $C \cdots H \cdots C$

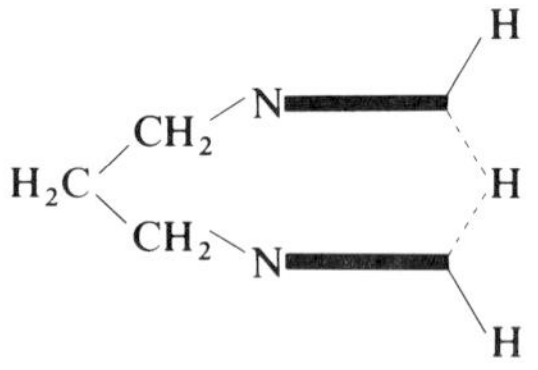

30

triad is bent. The same group of workers concluded that the transition state of a somewhat analogous *inter*molecular hydride transfer is linear (*133*).

A theoretical study of the transfer of hydride ion from cycloheptatriene and other compounds to cyclopropenium ion [Eq. (4-17)] uncovered no evidence

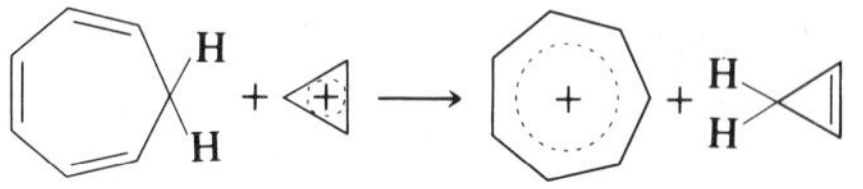

$$\text{(4-17)}$$

for the existence of triangular intermediate structures, and it was concluded that a linear or almost linear transition state was present in these reactions (*134–136*).

In summary, one can say that where linear transition states are readily attainable hydride transfers resemble proton transfers in showing some preference for the linear path, although lack of linearity is clearly no bar to high reactivity.

1. Systems with Hydride Bridges

It has long been customary to refer to the hydrogen links in diboranes as hydride bridges and to refer to 1,2-hydrogen migrations in carbonium ions as hydride shifts. Indeed, there are indications that symmetrical bridged species corresponding to the transition states of the latter reactions may have finite existences (e.g., **31**, the bridged form of the 2-butyl cation) (*137*). The atoms

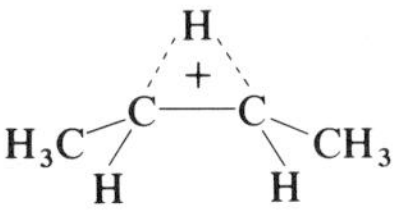

$$H_3C \overset{\displaystyle .H.}{\underset{H}{\overset{+}{C}}} \overset{+}{\underset{H}{C}} CH_3$$

31

that are bridged by hydrogen in such species are adjacent to one another, and the bridge need not be regarded as a hydride unit since it could, for example, be written as a proton embedded in a π system.

There are other carbocations, however, in which hydrogen bridging has been observed and in which there is compelling evidence that the hydrogen has hydridic character. These are the μ-hydrido-bridged cycloalkyl cations discovered by Sorensen, which can be generated by dissolving the appropriate alcohols or alkenes in $FSO_3H/SbF_5/SO_2ClF$ mixtures at low temperature [Eq. (4-18)] (*138–140*).

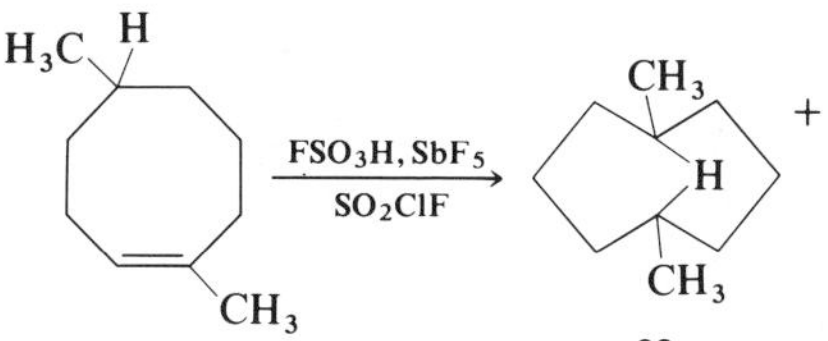

$$\xrightarrow[\text{SO}_2\text{ClF}]{\text{FSO}_3\text{H, SbF}_5} \qquad \text{(4-18)}$$

32

The evidence that **32** and other similarly generated species have symmetrical, bridged structures rather than being mixtures of equilibrating structures [as in Eq. (4-19)] is the following. The nmr spectrum shows that one of the protons in the ion has a very high field chemical shift (approximately $-6.3\ \delta$), which is not consistent with a normal carbonium ion structure but which is consistent with the presence of a single, shielded, "hydridic" hydrogen. This conclusion is buttressed by the magnitude of the $^{1}H-^{13}C$ coupling constant for this hydrogen and by the effect of deuterium substitution in the methyl groups (*141, 142*), which produces a temperature effect on the nmr spectrum that is opposite to that observed to be present in cations that are known to be rapidly equilibrating between two carbonium structures (*143, 144*), for example, those shown in Eq. (4-19).

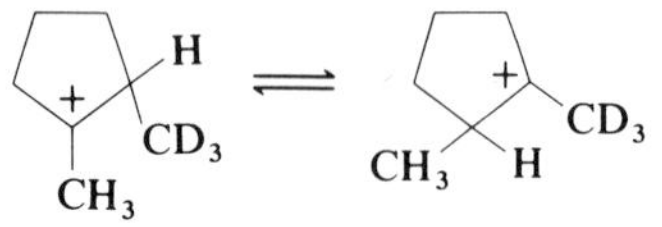

$$(4\text{-}19)$$

The dimethylcyclooctyl cation **32** is believed to have the boat–chair conformation shown in **33**, and the dimethylcyclodecyl cation **34**, which is somewhat less stable than **32**, is believed to have the conformation shown in **35**. It is interesting that X-ray crystallographic analysis of the structure of the hydrido-bridged diborane anion **36** shows it to have the same *cis*-decalin type

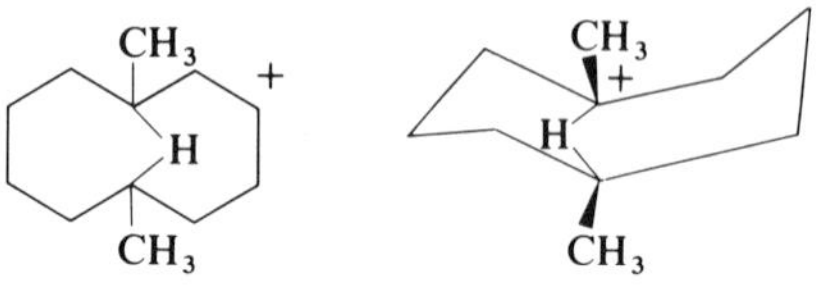

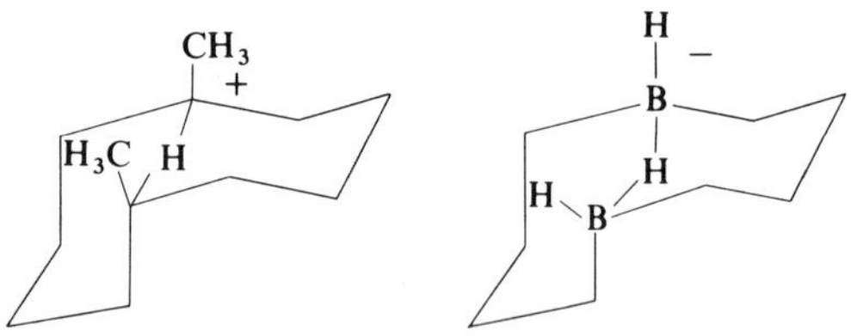

of structure as **35** (*145*). In the case of **36** the B—H—B bond angle is 140°, and the question arises as to the angle adopted by the corresponding C—H—C triad in **33**, **35**, and the other hydride-bridged cations. There is slight possibility of a crystallographic structure being obtained for the latter species since **32**, the most stable of the bridged cations, decomposes above $-60°C$. Theoretical studies have indicated a preference for a linear arrangement for the C—H—C unit in such systems, though there appear to be only small constraints on the location of the bridging hydrogen (*141, 146*). That is, it can be displaced with very little expenditure of energy. Furthermore, in many cases significant deviations from linearity doubtless result from the conformational requirements of the rings.

It is worth noting that the properties of the hydrido-bridged cations are not those expected of a protonated C—C σ bond (Chapter 3, Section VI) and, indeed, 9,10-dimethyldecalin in super acid does *not* generate the bridged species [Eq. (4-20)], whose absence can easily be ascertained by an examina-

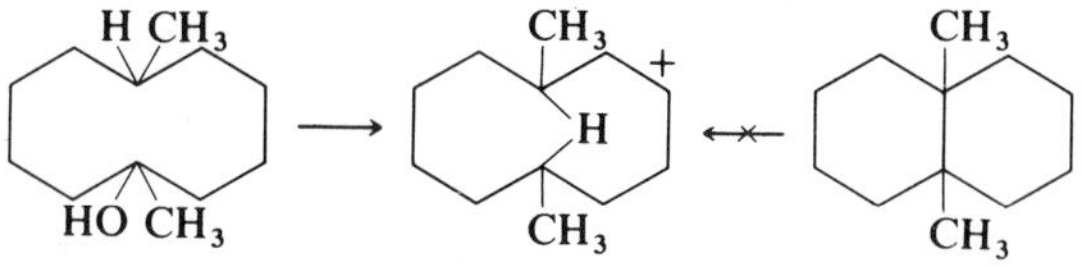

34
(4-20)

tion of the high-field region of the nmr spectrum where the hydridic proton absorbs.

2. Stereoelectronic Aspects of Hydride Transfer

An antiperiplanar arrangement of abstracted proton and leaving group is known to be favored for E2 reactions, and there is some evidence to suggest that hydride transfer is governed by similar factors. Thus, Deslongchamps explained the stereospecificity of the reaction shown in Eq. (4-21) on such grounds (*147*).

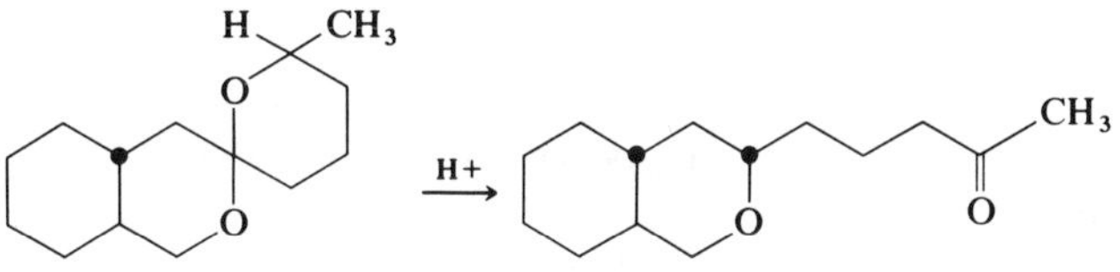

37
38
(4-21)

The spiroketal 37 opens in acidic solution to give an oxonium ion that undergoes an internal oxidation–reduction by means of hydride transfer to give 38, not the diastereomer 39 [Eq. (4-22)]. Note that in 38 the transferred hydrogen is antiperiplanar with respect to one of the lobes of the oxygen atom, whereas in 39 there is no such relationship. This stereochemical requirement can perhaps better be recognized if the reverse reaction is considered, with hydride as a leaving group. It would be reasonable to expect that the transition state formed from 38 in such a case would be of lower energy than one formed from 39. An alternative route to 39 exists in which the orbital alignment is satisfactory; it must produce a boat form as an intermediate, however, and hence it cannot compete with the route leading to 38.

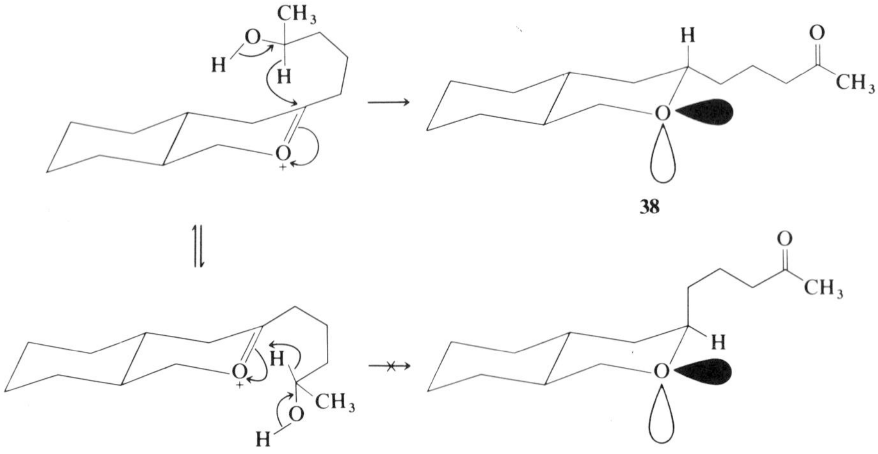

38

39

(4-22)

Orthoamides are effective hydride donors, and Wuest *et al.* have shown that they can reduce the proton to molecular hydrogen (*148, 149*). In practical terms this is made manifest by the pyrolysis of orthoamide salts to give guanidinium salts [Eq. (4-23)].

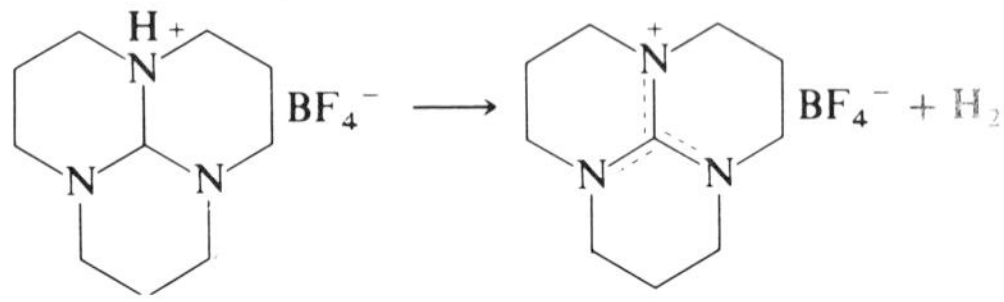

It is believed that the reaction proceeds by way of dissociation of the cation to orthoamide and tetrafluoroboric acid, followed by oxidation of the orthoamide by the Brønsted acid. The reaction is favored by stereoelectronic factors, as shown in **40**, where the lobes of the lone-pair orbitals on each of the three nitrogen atoms are antiperiplanar to the C—H σ orbital.

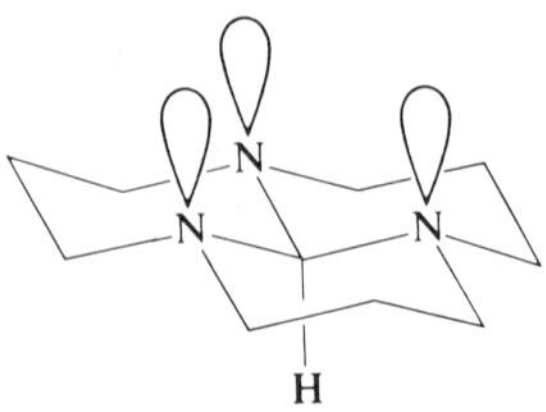

40

For the same reason the tristannaadamantane **41** is a much more effective hydride donor than is the open-chain compound **42**. In **41** all three C—Sn bonds are antiperiplanar to the central C—H bond (*149a*).

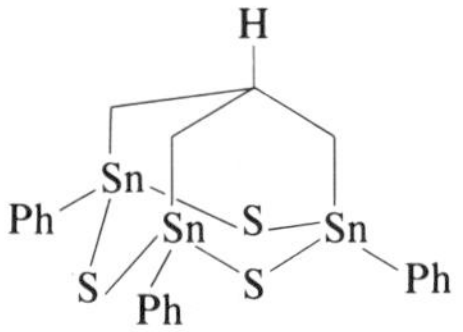

41 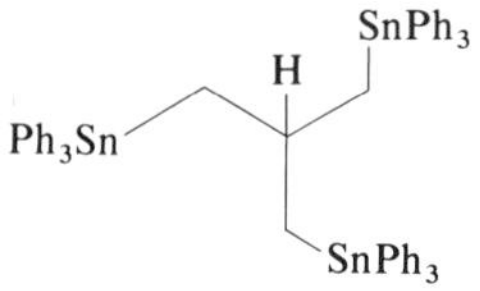   42

With regard to diastereotopic hydride transfer from carbon adjacent to oxygen there appears to be a kinetic preference for an axial site, a preference shared by hydrogen atom transfer but not by proton transfer (*150*).

III. Hydrogen Isotope Effects as a Criterion of Mechanism in Proton and Hydride Transfer

It has been customary to regard a "normal" deuterium isotope effect for cleavage of a C—H(D) bond as one that is in the range of 6 to 7 at room temperature. Soon after the discovery of deuterium in 1932 it was pointed out by Cremer and Polanyi that compounds containing deuterium should react more slowly than their protio analogues because of the difference in their zero-point energies (*151*). For C—H and C—D bonds this difference is ~ 1100 cal, the exact value being easily obtained for a particular compound from the absorption frequency in the infrared, commonly 2800–3000 cm^{-1} for a C–H stretch and 2100–2300 cm^{-1} for a C–D stretch. Knowing the precise value of the difference in zero-point stretching energies for a particular compound is not actually helpful since other factors are of much greater importance in determining the magnitude of an isotope effect in a particular reaction.

It was pointed out in Section II,B that the preferred geometry of hydride and proton transfers might not be the same and, as a consequence, their deuterium isotope effects might be expected to differ in magnitude. In particular, it has been suggested that transfer of a hydride (deuteride) should give rise to a smaller isotope effect than transfer of a proton. Do hydride transfer reactions have lower isotope effects than proton transfers or hydrogen atom transfers? This question cannot be answered definitively, despite several decades of work by many investigators and the determination of deuterium and tritium isotope effects for hundreds of reactions, although there does, indeed, seem to be a tendency for reactions considered to be hydride transfers on other grounds to have rather modest deuterium isotope effects, frequently in the range of 2 to 5 for k_H/k_D at room temperature (*27, 152–156*). Exceptions do exist, however, in which values are considerably greater than 6 to 7, the benchmark by which deuterium isotope effects have generally been judged. Furthermore, there are wide variations in the size of the isotope effects found in reactions that clearly involve proton transfer or, indeed, hydrogen atom transfer, and using the magnitude of a deuterium isotope effect as a criterion of mechanism is fraught with risk. Nonetheless, it seems fair to say that small isotope effects have been found more frequently in hydride than in proton or hydrogen atom processes.

It is a matter of historical interest that many of the early deuterium isotope effects that were measured involved proton transfer, for example, the ionization of nitromethane and the acid-catalyzed bromination of acetone (for which the rate-controlling step is proton removal), and these frequently turned out to be close to the "zero-point" value of 6 to 7 [Eqs. (4-24) and (4-25)] (157).

$$CD_3NO_2 + CH_3CO_2{}^- \longrightarrow CD_2NO_2{}^- + CH_3CO_2D \qquad k_H/k_D = 6.5, 25°C \quad (4\text{-}24)$$

$$\underset{\overset{\displaystyle\|}{CD_3CCD_3}}{\overset{\displaystyle O}{}} + Br_2 \xrightarrow{H^+} \underset{\overset{\displaystyle\|}{CD_3CCD_2Br}}{\overset{\displaystyle O}{}} + DBr \qquad k_H/k_D = 7.7, 25°C \quad (4\text{-}25)$$

Subsequently, it became clear that the magnitude of the isotope effect for these reactions can vary enormously, depending on the identity of the base that abstracts the proton and on the solvent. [The isotope effect also varies with temperature, but in most cases this is a systematic effect whose basis is well understood (158).] For example, changing the base in reaction (4-24) from acetate ion to hydroxide ion raises the isotope effect to 10.3 (159). Changes in the structure of the nitroalkane can also produce significant changes in the size of the isotope effect, particularly when hindered bases are used to effect proton removal. Thus, Lewis and Funderburk observed an isotope effect of 24.1 at 25°C for the acid–base reaction between 2-nitropropane and 2,6-dimethylpyridine (160, 161). There is little doubt that proton tunneling is involved here, and this is one of the two important variables that make it difficult to relate mechanism and isotope effect. The other uncertainty concerns the degree to which bending modes are altered in going from reactant to the transition state. Bending motions in general show much greater variation in frequency than do stretching motions, and it is difficult to make reasonable estimates of how these will change as the bond is broken. Furthermore, bending frequencies come in a less accessible part of the infrared spectrum than do stretching frequencies. [A complete analysis of vibrational changes accompanying isotopic substitution can sometimes be made for an equilibrium process involving compounds of limited complexity, and this has been done for the ionization of formic-d acid. Bell and Miller were able to calculate a value of 1.08 in this way for the ratio of the ionization constants of HCO_2H and DCO_2H in water, a value in excellent agreement with experimental data (162).]

Can the difference in zero-point energies of both bending *and* stretching vibrations vanish in going from reactants to transition state? Since these vibrations are orthogonal it is unrealistic to suppose so. Furthermore, as Bell pointed out, only one translational mode disappears in going to the transition state, corresponding to one real vibration in the reactants (*163*). That is not to say that both bending and stretching modes cannot be affected and that isotope effects greater than 6 or 7 must be attributed to something other than zero-point vibrational effects. However, while the total zero-point energy differences (stretching and bending) in a C—H(D) bond may correspond to a room temperature isotope effect as high as 20, it is hard to conceive of a process with the required geometrical parameters, other than a reaction such as methane pyrolysis where free hydrogen atoms presumably are formed.

Isotope Effect Maxima

When a proton (deuteron) is transferred from AH to :B to give A: and HB, the magnitude of the isotope effect is affected by the acid strengths of AH and BH. In particular, a maximum is frequently found in the isotope effect when AH and BH are of approximately equal strength (*164–169*). Two explanations for these maxima have been advanced: that of Westheimer (*164*) and that of Bell (*170, 171*). Westheimer considers the maxima to be a result of equal binding of the proton to two basic centers of equal strength, which, in turn, will have no zero-point energy associated with it. As a consequence, the isotope effect, which depends on the difference in zero-point energies of the reactant molecules diminishing as the transition state is formed, will be at a maximum under such circumstances. Bell, on the other hand, has attributed the isotope effect maximum to quantum mechanical tunneling, which should, indeed, be maximized by the free-energy difference between reactants and products being zero, that is, when the acids AH and BH are of equal strength. Regardless of the origin of the effect there is no doubt that the phenomenon is real (*172–175*) and that it can be present in hydrogen atom (*107, 176, 177*) and hydride (*16, 178*) transfers as well.

When a series of radicals reacts with *tert*-butyl mercaptan [Eq. (4-26)] deuterium isotope effects in the range of 1.9 to 6.7 at 25°C are found. The maximum rate occurs when the heat of reaction is near zero, that is, when the radicals R· and tBuS· are of comparable stability (*177*). In the case of hydride transfer from formate ion to a series of carbonium ions [Eq. (4.27)] the

$$tBuSD + R\cdot \longrightarrow tBuS\cdot + RD \qquad (4\text{-}26)$$

$$Ar_3C^+ + DCO_2^- \longrightarrow Ar_3CD + CO_2 \qquad (4\text{-}27)$$

deuterium isotope effects at 25°C are rather modest, as is so frequently observed to be the case with hydride transfers. They vary from 1.8 to 3.2, but here the maximum does not appear to coincide with the reaction whose overall free-energy change is zero (*16*). Similar results were found in the case of hydride transfer from triarylmethanes to a series of carbonium ions; in this case the deuterium isotope effect at 25°C varied from 1.7 to 4.8 (*178*).

IV. Competition among Proton, Hydrogen Atom, and Hydride Transfer Processes

A. General

The isomerization by base of acyloins [Eq. (4-28)] can be envisaged as taking place either by 1,2-hydride shifts [Eq. (4-29)] or by way of a prototropic (keto–enol) reaction involving enediol intermediates [Eq. (4-30)].

$$\text{RCHOH}-\overset{\overset{\text{O}}{\|}}{\text{C}}-\text{R}' \xrightarrow{\text{aq. HO}^-} \text{R}-\overset{\overset{\text{O}}{\|}}{\text{C}}-\text{CHOHR}' \qquad (4\text{-}28)$$

$$\text{RCHOH}-\overset{\overset{\text{O}}{\|}}{\text{C}}-\text{R}' + \text{HO}^- \rightleftharpoons \text{R}-\overset{\overset{\text{O}^-}{|}}{\underset{\underset{\text{H}}{|}}{\text{C}}}-\overset{\overset{\text{O}}{\|}}{\text{C}}-\text{R}' \overset{\text{H}_2\text{O}}{\rightleftharpoons} \text{R}-\overset{\overset{\text{O}}{\|}}{\text{C}}-\overset{\overset{\text{OH}}{|}}{\underset{\underset{\text{H}}{|}}{\text{C}}}-\text{R}' + \text{HO}^- \qquad (4\text{-}29)$$

$$\text{RCHOH}-\overset{\overset{\text{O}}{\|}}{\text{C}}-\text{R}' + \text{HO}^- \rightleftharpoons \text{R}-\overset{\overset{\text{OH}}{|}}{\text{C}}=\overset{\overset{\text{O}^-}{|}}{\text{C}}-\text{R}' + \text{H}_2\text{O} \rightleftharpoons \text{R}-\overset{\overset{\text{O}}{\|}}{\text{C}}-\overset{\overset{\text{OH}}{|}}{\text{CH}}-\text{R}' + \text{HO}^-$$

$$(4\text{-}30)$$

Activation of the substrate is accomplished by base in both routes, in one case by proton removal from the alcoholic hydroxyl group, and in the other by proton removal from the alcoholic carbon atom. It is generally agreed that the enediol route [Eq. (4-30)] is the favored one and, indeed, in some cases neutral enediols corresponding to the intermediates in Eq. (4-30) can be isolated (*123, 179*).

When one or more methylene units are interposed between the carbonyl and alcohol groups, ordinary enediols cannot be formed and in these cases the hydride shift route is used [Eq. (4-15)]. [Base-catalyzed homoenolization (*180–182*), which is the prototropic reaction of a hydrogen β to a carbonyl group, can occur under certain conditions, but such reactions are usually slow and are unlikely to compete successfully with the hydride shift route in the isomerization of β-hydroxy ketones. In this connection, most of the reactions discussed in this section, though apparently not homoenolization, are also subject to acid catalysis.]

A related reaction is the hydroxide-catalyzed conversion of α-keto-aldehydes to α-hydroxycarboxylate ions, for example, the rearrangement of phenylglyoxal to mandelate ion [Eq. (4-31)]. In this case an enediol intermediate is not possible, and the reaction proceeds by the hydride transfer route shown in Eq. (4-32). Hine and Koser have shown that the active intermediate is actually the dianion $C_6H_5COCHO_2^{2-}$ when the hydroxide ion concentration is greater than $\sim 0.003\ M$ (*183*). As expected in reactions of this sort, the aldehydic hydrogen is attached to the alcoholic carbon atom in the product and is not exchanged with the solvent pool (*184–186*).

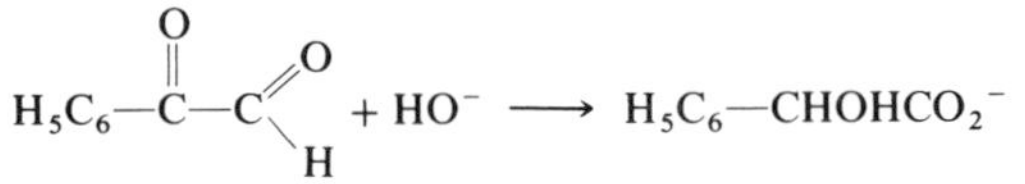

$$H_5C_6-\overset{\overset{\displaystyle O}{\|}}{C}-C\overset{\displaystyle O}{\underset{\displaystyle H}{\diagdown}} + HO^- \longrightarrow H_5C_6-CHOHCO_2^- \qquad (4\text{-}31)$$

$$H_5C_6-\overset{\overset{\displaystyle O}{\|}}{C}-C\overset{\displaystyle O}{\underset{\displaystyle H}{\diagdown}} + HO^- \rightleftharpoons H_5C_6-\overset{\overset{\displaystyle O}{(\!}}{C}-\overset{\overset{\displaystyle O^-}{)}}{\underset{\displaystyle H}{C}}-OH \longrightarrow H_5C_6-\overset{\overset{\displaystyle O^-}{|}}{\underset{\displaystyle H}{C}}-CO_2H \longrightarrow$$

$$C_6H_5CHOHCO_2^- \quad (4\text{-}32)$$

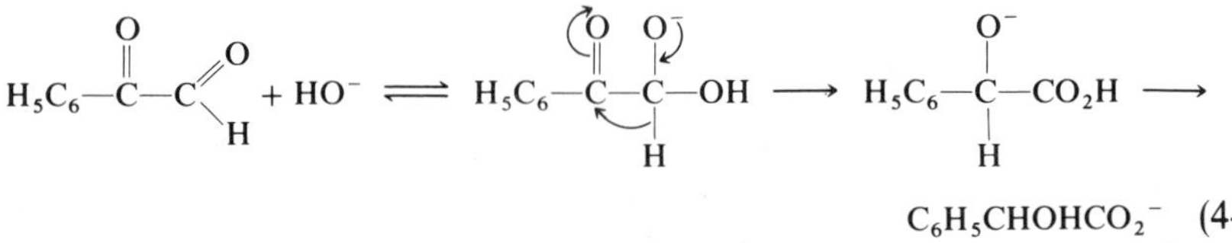

Franzen (*187, 188*) found that such isomerizations can be accomplished using much weaker bases than hydroxide ion and that 2-dialkylaminoethane-thiols are particularly efficient catalysts in this respect. It was at one time believed that this reaction took place by a hydride shift within the adduct formed by the addition of thiol to the aldehyde group of the substrate [Eq. (4-33a)] (*188, 189*). It was subsequently determined that exchange of the transferred hydrogen with solvent takes place (*190, 191*), and it appears that Eq. (4-33b), involving an enediol intermediate, is the preferred pathway (*192*).

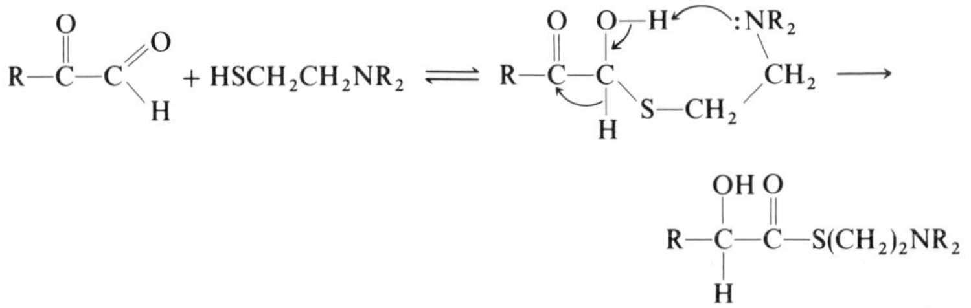

$$R-\overset{\overset{\displaystyle O}{\|}}{C}-C\overset{\displaystyle O}{\underset{\displaystyle H}{\diagdown}} + HSCH_2CH_2NR_2 \rightleftharpoons R-\overset{\overset{\displaystyle O}{\|}}{C}-\overset{\overset{\displaystyle O}{|}}{\underset{\displaystyle H}{\underset{\displaystyle S-CH_2}{C}}} \overset{H}{\underset{CH_2}{:NR_2}} \longrightarrow$$

$$R-\overset{\overset{\displaystyle OH}{|}}{\underset{\displaystyle H}{C}}-\overset{\overset{\displaystyle O}{\|}}{C}-S(CH_2)_2NR_2 \quad (4\text{-}33a)$$

$$R-\overset{\overset{\displaystyle O}{\|}}{C}-C\overset{\displaystyle O}{\underset{\displaystyle H}{\diagdown}} + HSCH_2CH_2NR_2 \rightleftharpoons R-\overset{\overset{\displaystyle O}{(\!}}{C}-\overset{\overset{\displaystyle OH}{|}}{\underset{\displaystyle H}{C}}-S\overset{CH_2}{\underset{\underset{R_2N}{CH_2}}{}} \longrightarrow$$

$$R-\overset{\overset{\displaystyle OH}{|}}{C}=\overset{\overset{\displaystyle OH}{|}}{C}-S(CH_2)_2NR_2 \longrightarrow R-\overset{\overset{\displaystyle OH}{|}}{\underset{\displaystyle H}{C}}-\overset{\overset{\displaystyle O}{\|}}{C}-S(CH_2)_2NR_2 \quad (4\text{-}33b)$$

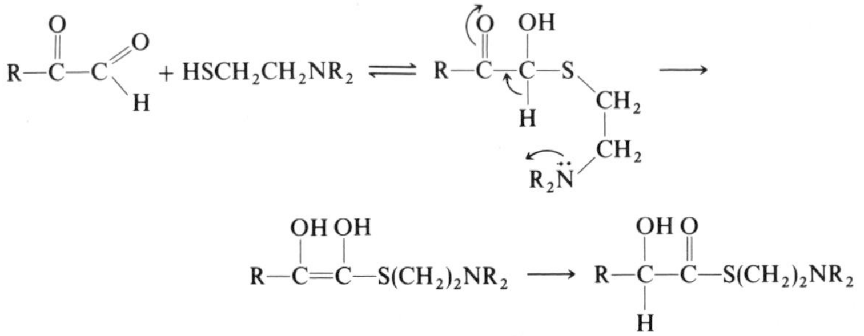

As with the acyloin isomerization [Eq. (4-28)] the two routes shown here differ with respect to which of the two hydrogen atoms in the —CHOH— unit is removed as a proton, the preferred enediol route proceeding by way of C—H bond rupture by base.

The glyoxalase-catalyzed isomerization of the thiohemiacetal of methyl-glyoxal to the thio ester of lactic acid [Eq. (4-34)] also appears to proceed

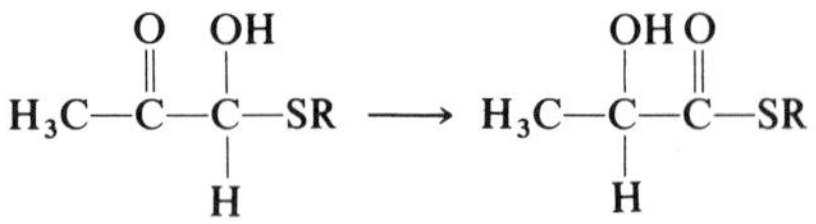

$$
\underset{\substack{|\\H}}{\overset{\substack{O\ OH\\||\ |}}{H_3C-C-C-SR}} \longrightarrow \underset{\substack{|\\H}}{\overset{\substack{OH\ O\\|\ ||}}{H_3C-C-C-SR}} \qquad (4\text{-}34)
$$

by way of the enediol pathway rather than by the hydride route (*191–194*).

The competition that may occur between hydrogen atom and hydride transfer is nicely illustrated by the work of Barclay and Dust (*195, 196*). They showed that the products formed in the decomposition of *N*-nitroso-2,4,6-tri-*tert*-butylacetanilide depend on the solvent used. Paradoxically, the nonpolar solvent benzene gives rise to products formed by an ionic route (hydride transfer), and the polar solvent triethylamine gives rise to products formed by a nonionic route (hydrogen atom transfer).

Direct competition between hydrogen atom and hydride transfer is frequently present in reactions in which a multiequivalent oxidant, such as manganese(VII), attacks a C—H bond (*197*). In some of these cases the experimental evidence will point clearly to one mechanism or the other. In other cases, for example, the oxidation of benzylamines to benzaldehydes by aqueous permanganate [Eq. (4-35)], the ambiguity is such as to suggest that

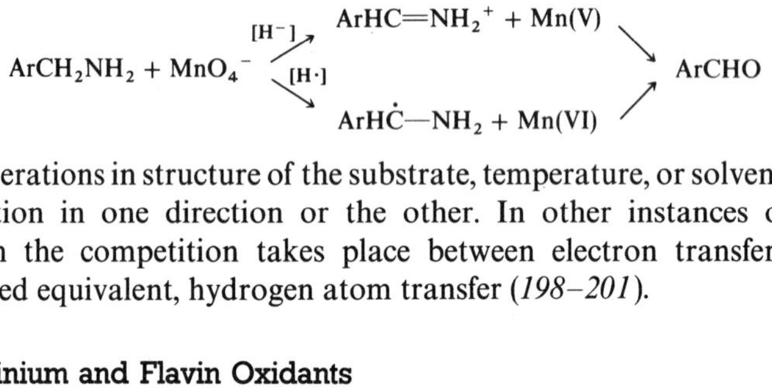

$$
ArCH_2NH_2 + MnO_4^- \underset{[H\cdot]}{\overset{[H^-]}{\rlap{\Big<}}} \quad \begin{array}{c} ArHC{=}NH_2^+ + Mn(V) \\[4pt] \\ ArH\dot{C}{-}NH_2 + Mn(VI) \end{array} \quad \rlap{\Big>} \quad ArCHO \qquad (4\text{-}35)
$$

minor alterations in structure of the substrate, temperature, or solvent may tip the reaction in one direction or the other. In other instances of amine oxidation the competition takes place between electron transfer and its protonated equivalent, hydrogen atom transfer (*198–201*).

B. Pyridinium and Flavin Oxidants

The principal coenzymes involved in the direct oxidation of organic substrates are the pyridinium coenzymes (e.g., NAD⁺, **43**) and the flavins (e.g., riboflavin, **44**), whose reactive positions are numbered in the formula **44**:

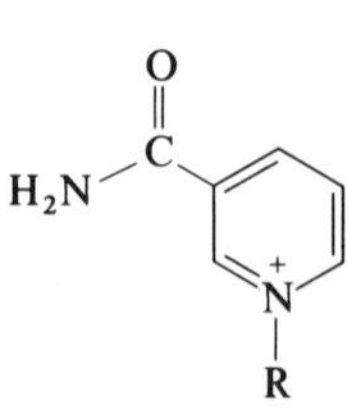

43 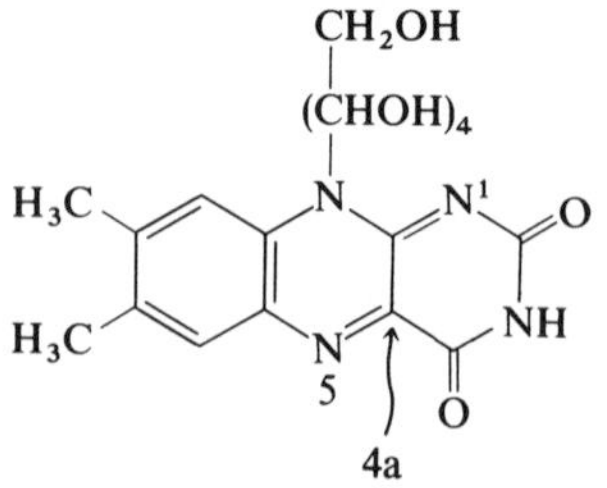   **44**

It has long been known that the conversion of NAD^+ to NADH in, for example, the oxidation of an alcohol is the result of direct transfer of hydrogen from the α-carbon atom of the alcohol to the 4 position of the pyridinium ring [Eq. (4-36)] (202, 203).

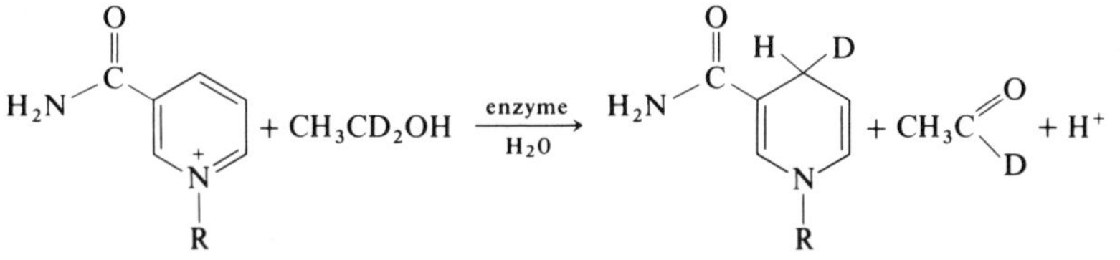

$$\text{NAD}^+ + CH_3CD_2OH \xrightarrow[\text{H}_2\text{O}]{\text{enzyme}} \text{NADH} + CH_3C\!\!\!\diagup^{O}_{\diagdown D} + H^+$$

(4-36)

Although a reaction path involving covalent adducts, analogous to the chromate ester oxidation mechanism (Section I), can be devised for such pyridinium oxidations (29, 204), it is generally agreed that the transfer of "hydride" occurs directly from substrate to C-4 of the pyridine ring (205–208). There is disagreement, however, about the timing of the transfer of the units that constitute "hydride," that is, a proton plus two electrons (156, 203, 209–214), although it appears that some reactions that had been considered to be possible multistep processes are, in fact, single-step hydride transfers (215–217).

Even if one assumes that hydride is transferred as a single unit there is still the question of the timing of the proton transfer that is required in most cases to complete the reduction. Thus, in the enzymatic reduction of aldehyde to alcohol by NADH there are indications that the proton transfer to carbonyl oxygen lags behind the hydride transfer to carbonyl carbon (218).

The situation with flavins is somewhat different. They readily form covalent adducts with various substrates, and plausible mechanisms can be constructed in which proton abstraction from a covalent adduct triggers the oxidation process, in a manner analogous to that of the classical chromate ester mechanism.

The molecular architecture of flavins is such that they can react in oxidation–reduction processes by a variety of reaction paths, some being 1-equivalent (one-electron) and others 2-equivalent (two-electron) (206, 207). They also differ from pyridinium coenzymes in having the capacity to react directly with molecular oxygen. The principal routes that have been suggested for the reactions of this versatile reagent are indicated in Scheme 4-1 by means of the structures of the intermediate species that would be formed in each route.

The highly conjugated isoalloxazine ring system of the flavins can add electrons with facility to form either radical-anions or dianions (Scheme 4-1),

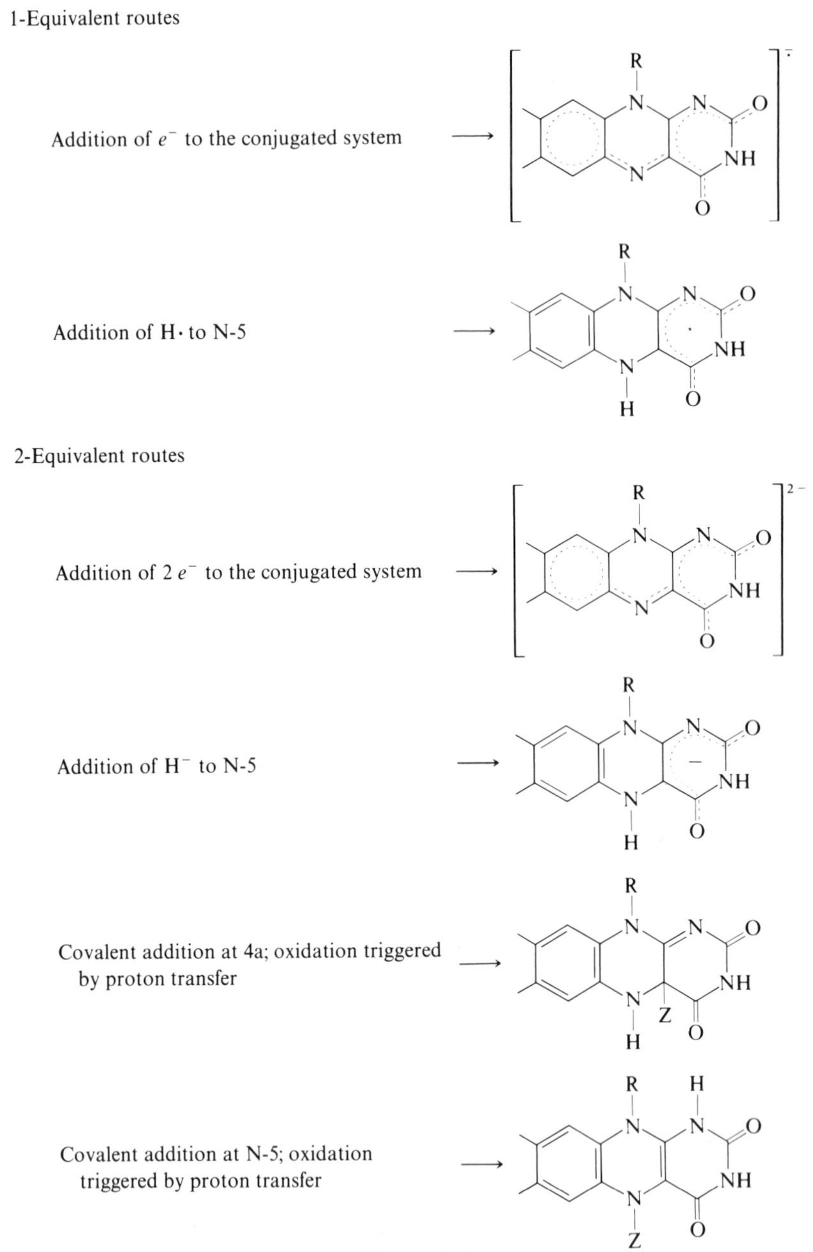

1-Equivalent routes

Addition of e^- to the conjugated system $\longrightarrow$

Addition of H· to N-5 $\longrightarrow$

2-Equivalent routes

Addition of 2 e^- to the conjugated system $\longrightarrow$

Addition of H$^-$ to N-5 $\longrightarrow$

Covalent addition at 4a; oxidation triggered by proton transfer $\longrightarrow$

Covalent addition at N-5; oxidation triggered by proton transfer $\longrightarrow$

Scheme 4-1

the state of protonation of these species depending, of course, on the acidity of the medium.

Of the possible mechanisms for 2-equivalent oxidation by flavins, other than concerted or consecutive transfer of two electrons, most attention has been directed at routes involving either covalent addition at 4a or N-5 (followed by bond reorganization that is triggered by proton transfer) or direct transfer of hydride to N-5. These alternatives will be considered in the sections that follow.

1. Oxidations in Flavin Adducts Triggered by Proton Transfer

A number of organic substrates that are oxidized by flavins are known to add to the isoalloxazine ring system, and it was suspected that these adducts produced the redox products (reduced flavin and oxidized substrate) by means of elimination reactions analogous to that shown in Eq. (4-2), that is, as processes triggered by proton transfer. Despite the attractiveness of such mechanisms and, indeed, the proof that such a mechanism does apply to the flavin oxidation of nitroalkanes, it appears that substrates of biological importance, in particular α-amino acids and α-hydroxy acids, do not follow this route in their oxidation by flavin coenzymes. Let us consider first the reaction of flavins with nitroethane.

Porter and Bright (*219, 220*) examined the kinetics of the conversion of nitroethane to acetaldehyde and nitrite by D-amino-acid oxidase, a flavin-containing enzyme [Eq. (4-37)].

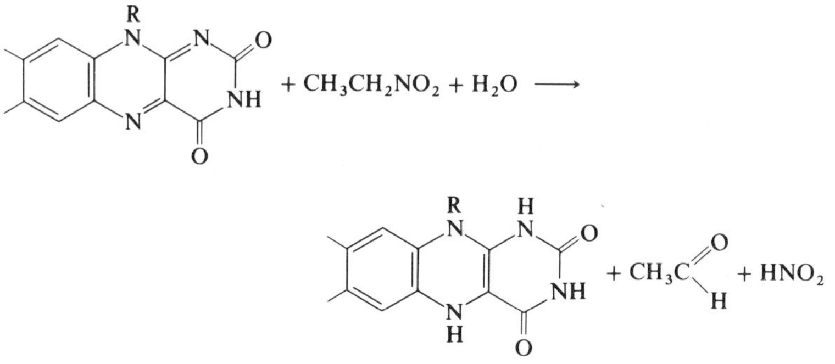

The reaction path was shown to be that in Scheme 4-2. Note that proton transfers have occurred in the first step, formation of the nitroethane anion, and that elimination of the flavin unit in its reduced form occurs in the final step.

Why is this pathway not used to oxidize alcohols (and hydroxy acids) or amines (or amino acids)? In some cases it is because the required carbanions cannot be formed; in others, the N-5 adducts have been shown not to lie on the

$$CH_3CH_2NO_2 + B: \longrightarrow CH_3\bar{C}HNO_2 + BH^+$$

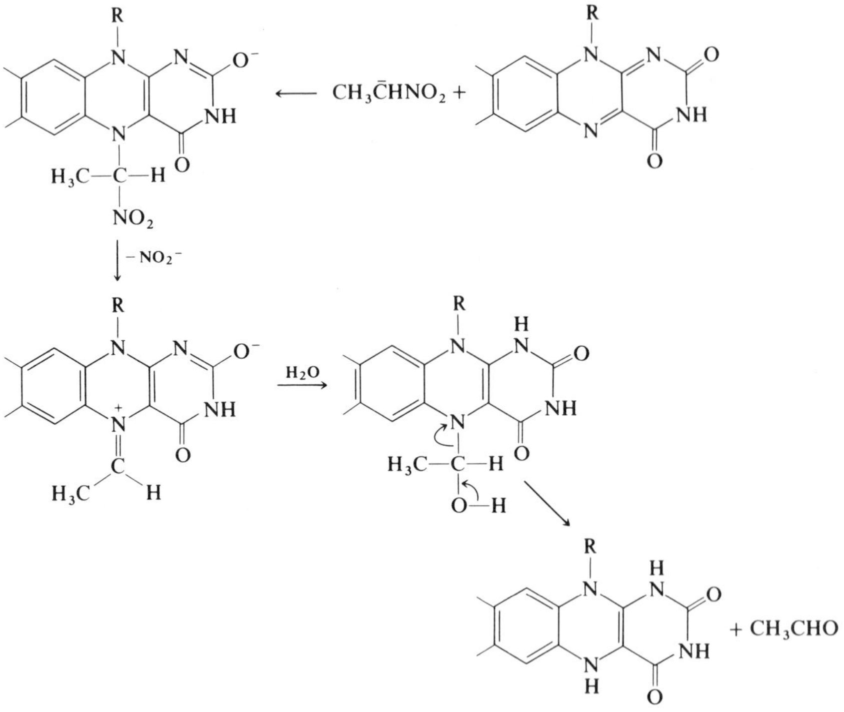

Scheme 4-2

reaction path. Bruice *et al.*, for example, showed that an adduct can be formed by starting with reduced flavin and pyruvic acid, that is, by going in the opposite direction. The only reaction of this adduct, however, is reversion to starting materials, which are converted to products (oxidized flavin and alcohol) by some other route (*221, 222*).

Covalent addition at position 4a, which is known to occur with relative ease in some cases, likewise does not appear to provide a route for oxidation of alcohol or amino groups (*222*).

2. Hydride Addition to Flavins

Flavins are 2-equivalent oxidants for a number of organic substrates, including the coenzyme NADH, the reduced form of NAD$^+$. The hydride-accepting properties of NAD$^+$ are well established, and it is not unreasonable to assume that NADH would be an effective hydride donor to flavin when these two coenzymes interact, as they do in the mitochondrial electron transport chain. This mechanism has been criticized on the basis of nitrogen being a poor site for hydride addition, and it is true that there are few instances in which such a process occurs in simple systems (29). Clearly, if hydride addition does indeed occur at N-5 it is because the latter is part of a conjugated system that stabilizes the negative charge. [Simultaneous protonation at N-1 to give 45 would not facilitate reaction, at least in neutral aqueous media, since the pK_{HA} of the N-1 position in 45 and related compounds is below 7.0 (223, 224).]

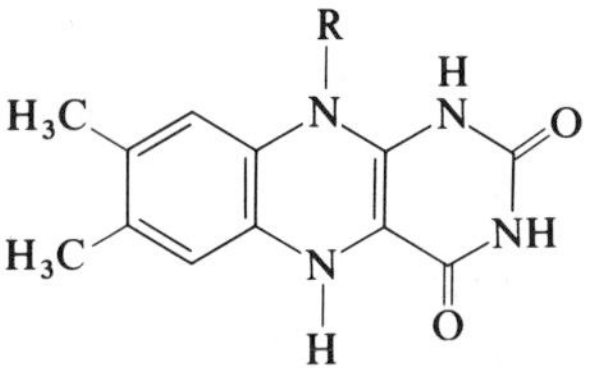

45

The use of isotopes had early shown that direct transfer of hydrogen, presumably as hydride, takes place to the C-4 position of NAD^+ and related compounds when they oxidize a variety of substrates (202, 203). Isotopes cannot be used in like fashion with the flavins, since hydrogen atoms attached to nitrogen undergo rapid exchange with hydroxylic media, the only conditions for which relevant conclusions regarding mechanism can be drawn. What at first sight seemed to be telling evidence in favor of the hydride mechanism in flavin redox reactions was the observation that a number of 5-deazaflavins, which contain a C—H unit at the 5 position instead of a nitrogen atom, do indeed undergo direct transfer of hydrogen to the 5 position when they act as oxidants (205). It has since become apparent, however, that the redox chemistry of 5-deazaflavins is more like that of the nicotinamide unit in NAD^+ than that of the isoalloxazine unit of flavins. Indeed, the lability of the N-5 proton that prevents isotopes from being used to trace its history may be part of the redox capability of the compound. Nonetheless, in the case of flavin reduction by dihydronicotinamides the weight of the evidence suggests that the reaction takes place by direct hydride transfer (215), although the question remains a vexed one.

3. Stepwise Hydride Transfer

Although the enzymatic reductions of pyridinium, flavin, and deazaflavin systems have been characterized here as hydride transfers, there are indications that a number of other reactions that appear to be simple hydride transfers may actually take place in a stepwise fashion, with an initial electron transfer being followed by either hydrogen atom transfer or by successive proton and electron transfers. Provided that these steps follow one another sufficiently closely that the intermediates cannot be intercepted by other reagents nor transient radicals detected (225, 226), it seems best to designate such processes simply as hydride transfers, as was done previously.

Clear evidence for the presence of intermediate oxidation states has been found in the case of the photo-induced reaction of benzophenone (46) with N-methylacridan (47). When this reaction, formally a hydride transfer, is conducted in benzene, it produces the ionic products 48 and 49 [Eq. (4-38)].

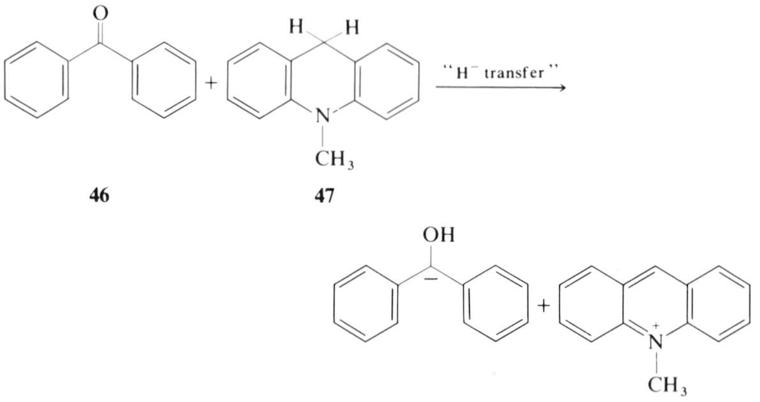

The transient electronic spectrum observed as this reaction takes place shows that the reaction actually occurs by initial formation of the diradical ion pair 50, followed by proton transfer to give 51 and 52 [Eq. (4-39)]. The latter

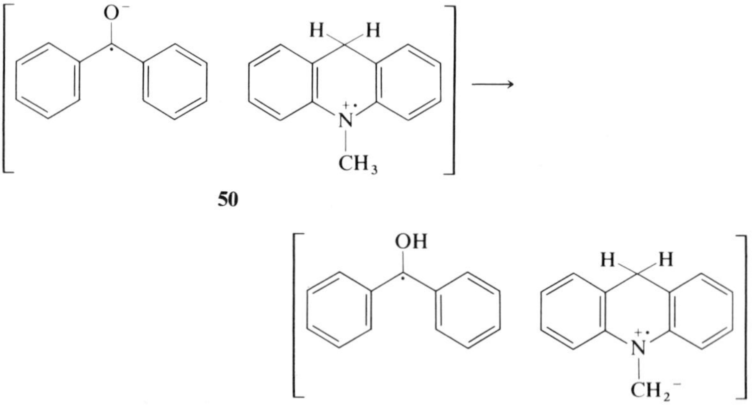

radical, which is also an ylid, undergoes a rapid [1,5]-sigmatropic shift to give 53, which immediately transfers an electron to 51 to give the products 48 and 49 (227, 228).

53

The proton and electron transfer steps that follow the initial electron transfer are extremely fast, being in the pico- and nanosecond time range. Even though this is a photochemical reaction there is evidence that, if the ground state process is also initiated by electron transfer, then the subsequent proton and electron transfers will be sequential as well. Because of the rapidity with which the subsequent steps follow the initial electron transfer and the inability of external reagents to intercept the transient intermediates, it seems likely that this and similar reactions will continue to be designated as hydride transfers. Finally, electron transfer in organic polymers and its coupling to proton transfer are matters of current interest (*229*).

References

1. F. Hibbert, *J. Chem. Soc., Chem. Commun.* p. 463 (1973).

2. A. Awwal and F. Hibbert, *J. Chem. Soc., Perkin Trans. 2* p. 1589 (1977).

3. P. B. Smith, J. L. Dye, J. Cheney, and J.-M. Lehn, *J. Am. Chem. Soc.* **103**, 6044 (1981).

4. F. Hibbert and K. P. P. Hunte, *J. Chem. Soc., Perkin Trans. 2* p. 1562 (1981).

5. N.-A. Bergman, Y. Chiang, and A. J. Kresge, *J. Am. Chem. Soc.* **100**, 5954 (1978).

6. M. M. Cox and W. P. Jencks, *J. Am. Chem. Soc.* **100**, 5956 (1978).

7. R. W. Alder, N. C. Goode, N. Miller, F. Hibbert, K. P. P. Hunte, and H. J. Robbins, *J. Chem. Soc., Chem. Commun.* p. 89 (1978).

8. W. van Gerresheim, C. Kruk, and J. W. Verhoeven, *Tetrahedron Lett.* **23**, 565 (1982).

9. R. S. Henry, F. G. Riddell, W. Parker, and C. I. F. Watt, *J. Chem. Soc., Perkin Trans. 2* p. 1549 (1976).

10. F. Hibbert, *J. Chem. Soc., Perkin Trans. 2* p. 1304 (1980).

11. R. J. P. Williams, *Chem. Soc. Rev.* p. 265 (1980).

12. R. Stewart, *Can. J. Chem.* **35**, 766 (1957).

13. R. Grinter and S. F. Mason, *Trans. Faraday Soc.* **60**, 889 (1969).

14. A. K. Colter, G. Saito, F. J. Sharom, and A. P. Hong, *J. Am. Chem. Soc.* **98**, 7833 (1976).

15. N. S. Isaacs, K. Javaid, and E. Rannala, *J. Chem. Soc., Perkin Trans. 2* p. 709 (1978).

16. R. Stewart and T. W. Toone, *J. Chem. Soc., Perkin Trans. 2* p. 1243 (1978).

17. D. Bethell, G. J. Hare, and P. A. Kearney, *J. Chem. Soc., Perkin Trans. 2* p. 684 (1981).

18. R. Stewart, "Oxidation Mechanisms, Applications to Organic Chemistry." Benjamin, New York, 1964.

19. M. Cohen and F. H. Westheimer, *J. Am. Chem. Soc.* **74**, 4387 (1952).

20. A. Leo and F. H. Westheimer, *J. Am. Chem. Soc.* **74**, 4383 (1952).

21. F. H. Westheimer and F. Novick, *J. Chem. Phys.* **11**, 506 (1943).

22. J. Rocek and J. Krupicka, *Collect. Czech. Chem. Commun.* **23**, 2068 (1958).

23. H. Kwart and J. H. Nickle, *J. Am. Chem. Soc.* **96**, 7572 (1974).

24. H. Kwart, *Suom. Kemistil. A* **34A,** 173 (1961).
25. J. Rocek, F. H. Westheimer, A. Eschenmoser, L. Moldovanyi, and J. Schreiber, *Helv. Chim. Acta* **45,** 2554 (1962).
26. R. Stewart and D. G. Lee, *Can. J. Chem.* **42,** 439 (1964).
27. R. Stewart, *in* "Isotopes in Organic Chemistry" (E. Buncel and C. C. Lee, eds.), Vol. 2, Chap. 7. Elsevier, Amsterdam, 1976.
28. H. O. House, "Modern Synthetic Reactions," p. 417. Benjamin, Menlo Park, 1972.
29. G. A. Hamilton, *in* "Progress in Bioorganic Chemistry" (E. T. Kaiser and F. J. Kézdy, eds.), Vol. 1, p. 83. Wiley, New York, 1971.
30. P. Schuster, G. Zundel, and C. Sandorfy, eds., "The Hydrogen Bond," Vols. I–III. North-Holland Publ., Amsterdam, 1976.
31. B. H. Robinson, *in* "Proton Transfer Reactions" (E. F. Caldin and V. Gold, eds.), Chap. 5. Chapman and Hall, London, 1975.
32. J. E. Crooks and B. H. Robinson, *Far. Symp. Chem. Soc.* **10,** 29 (1975).
33. S. L. Johnson and K. A. Rumon, *J. Phys. Chem.* **69,** 74 (1965).
34. J.-M. Leclerq, P. Dupuis, and C. Sandorfy, *Croat. Chim. Acta* **55,** 105 (1982).
35. L. Pauling, *J. Am. Chem. Soc.* **58,** 94 (1936).
36. M. D. Joesten and L. J. Schaad, "Hydrogen Bonding," p. 14. Dekker, New York, 1974.
37. D. Hadzi and S. Detoni, *in* "The Chemistry of Acid Derivatives" (S. Patai, ed.), Suppl. B, Part 1, Chap. 6. Wiley, Chichester, 1979.
38. D. Hadzi, *Z. Elektrochem.* **62,** 1157 (1958).
39. D. Hadzi, *J. Chem. Phys.* **34,** 1445 (1962).
40. T. Schaefer, S. R. Salman, T. A. Wildman, and P. D. Clark, *Can. J. Chem.* **60,** 342 (1982).
41. O. Kasende and Th. Zeegers-Huyskens, *J. Mol. Struct.* **75,** 201 (1981).
42. A. J. Kresge, G. W. Barry, K. R. Charles, and Y. Chiang, *J. Am. Chem. Soc.* **84,** 4343 (1962); see also ref. *43.*
43. J. De Taeye, G. Maes, and Th. Zeegers-Huyskens, *Bull. Soc. Chim. Belg.* **92,** 917 (1983).
44. I. Olovsson and P.-G. Jönsson, in ref. *30,* Vol. 2.
45. J. Roziere and J. M. Williams, *Inorg. Chem.* **15,** 1174 (1976).
46. D. J. Millen, *Croat. Chem. Acta* **55,** 133 (1982).
47. W. A. P. Luck, in ref. *30,* Vol. 2.
48. T. L. Brown, L. G. Butler, D. Y. Curtin, T. Hiyama, I. C. Paul, and R. B. Wilson, *J. Am. Chem. Soc.* **104,** 1172 (1982).
49. A. Novak, *Struct. Bonding (Berlin)* **18,** 177 (1974).
50. W. A. P. Luck, *Ber. Bunsenges. Phys. Chem.* **65,** 355 (1961); *Naturwissenschaften* **52,** 25 and 49 (1965).
51. G. Geiseler and S. Fruwert, *Z. Phys. Chem.* **26,** 111 (1960).
52. G. Geiseler, K. Quitsch, R. Gesemann, and H. J. Gesemann, *Z. Phys. Chem.* **35,** 10 (1962).
53. R. Yamdagni and P. Kebarle, *J. Am. Chem. Soc.* **95,** 3504 (1973); see also refs. *54* and *55.*
54. J. W. Larson, R. L. Clair, and T. B. McMahon, *Can. J. Chem.* **60,** 542 (1982).
55. J. M. Buschek, F. S. Jorgensen, and R. S. Brown, *J. Am. Chem. Soc.* **104,** 5019 (1982); E. A. Hillenbrand and S. Scheiner, *J. Am. Chem. Soc.* **106,** 6266 (1984) and refs. therein.
56. See, for example, P. A. Kollman and D. M. Hayes, *J. Am. Chem. Soc.* **103,** 2955 (1981).
57. P. G. Jonsson and A. Kvick, *Acta Crystallogr., Sect. B* **28B,** 1827 (1972).
58. See also G. Alagona, C. Ghio, and P. Kollman, *J. Am. Chem. Soc.* **105,** 5226 (1983); G. A. Jeffrey and J. Mitra, *J. Am. Chem. Soc.* **106,** 5546 (1984).
59. H. Shimanouchi and Y. Sasada, *Acta Crystallogr., Sect B* **29B,** 81 (1973).
60. L. M. Jackman, J. C. Trewella, and R. C. Haddon, *J. Am. Chem. Soc.* **102,** 2519 (1980).
61. A. O. Patil, D. Y. Curtin, and I. C. Paul, *J. Am. Chem. Soc.* **106,** 4010 (1984).
62. H. H. Limbach, J. Hennig, R. Kendrick, and C. S. Yannoni, *J. Am. Chem. Soc.* **106,** 4059 (1984).
63. G. A. Jeffrey and H. Maluszynska, *Int. J. Biol. Macromol.* **4,** 173 (1982).

64. R. Taylor, O. Kennard, and W. Versichel, *J. Am. Chem. Soc.* **106**, 244 (1984).

65. R. W. Wright and R. H. Marchessault, *Can. J. Chem.* **46**, 2567 (1968).

66. E. Fishman and T. L. Chen, *Spectrochim. Acta, Part A* **25A**, 1231 (1969).

67. See also M. Meot-Mer, *J. Am. Chem. Soc.* **106**, 278 (1984).

68. M. J. Kamlet and R. W. Taft, *J. Org. Chem.* **47**, 1734 (1982).

69. R. Taylor, O. Kennard, and W. Versichel, *J. Am. Chem. Soc.* **105**, 5761 (1983).

70. P. R. Rony, *J. Am. Chem. Soc.* **91**, 6090 (1969).

71. M. M. Cox and W. P. Jencks, *J. Am. Chem. Soc.* **103**, 572 (1981).

72. M. M. Cox and W. P. Jencks, *J. Am. Chem. Soc.* **103**, 580 (1981).

73. J. Spaulding, J. E. Stein, and J. E. Meany, *J. Phys. Chem.* **81**, 1359 (1977).

74. L. D. Amaral, K. Koehler, D. Bartenbach, T. Pletcher, and E. H. Cordes, *J. Am. Chem. Soc.* **89**, 3537 (1967).

75. P. Eugster and H. Zollinger, *Helv. Chim. Acta* **52**, 1985 (1969).

76. B. R. Glutz and H. Zollinger, *Helv. Chim. Acta* **52**, 1976 (1969).

77. J. L. Hogg, D. A. Jencks, and W. P. Jencks, *J. Am. Chem. Soc.* **99**, 4772 (1977).

78. M. F. Aldersley, A. J. Kirby, P. W. Lancaster, R. S. McDonald, and C. R. Smith, *J. Chem. Soc., Perkin Trans. 2* p. 1487 (1974).

79. J. Hine, M. S. Cholod, and R. A. King, *J. Am. Chem. Soc.* **96**, 835 (1974).

80. J. Hine, M. S. Cholod, and J. H. Jensen, *J. Am. Chem. Soc.* **93**, 2321 (1971).

81. R. D. Gandour, *Tetrahedron Lett.* p. 295 (1974).

82. S. Shinkai, N. Nakashima, and T. Kunitake, *Bull. Chem. Soc. Jpn.* **54**, 840 (1981).

83. Y. N. Lee and G. L. Schmir, *J. Am. Chem. Soc.* **101**, 3026 (1979).

84. R. P. Bell, B. G. Cox, and J. B. Henshall, *J. Chem. Soc., Perkin Trans. 2* p. 1232 (1972).

85. R. P. Bell and B. A. Timimi, *J. Chem. Soc., Perkin Trans. 2,* p. 1518 (1973).

86. E. T. Harper and M. L. Bender, *J. Am. Chem. Soc.* **87**, 5625 (1965).

87. J. K. Coward and T. C. Bruice, *J. Am. Chem. Soc.* **91**, 5339 (1969).

88. R. P. Bell and M. A. D. Fluendy, *Trans. Faraday Soc.* **59**, 1623 (1963).

89. F. M. Menger, *Tetrahedron* **39**, 1013 (1983).

90. Ref. *89*, p. 1032.

91. F. M. Menger, J. F. Chow, H. Kaiserman, and P. C. Vasquez, *J. Am. Chem. Soc.* **105**, 4996 (1983).

92. P. Deslongchamps, "Stereoelectronic Effects in Organic Chemistry." Pergamon, Oxford, 1983.

93. E. J. Corey and R. A. Sneen, *J. Am. Chem. Soc.* **78**, 6269 (1956).

94. The stereochemistry of eliminations is reviewed in refs. *95* and *96* and is analyzed in terms of frontier orbitals in ref. *97*.

95. S. Wolfe, *Acc. Chem. Res.* **5**, 102 (1972).

96. J. Sicher, *Angew. Chem., Int. Ed. Engl.* **11**, 200 (1972).

97. R. D. Bach, R. C. Badger, and T. J. Lang, *J. Am. Chem. Soc.* **101**, 2845 (1979).

98. H. A. Rachid, C. Larrieu, M. Chaillet, and J. Elguero, *Tetrahedron* **39**, 1307 (1983).

99. S. Wolfe, H. B. Schlegel, I. G. Czismadia, and F. Bernardi, *Can. J. Chem.* **53**, 3365 (1975).

100. R. R. Fraser and P. J. Champagne, *J. Am. Chem. Soc.* **100**, 657 (1978).

101. See refs. given in ref. *100*.

102. A. R. Fersht and A. J. Kirby, *J. Am. Chem. Soc.* **89**, 4853 and 4857 (1967).

103. R. P. Bell and B. A. Timiwi, *J. Chem. Soc., Perkin Trans. 2* p. 1518 (1973).

104. A. J. Kirby, *in* "Advances in Physical Organic Chemistry" (V. Gold and D. Bethell, eds.), Vol. 17, p. 183. Academic Press, London, 1980.

105. A. J. Kirby and G. J. Lloyd, *J. Chem. Soc., Perkin Trans. 2* p. 1762 (1976).

106. C. F. Bernasconi, S. A. Hibdon, and S. E. McMurry, *J. Am. Chem. Soc.* **104**, 3459 (1982).

107. E. S. Lewis, *in* "Isotopes in Organic Chemistry" (E. Buncel and C. C. Lee, eds.), Vol. 2, p. 134. Elsevier, Amsterdam, 1976.

108. B. M. Gimarc, "Molecular Structure and Bonding," p. 27. Academic Press, New York, 1979.

109. See also P. J. Brown and E. F. Hayes, *J. Chem. Phys.* **55**, 922 (1971).

110. H. Kwart, *Acc. Chem. Res.* **15**, 401 (1982) and refs. therein; see, however, B. Anhede and N.-Å. Bergman, *J. Am. Chem. Soc.* **106**, 7634 (1984).

111. H. Kwart, M. W. Brechbiel, R. M. Acheson, and D. C. Ward, *J. Am. Chem. Soc.* **104**, 4671 (1982).

112. A. A. Dzakpasu, S. E. V. Phillips, J. R. Scheffer, and J. Trotter, *J. Am. Chem. Soc.* **98**, 6049 (1976).

113. J. R. Scheffer, and A. A. Dzakpasu, *J. Am. Chem. Soc.* **100**, 2163 (1978).

114. P. J. Wagner, *in* "Molecular Rearrangements in Ground and Excited States" (P. de Mayo, ed.), Chap. 20. Wiley (Interscience), New York, 1980.

115. W. G. Dauben, L. Salem, and N. J. Turro, *Acc. Chem. Res.* **8** 41 (1975).

116. S. Ariel, V. Ramamurthy, J. R. Scheffer, and J. Trotter, *J. Am. Chem. Soc.* **105**, 6959 (1983).

117. W. K. Appel, Z. Q. Jiang, J. R. Scheffer, and L. Walsh, *J. Am. Chem. Soc.* **105**, 5354 (1983).

118. M. A. Winnick, *Acc. Chem. Res.* **10**, 173 (1977).

119. E. S. Lewis and M. C. R. Symons, *Q. Rev. Chem. Soc.* (*London*) **12**, 230 (1958).

120. M. F. Hawthorne and E. S. Lewis, *J. Am. Chem. Soc.* **80**, 4296 (1958).

121. G. A. Olah, Y. K. Mo, and J. A. Olah, *J. Am. Chem. Soc.* **95**, 4939 (1973).

122. C: Djerassi, *Org. React.* (*N.Y.*) **6**, 207 (1951).

123. C. J. Collins and J. F. Eastham, *in* "The Chemistry of the Carbonyl Group" (S. Patai, ed.), p. 778. Wiley (Interscience), New York, 1966.

124. W. Acklin and V. Prelog, *Helv. Chim. Acta* **42**, 1239 (1959).

125. E. W. Warnhoff, *Can. J. Chem.* **55**, 1635 (1977).

126. A. Kasal and A. Trka, *Collect. Czech. Chem. Commun.* **42**, 1389 (1977).

127. J. M. Shepherd, D. Singh, and P. Wilder, *Tetrahedron Lett.* p. 2743 (1974).

128. S. Danishefsky, M. Hirama, N. Fritsch, and J. Clardy, *J. Am. Chem. Soc.* **101**, 7013 (1979).

129. G.-A. Craze and I. Watt, *J. Chem. Soc., Perkin Trans. 2* p. 175 (1981).

130. R. Cernik, G.-A. Craze, O. S. Mills, and I. Watt, *J. Chem. Soc., Perkin Trans. 2* p. 361 (1982).

131. H. B. Burgi, J. M. Lehn, and G. Wipff, *J. Am. Chem. Soc.* **96**, 1956 (1974).

132. W. van Gerresheim, C. Kruk, and J. W. Verhoeven, *Tetrahedron Lett.* **23**, 565 (1982).

133. W. van Gerresheim and J. W. Verhoeven, *J.R. Neth. Chem. Soc.* **102**, 339 (1983); see also S. M. van der Kerk, W. van Gerresheim, and J. W. Verhoeven, *J.R. Neth. Chem. Soc.* **103**, 143 (1984).

134. M. C. Donkersloat and H. M. Buck, *J. Am. Chem. Soc.* **103**, 6549 (1981).

135. R. H. A. M. Brounts and H. M. Buck, *J. Am. Chem. Soc.* **105**, 1284 (1983).

136. See also N. E. Okazawa and T. S. Sorensen, *Can. J. Chem.* **60**, 2180 (1982).

137. J. J. Dannenberg, B. J. Goldberg, J. K. Barton, K. Dill, D. H. Weinwurzel, and M. O. Longas, *J. Am. Chem. Soc.* **103**, 7764 (1981).

138. R. P. Kirchen and T. S. Sorensen, *J. Chem. Soc., Chem. Commun.* p. 769 (1978).

139. R. P. Kirchen, T. S. Sorensen, and K. Wagstaff, *J. Am. Chem. Soc.* **100**, 6761 (1978).

140. R. P. Kirchen, K. Ranganayakulu, B. P. Singh, and T. S. Sorensen, *Can. J. Chem.* **59**, 2173 (1981).

141. R. P. Kirchen, K. Ranganayakulu, A. Rauk, B. P. Singh, and T. S. Sorensen, *J. Am. Chem. Soc.* **103**, 588 (1981).

142. R. P. Kirchen, N. Okazawa, K. Ranganayakulu, A. Rauk, and T. S. Sorensen, *J. Am. Chem. Soc.* **103**, 597 (1981).

143. M. Saunders, L. Telkowski, and M. R. Kates, *J. Am. Chem. Soc.* **99**, 8070 (1977).

144. M. Saunders and M. R. Kates, *J. Am. Chem. Soc.* **99**, 8071 (1977).

145. D. Saturnino, M. Yamauchi, W. R. Clayton, R. W. Nelson, and S. G. Shore, *J. Am. Chem. Soc.* **97**, 6063 (1975).

146. W. J. Hehre, R. F. Stewart, and J. A. Pople, *J. Chem. Phys.* **51,** 2657 (1969).
147. P. Deslongchamps, D. D. Rowan, and N. Pothier, *Can. J. Chem.* **59,** 2787 (1981).
148. J. M. Erhardt and J. D. Wuest, *J. Am. Chem. Soc.* **102,** 6363 (1980).
149. J. M. Erhardt, E. R. Grover, and J. D. Wuest, *J. Am. Chem. Soc.* **102,** 6365 (1980).
149a. Y. Ducharme, S. Latour, and J. D. Wuest, *J. Am. Chem. Soc.* **106,** 1499 (1984).
150. S. Wolfe, A. Stolow, and L. A. LaJohn, *Can. J. Chem.* **62,** 1470 (1984) and refs. therein.
151. E. Cremer and M. Polanyi, *Z. Phys. Chem. B* **19B,** 443 (1932).
152. R. Stewart, K. C. Teo, and L. K. Ng, *Can. J. Chem.* **58,** 2497 (1980).
153. G. W. Cowell, A. Ledwith, A. C. White, and H. J. Woods, *J. Chem. Soc. B* p. 227 (1970).
154. P. Muller and J. Rocek, *J. Am. Chem. Soc.* **94,** 2716 (1972).
155. P. Muller, *Helv. Chim. Acta* **56,** 1243 (1973).
156. R. Srinivasan, T. T. Medary, H. F. Fisher, D. J. Norris, and R. Stewart, *J. Am. Chem. Soc.* **104,** 807 (1982).
157. O. Reitz, *Z. Phys. Chem. A* **175A,** 257 (1936); **179A,** 119 (1937).
158. K. B. Wiberg, *Chem. Rev.* **55,** 713 (1955); for an example of an unusual temperature-dependent isotope effect, see H. F. Koch and A. S. Koch, *J. Am. Chem. Soc.* **106,** 4536 (1984).
159. R. P. Bell and D. M. N. Goodall, *Proc. R. Soc., London, Ser. A* **294A,** 273 (1966).
160. E. S. Lewis and L. F. Funderburk, *J. Am. Chem. Soc.* **89,** 2322 (1967).
161. L. F. Funderburk and E. S. Lewis, *J. Am. Chem. Soc.* **86,** 2531 (1964).
162. R. P. Bell and W. B. T. Miller, *Trans. Faraday Soc.* **59,** 1147 (1963).
163. R. P. Bell, *Chem. Soc. Rev.* p. 513 (1974).
164. F. H. Westheimer, *Chem. Rev.* **61,** 265 (1961).
165. J. Bigeleisen, *Pure Appl. Chem.* **8,** 217 (1964).
166. R. A. More O'Ferrall *in* "Proton Transfer Reactions" (E. F. Caldin and V. Gold, eds.), Chap. 8. Chapman and Hall, London, 1975.
167. R. P. Bell and B. G. Cox, *J. Chem. Soc. B* p. 194 (1970); see, however, ref. *168.*
168. B. Anhede, L. Baltzer, and N.-A. Bergman, *Acta Chem. Scand., Ser. A* **36A,** 39 (1982).
169. L. Melander and W. H. Saunders, "Reaction Rates of Isotopic Molecules." Wiley (Interscience), New York, 1980.
170. R. P. Bell, W. H. Sachs, and R. L. Tranter, *Trans. Faraday Soc.* **67,** 1995 (1967).
171. R. P. Bell, "The Tunnel Effect in Chemistry." Chapman and Hall, London, 1980; see also W. Siebrand, T. A. Wildman, and M. Z. Zgierski, *J. Am. Chem. Soc.* **106,** 4083 and 4089 (1984).
172. A. J. Kresge, *Discuss. Faraday Soc.* **39,** 49 (1965).
173. P. J. Smith *in* "Isotopes in Organic Chemistry" (E. Buncel and C. C. Lee, eds.), Vol. 2, Chap. 6. Elsevier, Amsterdam, 1976.
174. J. L. Longridge and F. A. Long, *J. Am. Chem. Soc.* **89,** 1292 (1967).
175. D. J. McLennan and R. J. Wong, *Tetrahedron Lett.* **28,** 2887 (1972).
176. E. S. Lewis and M. M. Butler, *J. Chem. Soc., Chem. Commun.* p. 941 (1971).
177. W. A. Pryor and K. G. Kneipp, *J. Am. Chem. Soc.* **93,** 5584 (1971).
178. D. Bethell, G. J. Hare, and P. A. Kearney, *J. Chem. Soc., Perkin Trans. 2* p. 684 (1981).
179. B. Eistert and H. Munder, *Chem. Ber.* **91,** 1415 (1958).
180. A. Nickon and J. L. Lambert, *J. Am. Chem. Soc.* **84,** 4604 (1962); **88,** 1905 (1966).
181. D. H. Hunter, J. B. Stothers, and E. W. Warnhoff, *in* "Rearrangements in Ground and Excited States" (P. de Mayo, ed.), Vol. 1, Chap. 6. Academic Press, New York, 1980.
182. A. K. Cheng, A. K. Ghosh, I. Sheepy, and J. B. Stothers, *Can. J. Chem.* **59,** 3379 (1981).
183. J. Hine and G. F. Koser, *J. Org. Chem.* **36,** 3591 (1971).
184. H. Fredenhagen and K. F. Bonhoeffer, *Z. Phys. Chem. A* **181,** 379 (1938).
185. P. Salomaa, *Acta Chem. Scand.* **10,** 311 (1956).
186. W. E. Doering, T. I. Taylor, and E. F. Schoenewaldt, *J. Am. Chem. Soc.* **70,** 455 (1948).

187. V. Franzen, *Chem. Ber.* **88,** 1361 (1955).

188. V. Franzen, *Chem. Ber.* **90,** 623 (1957).

189. H. R. Mahler and E. H. Cordes, "Biological Chemistry," p. 340. Harper and Row, New York, 1966.

190. S. S. Hall, A. M. Doweyko, and F. Jordan, *J. Am. Chem. Soc.* **98,** 7460 (1976).

191. S. S. Hall, A. M. Doweyko, and F. Jordan, *J. Am. Chem. Soc.* **100,** 5934 (1978).

192. T. Okuyama, S. Komoguchi, and T. Fueno, *J. Am. Chem. Soc.* **104,** 2582 (1982).

193. E. Racker, *J. Biol. Chem.* **190,** 685 (1951).

194. R. V. J. Chari and J. W. Kozarich, *J. Biol. Chem.* **256,** 9785 (1981).

195. L. R. C. Barclay and J. M. Dust, *Can. J. Chem.* **60,** 607 (1982).

196. L. R. C. Barclay, A. G. Briggs, W. E. Briggs, J. M. Dust, and J. A. Gray, *Can. J. Chem.* **57,** 2172 (1979).

197. M. M. Wei and R. Stewart, *J. Am. Chem. Soc.* **88,** 1974 (1966).

198. D. H. Rosenblatt, G. T. Davis, L. A. Hull, and G. D. Forberg, *J. Org. Chem.* **33,** 1649 (1968).

199. L. H. Hull, G. T. Davis, D. H. Rosenblatt, H. K. R. Williams, and R. C. Weglein, *J. Am. Chem. Soc.* **89,** 1163 (1967).

200. J. R. Lindsay Smith and L. A. V. Mead, *J. Chem. Soc., Perkin Trans. 2* p. 206 (1973).

201. E. P. Burrows and D. H. Rosenblatt, *J. Org. Chem.* **48,** 992 (1983).

202. F. H. Westheimer, H. F. Fisher, E. E. Conn, and B. Vennesland, *J. Am. Chem. Soc.* **73,** 2403 (1951).

203. R. H. Abeles, R. F. Hutton, and F. H. Westheimer, *J. Am. Chem. Soc.* **79,** 712 (1957).

204. K. Wallenfels and D. Hofmann, *Tetrahedron Lett.* **4,** 151 (1962).

205. C. Walsh, F. Jacobson, and C. C. Ryerson, *Adv. Chem. Ser. No. 191* p. 123 (1980).

206. C. Walsh, *Acc. Chem. Res.* **13,** 148 (1980).

207. C. T. Suckling, *Chem. Soc. Rev.* **13,** 97 (1984).

208. G. Blankenhorn and E. G. Moore, *J. Am. Chem. Soc.* **102,** 1092 (1980).

209. J. J. Steffens and D. M. Chipman, *J. Am. Chem. Soc.* **93,** 6694 (1971).

210. L. C. Kurz and C. Frieden, *Biochemistry* **16,** 5207 (1977).

211. D. J. Creighton, J. Hajdu, and D. S. Sigman, *J. Am. Chem. Soc.* **98,** 4619 (1976).

212. F. M. Martens, J. W. Verhooven, R. A. Gase, U. K. Pandit, and Th. J. de Boer, *Tetrahedron* **34,** 443 (1978).

213. A. Ohno, T. Shio, H. Yamamoto, and S. Oka, *J. Am. Chem. Soc.* **103,** 2045, (1981).

214. I. MacInnes, D. C. Nonhebel, S. T. Orszulik, and C. J. Suckling, *J. Chem. Soc., Chem. Commun.* p. 121 (1982).

215. M. F. Powell and T. C. Bruice, *J. Am. Chem. Soc.* 105, 1014 (1983).

216. A. van Laar, H. J. van Ramesdonk, and J. W. Verhoeven, *J. R. Neth. Chem. Soc.* **102,** 157 (1983).

217. D. Ostovic, R. M. G. Roberts, and M. M. Kreevoy, *J. Am. Chem. Soc.* **105,** 7629 (1983).

218. R. G. Morris, G. Saliman, and M. F. Dunn, *Biochemistry* **19,** 725 (1980).

219. D. J. T. Porter and H. J. Bright, *J. Biol. Chem.* **252,** 4361 (1977).

220. D. J. T. Porter, J. G. Voet, and H. J. Bright, *J. Biol. Chem.* **248,** 4400 (1973).

221. R. F. Williams, S. S. Shinkai, and T. C. Bruice, *J. Am. Chem. Soc.* **99,** 921 (1977).

222. T. C. Bruice, *Adv. Chem. Ser. No. 191* p. 89 (1980).

223. T. C. Bruice and Y. Yano, *J. Am. Chem. Soc.* **97,** 5263 (1975).

224. L.-C. Yuan and T. C. Bruice, *J. Am. Chem. Soc.* **106,** 1530 (1984).

225. See, for example, E. C. Ashby, A. B. Goel, and R. N. DePriest, *J. Am. Chem. Soc.* **102,** 7779 (1980).

226. See also D. L. Kleyer and T. H. Koch, *J. Am. Chem. Soc.* **105,** 5911 (1983).

227. K. S. Peters, E. Pang, and J. Rudzki, *J. Am. Chem. Soc.* **104,** 5535 (1982).

228. L. E. Manring and K. S. Peters, *J. Am. Chem. Soc.* 105, 5708 (1983).

229. G. Inzelt, J. Q. Chambers, J. F. Kinstle, and R. W. Day, *J. Am. Chem. Soc.* **106,** 3396 (1984).

5

Alternative Sites
of Protonation
and Deprotonation

I. Introduction

Many organic acids and bases have more than one site for removal or addition of a proton, and determining which of these is the principal locus of reaction is not always straightforward. In some cases a markedly acidic and a markedly basic group are present in the same molecule (e.g., amino acids), and in such amphiprotic systems there is no difficulty in locating the reactive sites. Alternative sites of deprotonation now appear in the cation, however, and alternative sites of protonation now appear in the anion. For purposes of discussion it will be convenient to consider three separate categories: alternative sites of protonation, alternative sites of deprotonation, and amphiprotic systems that form zwitterions.

II. Alternative Sites of Protonation

Molecules with two or more reactive centers are called *ambident*. There are two types of ambident base, that is, systems in which alternative sites of protonation exist. The first is exemplified by the amide, carboxyl, and amidine units; each of these is a single functional group that contains more than one atom capable of bonding to a proton. The second type comprises molecules that contain two or more separate basic functional groups. These two classes of ambident base are considered in the following sections.

A. Ambident Functional Groups

This subject was extensively reviewed by Liler in 1975 (*1*), and the following discussion therefore only touches on the highlights of the earlier work. In addition, the question of the site of protonation in amides, in particular, was reviewed by Homer and Johnson in 1971 (*2*). It is interesting that opposite conclusions were drawn in these two reviews with regard to the question of N-versus O-protonation of amides. Homer and Johnson, like the vast majority of workers in the field (3, 4), concluded that O-protonation occurs with amides [Eq. (5-1)]. Liler, on the other hand, favors N-protonation for aqueous solutions [Eq. (5-2)] and is the principal protagonist of this school of thought.

$$R-C\overset{\displaystyle O}{\underset{\displaystyle NH_2}{<}} + H^+ \rightleftharpoons R-C\overset{\displaystyle OH}{\underset{\displaystyle NH_2}{<}}+ \qquad (5\text{-}1)$$

$$R-C\overset{\displaystyle O}{\underset{\displaystyle NH_2}{<}} + H^+ \rightleftharpoons R-C\overset{\displaystyle O}{\underset{\displaystyle \overset{+}{N}H_3}{<}} \qquad (5\text{-}2)$$

What sort of evidence has been adduced to support the conclusion that amides are protonated at oxygen, and why has it not been universally accepted? The most telling evidence comes from nmr studies, and there is now virtually universal agreement that the O-protonated form is favored in concentrated acid solutions and in the gas phase. The fact that there is considerable resonance stabilization in this form makes it intuitively attractive as well.

The nmr spectra of amides in concentrated sulfuric acid and in moderately concentrated (72%) perchloric acids are consistent only with the O-protonated form (5, 6), as are the low-temperature nmr spectra of formamide, acetamide, benzamide, and their N-methyl and N,N-dimethyl derivatives in fluorosulfuric acid (7, 8). In the latter cases the absorption of the proton attached to the carbonyl oxygen atom can be clearly seen.

Salts of many amides can be prepared, and a number of tiese have been the object of infrared and Raman studies. Although many of these studies, like so many others dealing with amide protonation, are ambiguous to some extent (1, 2), definitive results that favor O-protonation have been obtained in a few cases. For example, earlier infrared work that appeared to prove the presence of a carbonyl group in amide salts and that would be consistent only with N-protonation was shown by the use of ^{18}O-substituted salts to be based on a false assignment (9, 10).

Does the site of protonation of amides switch from oxygen in concentrated acid to nitrogen in dilute acid, as Liler contends? The two principal lines of investigation that have been followed in recent years in attempts to resolve this question are (a) nmr studies, including rates of exchange of the amido protons (1, 6, 12–15), and (b) studies of the small changes observed in the ultraviolet (and nmr) spectra of the cations as the acid concentration is changed (15–17).

Exchange of amido protons with protic solvents is catalyzed by both acids and bases. Perrin and Johnston (12) showed that the acid-catalyzed reaction has two routes, one via the N-protonated form and one via the O-protonated form. Amides with electron-withdrawing substituents exchange predominantly through the O-protonated form, while many others react via the N-protonated form. They concluded, however, that the latter cations must be very strong acids, which means that they are minor components in the tautomeric equilibrium shown in Eq. (5-3). That is, reaction proceeds through

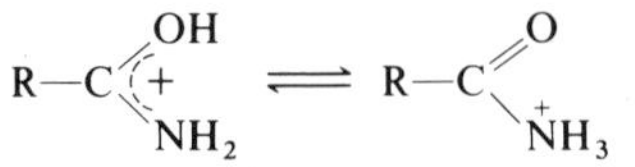

$$R-C\overset{OH}{\underset{NH_2}{\langle}}+ \rightleftharpoons R-C\overset{O}{\underset{NH_3}{\diagdown}} \qquad (5\text{-}3)$$

the minor component of the equilibrium, the N-protonated form. Such a result would be consistent with the value of -7 or so that Fersht calculated for the pK_{BH^+} of N-protonated amides (18).

Protonation of amides usually causes modest but significant changes in their ultraviolet and nmr spectra. For most amides this takes place in the region of 15 to 50% sulfuric acid by weight. As solutions of amide cation are made still more acidic, the ultraviolet and nmr spectra of most amides continue to undergo changes, although these are almost always quite small. Liler interpreted these changes as a medium effect on the equilibrium shown in Eq. (5-3) with the N-protonated forms, which will be strongly hydrogen-bonded to water, being displaced by the O-protonated forms in concentrated acid (15–17). The latter ions possess considerable charge delocalization, and it is argued that they will be less affected by the removal of water from the medium than will the former. Accordingly, the equilibrium will be tipped from the N- to the O-protonated form as the concentration of acid becomes high.

The alternative explanation, and the one that is generally accepted, is that the observed medium effect is not on the equilibrium but on the spectra of the O-protonated forms. Until it is shown that the latter explanation is untenable by, for example, studying model compounds such as **1** or **2** in which alkyl

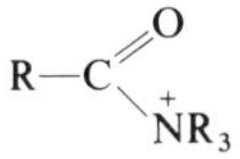

$$R-C\overset{O}{\underset{NR_3}{\diagdown}} \qquad R-C\overset{OR}{\underset{NR_2}{\langle}}+$$

$$\textbf{1} \qquad\qquad \textbf{2}$$

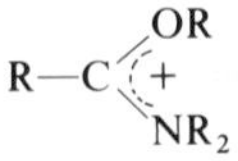

groups immobilize the structures, it seems likely that the O-protonated form will continue to be regarded as the major cationic form of amides under all conditions.

It should be noted that amido groups in which resonance is prevented by geometrical constraints are much stronger bases than ordinary amides. For example, 2,2-dimethylquinuclidone-6 has a pK_{BH^+} of 5.33 (19), undoubtedly as a result of protonation taking place on nitrogen to give the cation **3**. The

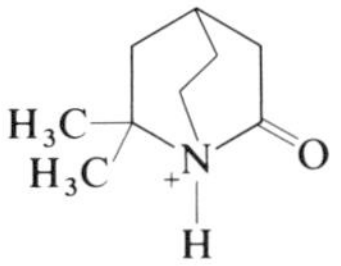

3

loss of resonance in both the amide and the O-protonated cation has the effect of sharply increasing the basicity of the amino group at the expense of the carbonyl group.

Whereas thioamides are protonated on sulfur, sulfonamides are protonated on nitrogen (Chapter 3, Section VII), as apparently are phosphinamides (20).

Other ambident groups include carboxyl and amidino. There is no doubt whatever that the amidino group is protonated at the imino nitrogen to give the resonance-stabilized cation [Eq. (5-4)], and it has become clear that the carboxyl group gives the resonance-stabilized ion as well [Eq. (5-5)]. There

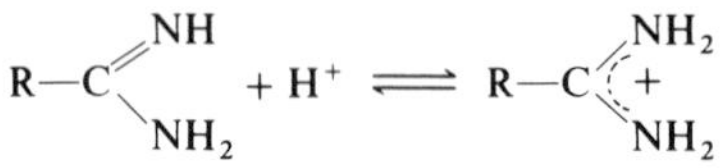

$$
R-C{\overset{NH}{\underset{NH_2}{\big<}}} + H^+ \rightleftharpoons R-C{\overset{NH_2}{\underset{NH_2}{\big<}}} + \qquad (5\text{-}4)
$$

$$
R-C{\overset{O}{\underset{OH}{\big<}}} + H^+ \rightleftharpoons R-C{\overset{OH}{\underset{OH}{\big<}}} + \qquad (5\text{-}5)
$$

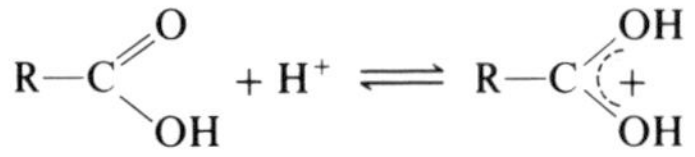

had earlier been some doubt on this score since it had been assumed that the intrinsic basicity of a hydroxyl group was very much greater than that of a carbonyl group. We now suspect that most aldehydes and ketones are not much weaker than the analogous alcohols (Chapter 3, Section VIII,A), and when one considers the large degree of resonance that must be present in the carbonyl-protonated form, there is little doubt that it is the strongly favored cation. Furthermore, there is evidence that the charge can be delocalized to the aromatic ring when R is aryl, and this is consistent with carbonyl protonation (21).

B. Other Ambident Bases

Enolate anions are ambident species that are analogous to neutral amides in the sense that there are two sites of protonation and the equilibrium constant K_T linking the two tautomeric protonated forms is a function of the acidity constants of these two species, that is, K_{HA}^{keto} and K_{HA}^{enol} [Eq. (5-6)].

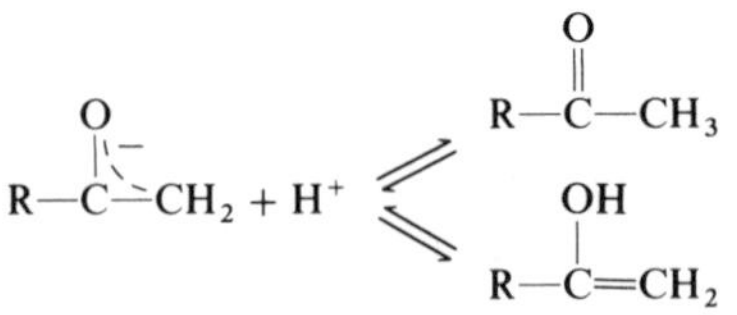

$$
R-\overset{O^-}{\underset{}{C}}-CH_2 + H^+ \underset{}{\overset{}{\big<}}
\begin{array}{l}
R-\overset{O}{\overset{\|}{C}}-CH_3 \\[4pt]
\overset{OH}{\underset{}{\big|}} \\
R-C{=}CH_2
\end{array}
\qquad (5\text{-}6)
$$

It can be readily shown that K_T is equal to $K_{HA}^{keto}/K_{HA}^{enol}$, where $K_T = $ [enol]/[keto]. If, as is often the case, the keto form is the more stable of the two ($K < 1$), then the ketone is the weaker acid by the amount of the tautomeric equilibrium constant, that is, $K_{HA}^{keto} = K_T K_{HA}^{enol}$. This is a general phenomenon; the minor component of a pair of compounds in equilibrium with a common anion is the stronger acid. Thus, the aci forms of nitroalkanes are stronger acids than the nitro forms, and the keto forms of phenols are stronger acids than the phenolic forms.

Stronger acid	Weaker acid

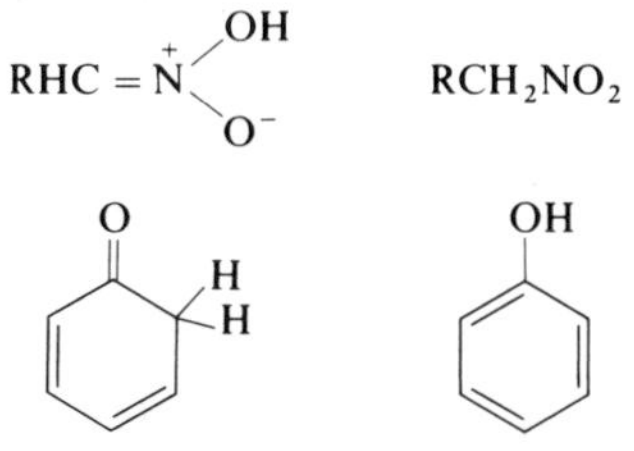

The rates at which acids ionize do not necessarily follow the order of their equilibrium acidities; 2,4-cyclohexadienone cannot be expected to lose a proton to a base as rapidly as does its weaker isomer phenol.

With regard to bases the minor component of a pair of neutral compounds that are protonated to give a common cation is the stronger base. The ratio of base strengths is given by the tautomeric constant linking the two neutral compounds as, for example, in Eq. (5-7); in water the hydroxypyridine is the

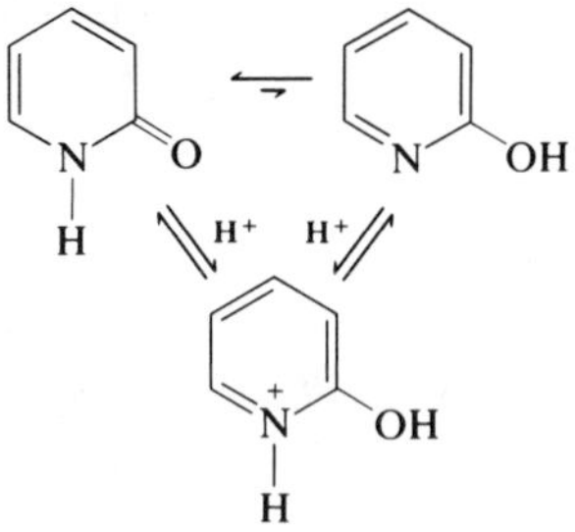

$$(5\text{-}7)$$

minor tautomer and the stronger base, and the α-pyridone is the major tautomer and the weaker base (22–24). The macroscopic pK_{BH^+} for the tautomeric mixture, of course, a composite value that reflects the proportion of the two compounds in the mixture (Section IV).

The neutral tautomeric pair shown in Eq. (5-7) is also in equilibrium with a common anion **4**, the minor tautomer being the stronger acid of the two, as

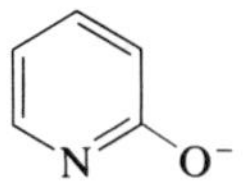

4

well as being the stronger base. It may seem paradoxical that protonation of the anion takes place preferentially at nitrogen rather than at anionic oxygen, whereas the molecule so produced prefers to be protonated at neutral oxygen rather than at nitrogen.

The acid–base equilibrium of the oxy derivatives of the species shown in Eq. (5-7) is given in Eq. (5-8), where the tautomeric mixture (*25, 26*) gives

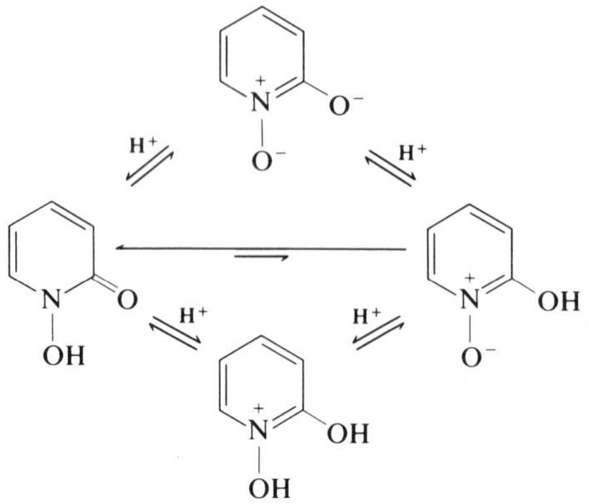

(5-8)

a common cation on protonation and a common anion on deprotonation. Again, the minor tautomer is the stronger acid and stronger base. (Tautomers are structural isomers in a mobile equilibrium; the term *protomer* is sometimes used for tautomers in which the mobile group is hydrogen. See ref. *27* and references therein for accounts of the energetics of protomerization of various heterocyclic systems.)

The 2-, 3-, and 4-aminopyridines each have two sites of protonation, with the ring nitrogen being the preferred site in all cases. The proportion of the two protonated forms in such cases can be calculated with some accuracy with the help of the Hammett equation (*28–30*), as can be illustrated for the case of 4-aminopyridine [Eq. (5-9)].

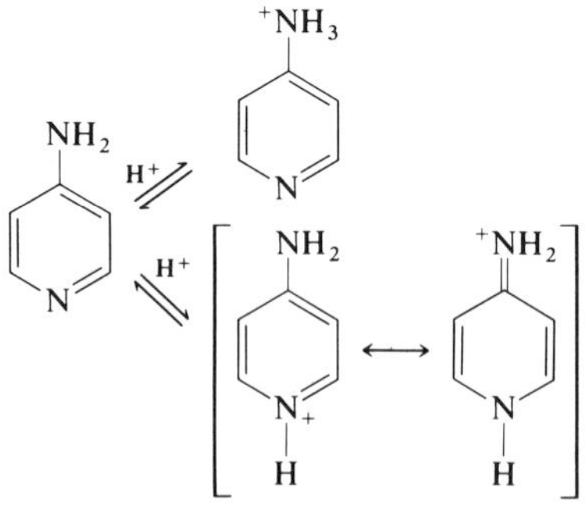

(5-9)

The pK_{BH^+} for amino-protonated 4-aminopyridine is given by Eq. (5-10),

$$pK_{BH^+}^{amino} = pK_{BH^+}^{aniline} - \rho_{NH_2}\sigma_N \qquad (5\text{-}10)$$

where ρ_{NH_2}, the reaction constant for anilinium deprotonation, has the value 2.88; σ_N, the substituent constant for a 4-aza substituent, has the value 0.83; and the pK_{BH^+} of aniline is 4.58 (28).

The pK_{BH^+} thus calculated for the amino group of 4-aminopyridine is 2.19, compared with the experimental value of 9.11 (31), which clearly must refer to protonation at the nitrogen atom of the ring to give the resonance-stabilized cation. Note that the more basic center, which is of course the preferred site of protonation, gives rise to the cation that is the major tautomer and weaker acid. The difference of almost seven pK units for protonation at the two sites shows that the concentration in water of the ring-protonated form is 10^7 times greater than that of the amino-protonated form, regardless of the acidity of the solution.

Numerous examples of calculations of the sort shown have been given by Perrin et al. (28).

2-Aminopyridine is ~ 2.4 pK units weaker as a base than 4-aminopyridine and, indeed, is not very much more basic than the 3-aminopyridine isomer, in which the charge cannot be formally delocalized to the amino group [Eqs. (5-11) and (5-11a)].

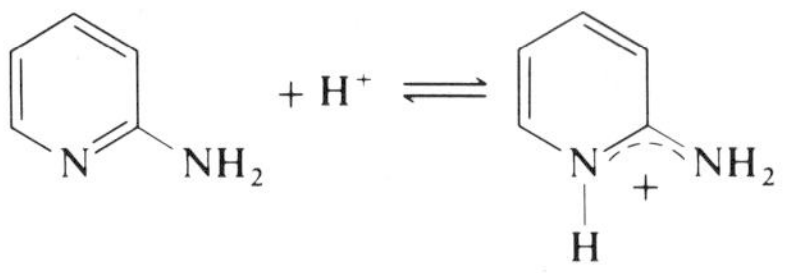

$$pK_{BH^+} = 6.71 \ (32) \qquad (5\text{-}11)$$

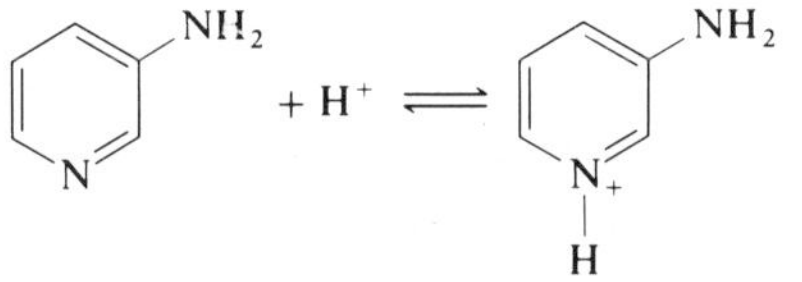

$$pK_{BH^+} = 6.03 \ (32) \qquad (5\text{-}11a)$$

Other methods for determining the site of protonation in ambient bases involve spectroscopic studies of the cations, particularly comparisons with model systems in which the tautomeric proton is replaced by an immobile alkyl group (33).

There are three isomeric aminopyrimidines, and they resemble the amino-pyridines in being protonated at ring nitrogen [Eqs. (5-12)–(5-14)]. In the

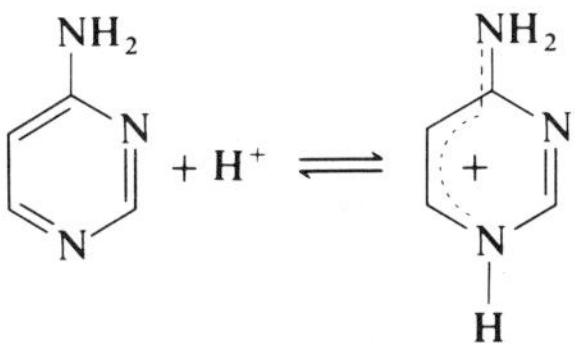

$$pK_{BH^+} = 5.60 \ (34) \qquad (5\text{-}12)$$

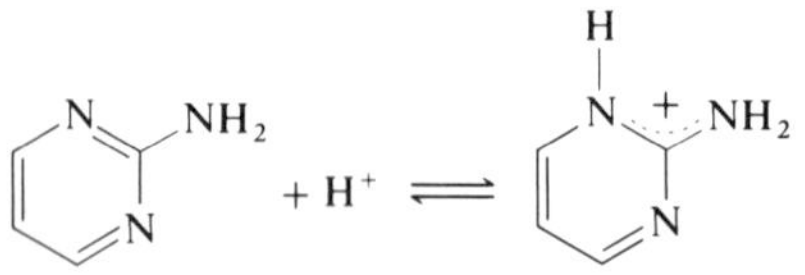

$$\text{(structure)} + H^+ \rightleftharpoons \text{(protonated structure)} \qquad pK_{BH^+} = 3.44\ (34) \qquad (5\text{-}13)$$

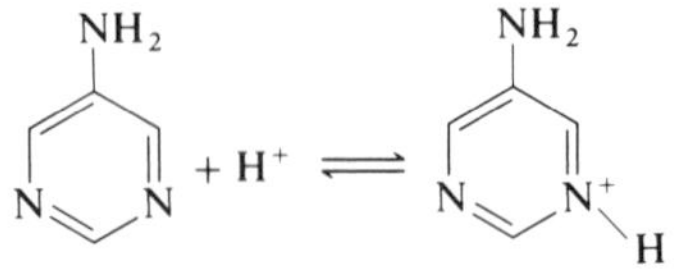

$$\text{(structure)} + H^+ \rightleftharpoons \text{(protonated structure)} \qquad pK_{BH^+} = 2.60\ (35) \qquad (5\text{-}14)$$

case of the 4-amino compound, which is unsymmetrical, it is N-1 and not N-3 that is protonated first [Eq. (5-12)]. The conjugate acid of 5-aminopyrimidine, the weakest of the three bases, has a formally localized charge [Eq. (5-14)], but this form is nonetheless more stable than the tautomeric form having the proton on the amino group. The pK_{BH^+} of the latter can be calculated using the Hammett equation and the methods described earlier to be 0.4, making the amino group more than two pK units weaker as a base than the ring nitrogen. The amino-protonated form is therefore present to the extent of less than 1% of the ring-protonated form in aqueous solution.

The 4-aminopyrimidine unit is present in thiamine, vitamin B_1; here, too, the N-1 position is the site of protonation to give the conjugate acid **5** (*36*),

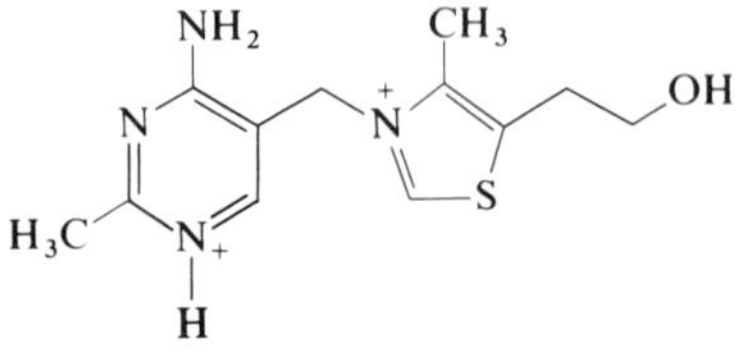

5

which has a pK_{HA} of 4.93 (*37*). The presence of the additional positive charge in the molecule doubtless accounts for thiamine being a somewhat weaker base than 4-aminopyrimidine.

In neutral solution thiamine exists almost exclusively as the monocation, but at slightly higher pH appreciable quantities of neutral pseudobase are formed by hydroxide addition to the carbon atom that is flanked by nitrogen and sulfur in the thiazole ring (*38*). [*Pseudobase* is the term used, particularly in heterocyclic chemistry, for adducts formed between cations and hydroxide ion (*39*).]

The site of protonation of a large number of naturally occurring pyrimidines has been reviewed by Kwiatkowski and Pullman (*40*).

Listed in Table 5-1 are a number of ambient heterocyclic systems with the preferred positions of successive protonations being shown. For simplicity's sake charge delocalization in the ions is not shown. Further examples of

Table 5-1

PRINCIPAL MONO- AND DIPROTONATION SITES IN SOME AMBIDENT HETEROCYCLIC SYSTEMS

Name	Structure	Principal monoprotonated form	Principal diprotonated form	Ref.
2-Aminopyrimidine	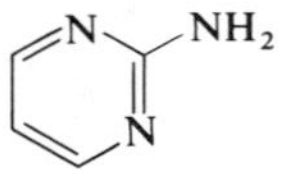	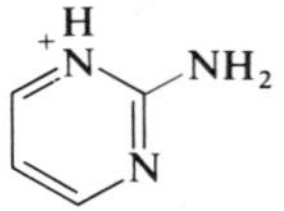	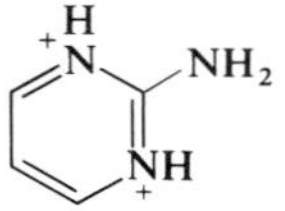	34
4-Aminopyrimidine	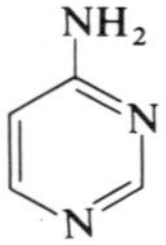	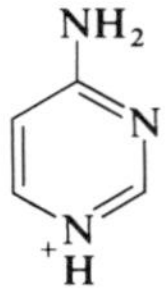	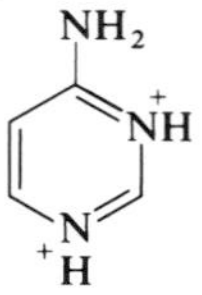	34, 66
2,4-Diaminopyrimidine	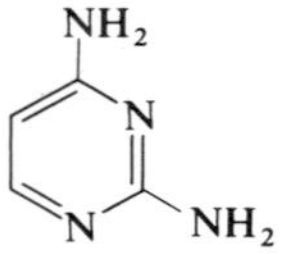	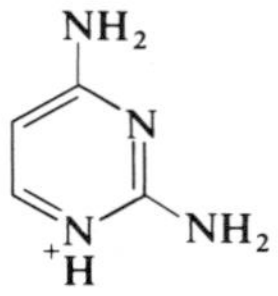	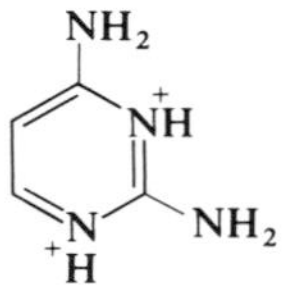	71[b]
3-Aminopyridazine	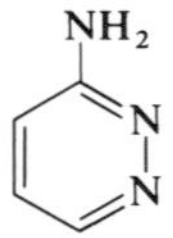	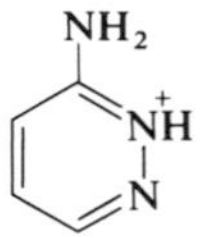	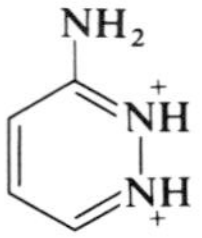	69

(cont.)

Table 5-1 (*cont.*)

Name	Structure	Principal monoprotonated form	Principal diprotonated form	Ref.
1-Methyl-2-imino-1,2-dihydropyrimidine	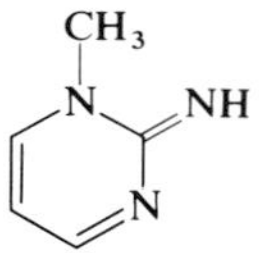	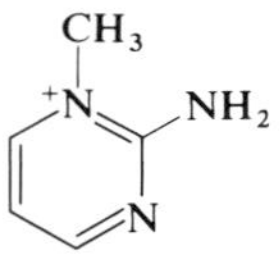	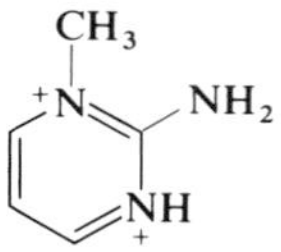	70
2,4,6-Triamino-pyrimidine	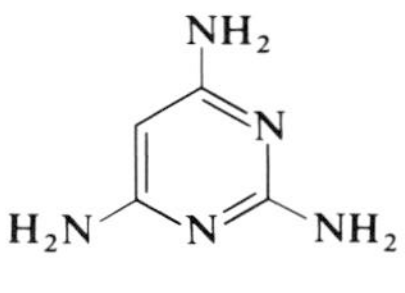	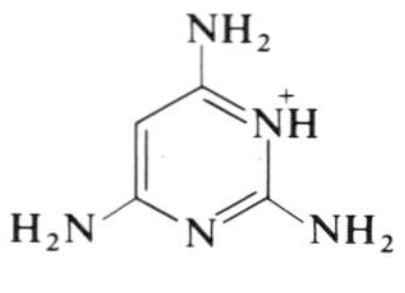	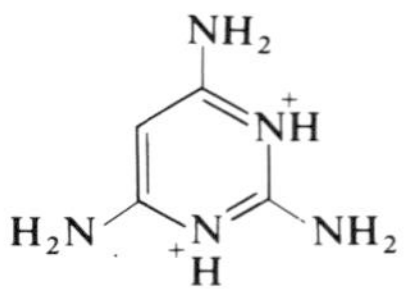	71
Melamine	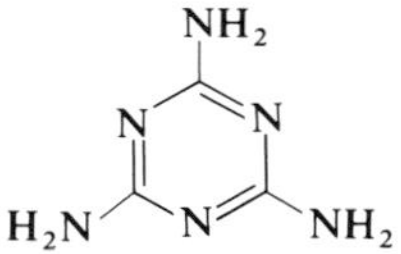	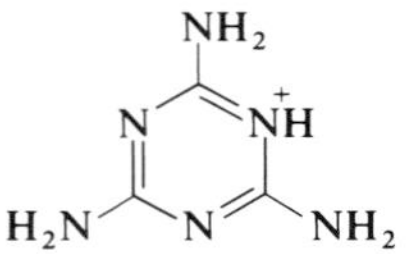	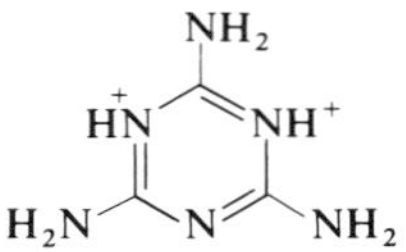	72, 73
4,5-Diaminoacridine	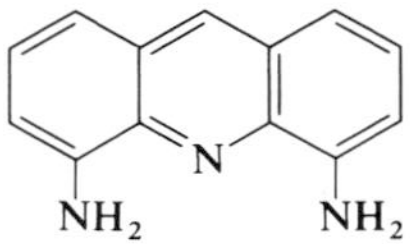	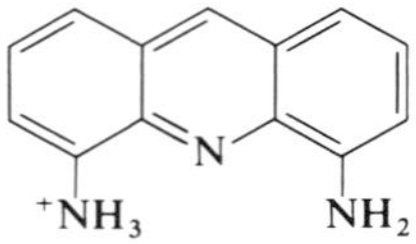	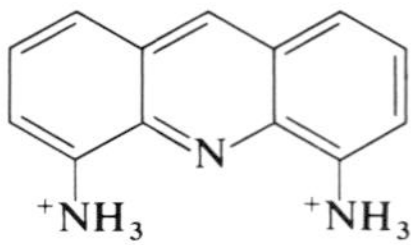	74
Uracil	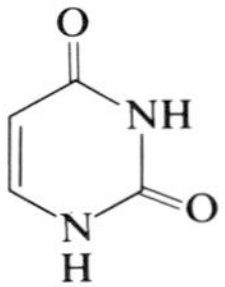	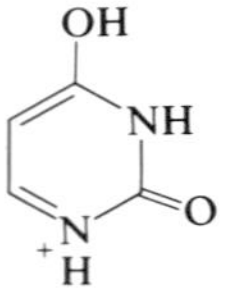	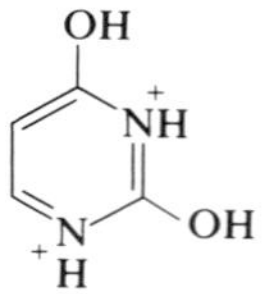	40, 67

194

Compound				Ref.
Cytosine	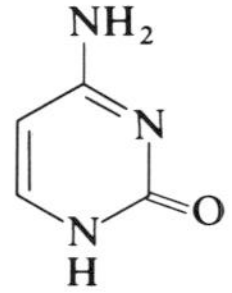	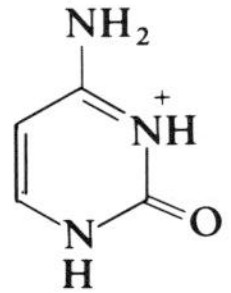	—	40
2-Pyrimidone (anion)	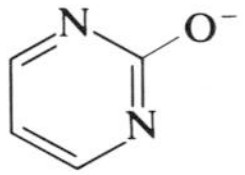	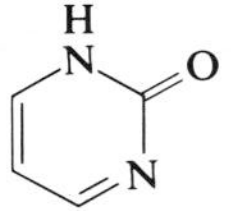	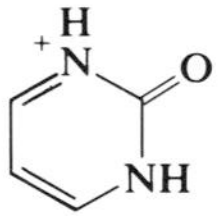	67, 70
4-Pyrimidone (anion)	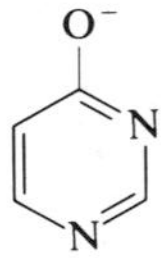	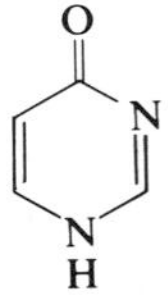	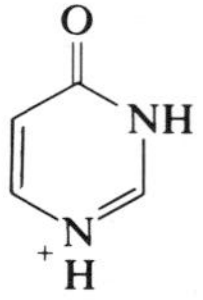	67
Glutaconimide (anion)	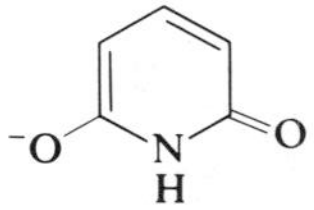	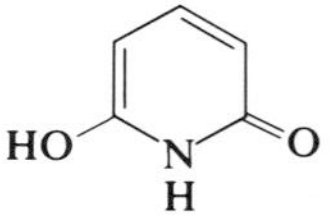	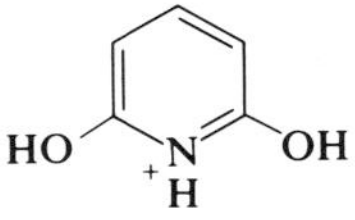	75, 76
3-Hydroxypyridine 1-oxide (anion)	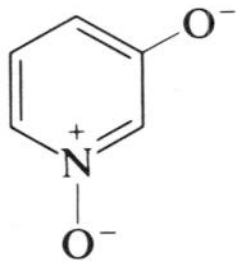	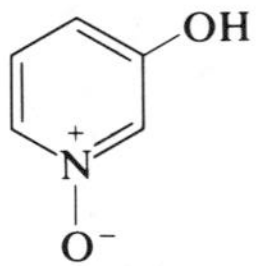	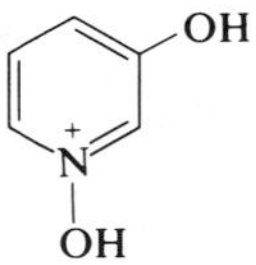	75

(cont.)

Table 5-1 (*cont.*)

Name	Structure	Principal monoprotonated form	Principal diprotonated form	Ref.
Guanine (anion)	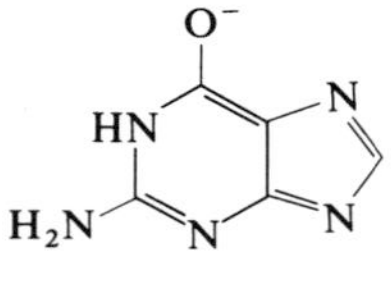	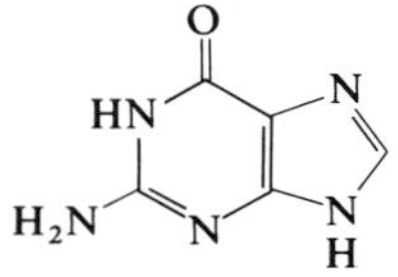	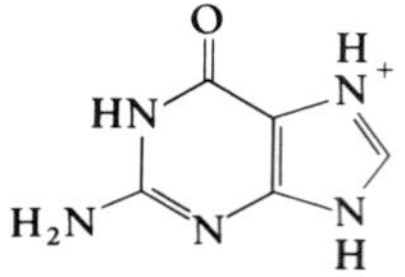	75, 76
9-Methylguanine (anion)	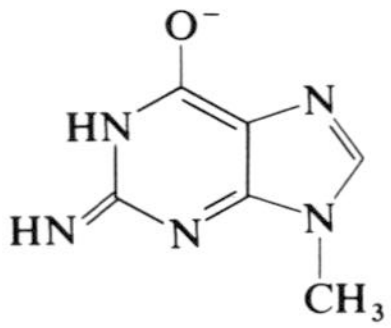	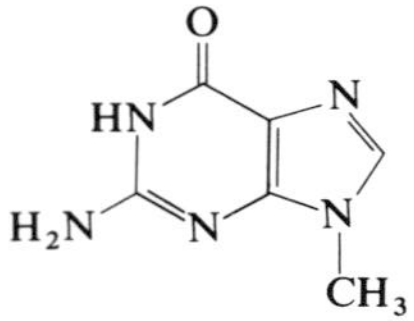	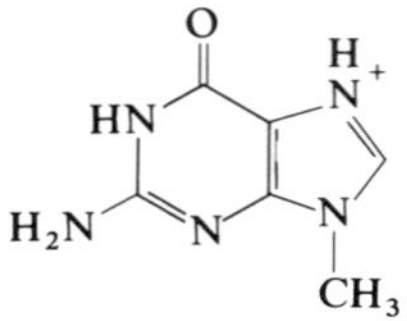	73, 75
3,5-Diazaindole (anion)	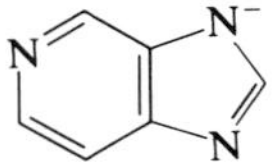	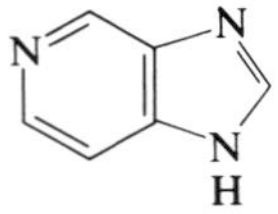	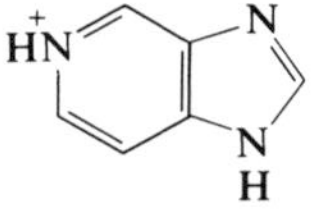	73
3-Cinnoline (anion)	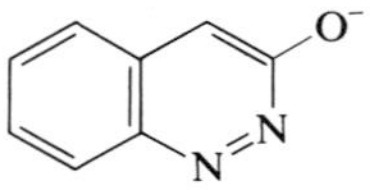	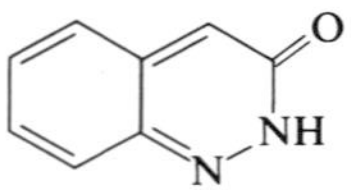	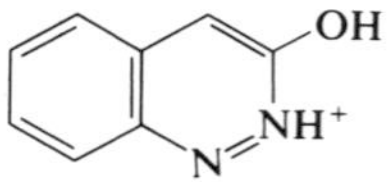	75

[a] In some cases only one of the protonated structures is given in the cited references, and in these cases the structure of the other form has been inferred.

[b] Reference 68 gives the ratio of 1- to 3-monoprotonated ions as 82:18.

ambident behavior in heterocyclic bases can be found in the accounts of Liler (1), Elguero et al. (29), and elsewhere.

Competition between C- and N-protonation occurs in a number of important nitrogen heterocycles containing five-membered rings. Pyrrole is protonated preferentially, though somewhat reluctantly, on carbon [Eq. (5-15)]. It has a pK of -3.8, which corresponds to being half-protonated in

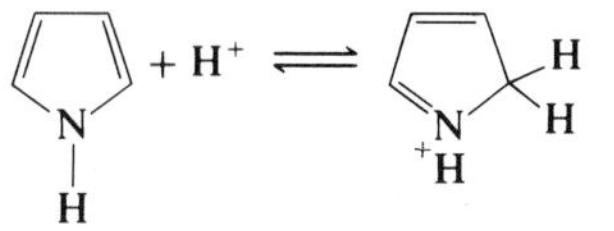

$$\text{(structure)} + \text{H}^+ \rightleftharpoons \text{(structure)} \tag{5-15}$$

38% sulfuric acid (Chapter 3, Section VI). The low proton affinity of nitrogen is due to its lone pair of electrons being part of the aromatic sextet of the ring.

Two diazoles, imidazole and pyrazole [Eqs. (5-16) and (5-17), respectively],

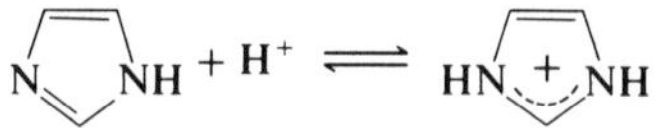

$$\text{(structure)} + \text{H}^+ \rightleftharpoons \text{(structure)} \tag{5-16}$$

$$\text{(structure)} + \text{H}^+ \rightleftharpoons \text{(structure)} \tag{5-17}$$

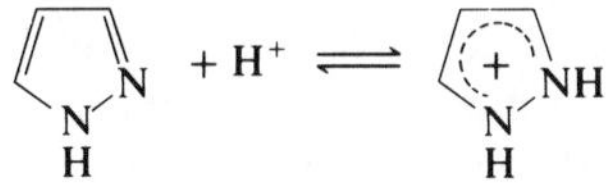

are both protonated on nitrogen, with imidazole, in particular, being much more basic than pyrrole; they have pK_{BH^+} values of 6.99 and 2.48, respectively (3?). In neither case does the added proton make use of the lone pair of electrons that is required for aromaticity; that is, protonation occurs in each case at the nitrogen atom that has its lone pair in the plane of the ring.

The dihydropyrazoles (pyrazolines) are preferentially protonated at either carbon or nitrogen, depending on the presence of substituents and, particularly, on the position of the double bond in the ring (41–44). 2-Pyrazoline is protonated at nitrogen [Eq. (5-18)], whereas 3-pyrazoline, judging from the behavior of its alkyl derivatives, is protonated at carbon [Eq. (5-19)]. In

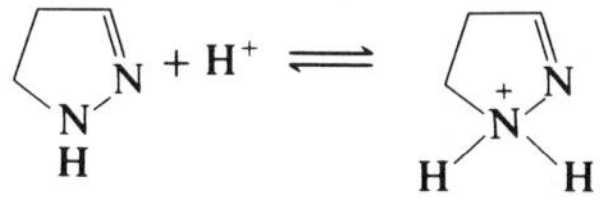

$$\text{(structure)} + \text{H}^+ \rightleftharpoons \text{(structure)} \tag{5-18}$$

$$\text{(structure)} + \text{H}^+ \rightleftharpoons \text{(structure)} \tag{5-19}$$

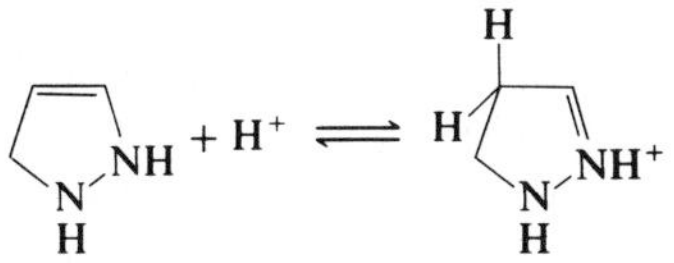

the latter case, as with enamines and vinyl ethers in general, rapid protonation at nitrogen or oxygen is followed by isomerization to the more stable carbon-protonated forms (45–47).

Purine contains both an imidazole and a pyrimidine ring; it exists in water as almost equal amounts of the tautomers **6** and **7**, with protonation taking place exclusively at the N-1 position to give the delocalized ions **8** and **9** in comparable amounts (*48*, *48a*).

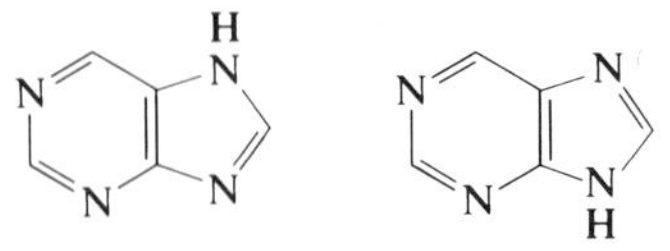

The pK_{BH^+} of purine is 2.39 (*49*), meaning that it is more basic than pyrimidine ($pK_{BH^+} = 1.2$) but much less basic than imidazole ($pK_{BH^+} = 6.99$). It is clear that the fused imidazo ring has a base-strengthening effect on the pyrimidine ring and that the latter has a very large base-weakening effect on the imidazole ring.

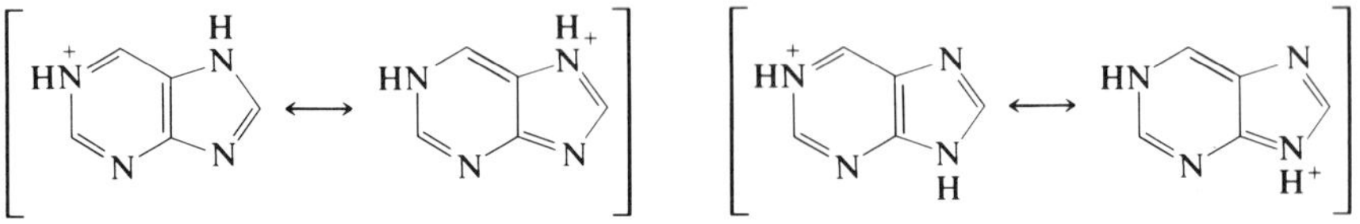

Polyamino- and polyhydroxyarenes may undergo C-protonation even in solutions that are only mildly acidic, provided that sufficient conjugation is present (Chapter 3, Section VI). In general, C-protonation occurs readily only when the positive charge can be delocalized by resonance, preferably to nitrogen atoms elsewhere in the molecule. (See ref. *50* for an unusual case of a monoaminoarene undergoing C-protonation in solution.) N-Protonation, on the other hand, may give rise to cations in which the charge is either localized or delocalized, depending on the structural features of the molecule. Since a localized charge will interact strongly with neighboring molecules of solvent we can expect to find the stabilization of ions with such charges to be strongly medium dependent, unlike the situation with ions that have extensive charge delocalization. Thus, if an ambident base can be protonated to give a cation with a localized charge and a second with a delocalized charge the tautomeric equilibrium will tend to shift toward the latter as the polarity or hydrogen bond acceptor properties of the medium decrease (*51*). If the delocalized tautomeric cation is favored in aqueous solution, as is the case with the conjugate acids of pyrrole, indole, 1,3,5-triaminobenzene, amides (probably), carboxylic acids, γ-pyrones, vinyl ethers, and others, one can be reasonably sure that it will also be the dominant form in other media, and particularly in the gas phase, where resonance is the only stabilizing feature available to an isolated molecule or ion.

Monosubstituted hydrazines have two different basic sites; infrared work shows that the terminal group is preferentially protonated (54), as do studies of the effect of ring substitution. The Hammett ρ value is close to that for substituted benzylamines and far from the value for anilines [Eqs. (5-20)–(5-22)] (55).

$$ArNHNH_2 + H^+ \rightleftharpoons ArNHNH_3^+ \qquad \rho = -1.17 \qquad (5\text{-}20)$$

$$ArCH_2NH_2 + H^+ \rightleftharpoons ArCH_2NH_3^+ \qquad \rho = -1.06 \qquad (5\text{-}21)$$

$$ArNH_2 + H^+ \rightleftharpoons ArNH_3^+ \qquad \rho = -2.88 \qquad (5\text{-}22)$$

Furthermore, phenylhydrazine ($pK_{BH^+} = 7.95$) (56) is more basic than aniline ($pK_{BH^+} = 4.60$), and this does not square with protonation at the interior nitrogen atom, since the neutral NH_2 substituent should have a small base-weakening effect on its neighbor, judging from its σ^* value (28, 57) and from the fact that hydrazine ($pK_{BH^+} = 8.07$) (58) is a weaker base than ammonia ($pK_{BH^+} = 9.24$). Provided that NH_2 is not base strengthening one can see from these pK values that the concentration of $C_6H_5\overset{+}{N}H_2NH_2$ must be less than 0.1% of the concentration of $C_6H_5NHNH_3^+$.

Alkylation of hydrazine causes a continuous decrease in base strength (58, 59) that is difficult to explain using standard substituent-effect arguments, although it has been argued that protonation occurs preferentially at the more alkylated nitrogen atom (60).

A substituted azobenzene dissolved in sulfuric acid will give rise to two protonated forms, the distribution of which will depend on the nature of the substituent [Eq. (5-23)].

$$ArN{=}NC_6H_5 + H^+ \rightleftharpoons Ar\overset{H}{N}{=}\underset{+}{N}C_6H_5 + ArN{=}\underset{+}{\overset{H}{N}}C_6H_5 \qquad (5\text{-}23)$$

In the case of a 4-amino substituent, protonation can be achieved in the aqueous region ($pK_{BH^+} = 2.8$) (32). The equilibrium in the cation is between the species shown below, with the charge-localized ammonium form (10) being slightly favored over the resonance-stabilized azonium form (11) in water (61).

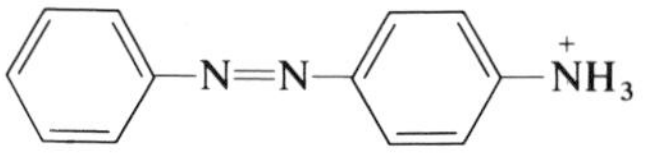

10

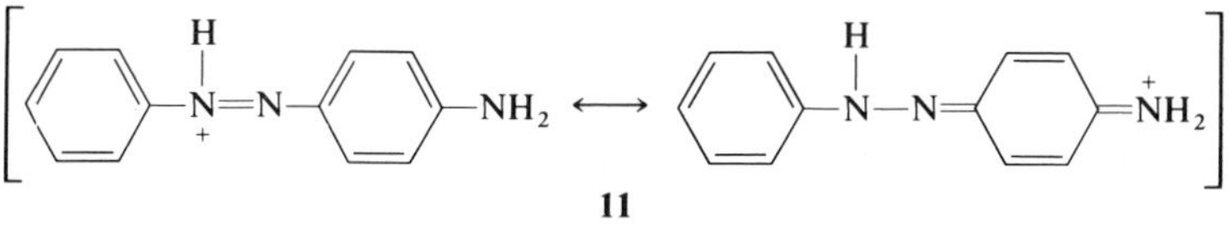

11

The factors affecting competition between carbon and oxygen protonation in ketenes have been analyzed by Nguyen and Hegarty (65a).

III. Alternative Sites of Deprotonation

In the previous section alternative sites of protonation of bases was considered and, although considering alternative sites of deprotonation of acids often amounts to examining the other side of the same coin, there are enough points of practical difference in the two to make the distinction worthwhile. Thus, there are many molecules of importance in which only a single deprotonation is of consequence but in which the species so produced have two protonation sites. Similarly, there are molecules with more than one likely site for proton loss but in which the anions so produced have only a single protonation site. An example of a reaction of the first sort is the deprotonation of nitromethane, which has only one kind of hydrogen, to form the ambident acinitronate anion. A reaction of the second sort is the ionization of an unsymmetrical dicarboxylic acid, which has two kinds of ionizable hydrogen, to form a nonambident monocarboxylate ion. Amino acids are examples of compounds that share some of both characteristics, and they are discussed in Section IV.

As seen in Chapter 2 many of the important functional groups of organic chemistry occupy distinctly different parts of the pK spectrum, and unless special features are present to stabilize their anions the choice of deprotonation site for a compound containing more than one of the following groups can be taken directly from the indicated order of precedence:

$$RSO_3H, RPO_3H_2, RCO_2H, ArOH, ROH, RNH_2, RCH_3$$

Intramolecular inversions of acid strength of these functional groups are not difficult to find, of course, 2,4,6-trinitrophenol being much stronger than acetic acid, for example. It is rare, however, to find such inversions for groups in the same molecule, at least in water and similar solvents.

The interplay between successive ionizations of polyprotic acids has been most closely examined in the case of polycarboxylic acids and aminocarboxylic acids. The dissociation of a dicarboxylic acid is shown in Eq. (5-24), where the symbols a through d indicate the relevant microscopic equilibrium constants, that is,

$$a = \left[C\begin{matrix} CO_2^- \\ CO_2H \end{matrix} \right] [H^+] \bigg/ \left[C\begin{matrix} CO_2H \\ CO_2H \end{matrix} \right]$$

and so on. (In the interests of simplicity the usual double arrow is replaced by a single arrow and the symbol K is omitted when microequilibria are shown.)

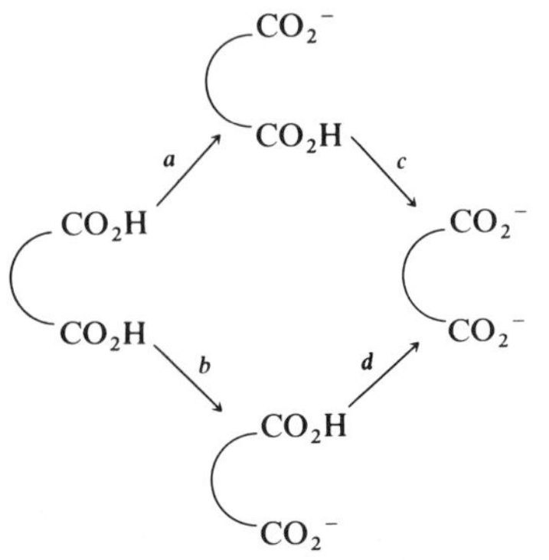

$$(5-24)$$

The macroscopic dissociation constants K_1 and K_2 are usually determined by potentiometry using the glass electrode. They are related to the microscopic values by Eqs. (5-25) and (5-26) (77, 78).

$$K_1 = a + b \qquad (5-25)$$

$$1/K_2 = 1/c + 1/d \qquad (5-26)$$

In the case of symmetrical diprotic acids such as succinic acid, $a = b$ and $c = d$. Consequently, Eqs. (5-25) and (5-26) reduce to Eqs. (5-27) and (5-28), and the microscopic values become immediately available from the experimental quantities K_1 and K_2.

$$K_1 = 2a \qquad (5-27)$$

$$K_2 = c/2 \qquad (5-28)$$

In the case of unsymmetrical diprotic acids, where $a \neq b$ and $c \neq d$, the microscopic values cannot be obtained from measurements of K_1 and K_2 alone. In such cases model compounds may be of help. For example, the dissociation constant of the monoalkyl esters can be taken as approximating the values of a and b (79, 80). Analogous models for the species involved in the second dissociation, represented by c and d, do not exist but they are not actually necessary since there is a relationship between the four microscopic dissociation constants [Eq. (5-29)] that enables us to calculate c and d,

$$ac = bd \qquad (5-29)$$

provided that valid estimates of the equilibrium constants a and b have been made. Indeed, only one of a and b need be independently determined since the other becomes immediately available from the relationship $K_1 = a + b$. Combining Eqs. (5-25), (5-26), and (5-29) gives Eq. (5-30), which enables us to calculate c and, in turn, d.

$$c = K_2(1 + b/a) \qquad (5-30)$$

This procedure is illustrated using 2,2-diphenylhexanedioic acid and its two monomethyl esters (**12**, **13**, and **14**, where ϕ = phenyl), whose macroscopic dissociation constants were determined in water at 25°C by le Moal (*78, 81*).

$$C\phi_2CO_2H$$
$$|$$
$$(CH_2)_3CO_2H$$

12
$K_1 = 6.80 \times 10^{-5}$
$K_2 = 4.00 \times 10^{-6}$

$$C\phi_2CO_2H$$
$$|$$
$$(CH_2)_3CO_2Me$$

13
$K_{HA} = 5.50 \times 10^{-5}$

$$C\phi_2CO_2Me$$
$$|$$
$$(CH_2)_3CO_2H$$

14
$K_{HA} = 1.44 \times 10^{-5}$

Setting a and b of **12** equal to K_{HA} of **13** and K_{HA} of **14**, respectively, and using Eqs. (5-29) and (5-30) gives the following microscopic dissociation constants for the species **12**, **15**, and **16**:

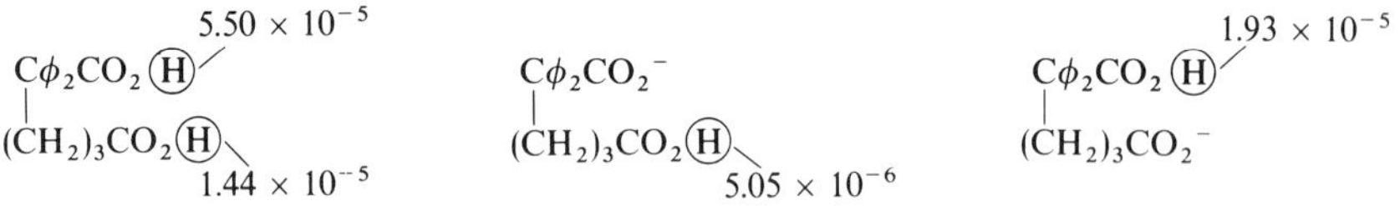

$$5.50 \times 10^{-5}$$
$$C\phi_2CO_2\text{(H)}$$
$$|$$
$$(CH_2)_3CO_2\text{(H)}$$
$$1.44 \times 10^{-5}$$

12

$$C\phi_2CO_2^-$$
$$|$$
$$(CH_2)_3CO_2\text{(H)}$$
$$5.05 \times 10^{-6}$$

15

$$1.93 \times 10^{-5}$$
$$C\phi_2CO_2\text{(H)}$$
$$|$$
$$(CH_2)_3CO_2^-$$

16

A check on the reasonableness of the basic assumption that **13** and **14** are suitable models for the two carboxyl units in **12** can be made by inspection, since K_1 of **12** must be equal to the sum of a and b, that is, to the sums of the K_{HA} values of the model compounds. In the case of these acids there is satisfactory agreement (6.80×10^{-5}, compared with 6.94×10^{-5}), as there is for most, but not all, of the other dicarboxylic systems studied by le Moal.

The concentrations of the microscopic species can be calculated as follows. The general equation is shown in Eq. (5-31), where a through d are the relevant

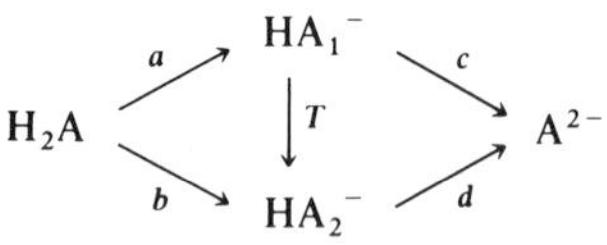

$$H_2A \begin{array}{c} \xrightarrow{a} HA_1^- \xrightarrow{c} \\ \downarrow T \qquad A^{2-} \\ \xrightarrow{b} HA_2^- \xrightarrow{d} \end{array} \tag{5-31}$$

equilibrium constants as used earlier and T represents the tautomeric equilibrium constant linking the two isomeric monoanions [Eq. (5-32)].

$$T = [HA_2^-]/[HA_1^-] = b/a = c/d \tag{5-32}$$

The concentrations of H_2A and A^{2-} and the total monoanion concentration can be calculated from the pH of the solution, provided that K_1 and

K_2, the macroscopic constants, are known; see Eqs. (5-33)–(5-35) (82), where [C] stands for the total concentration of substrate.

$$[H_2A] = \frac{[C]}{1 + K_1/[H^+] + K_1K_2/[H^+]^2} \tag{5-33}$$

$$[HA_1^-] + [HA_2^-] = \frac{[C]}{1 + [H^+]/K_1 + K_2/[H^+]} \tag{5-34}$$

$$[A^{2-}] = \frac{[C]}{1 + [H^+]/K_2 + [H^+]^2/K_1K_2} \tag{5-35}$$

Provided that the equilibrium microconstants are known the concentration of the individual monoanions can be obtained by combining Eqs. (5-32) and (5-34) to give Eqs. (5-35) and (5-36).

$$[HA_1^-] = \frac{[C]}{(1 + [H^+]/K_1 + K_2/[H^+])(1 + b/a)} \tag{5-36}$$

$$[HA_2^-] = \frac{[C]}{(1 + [H^+]/K_1 + K_2/[H^+])(1 + a/b)} \tag{5-37}$$

The concentration of the monoanions reaches a maximum at a pH that is midway between pK_1 and pK_2. The midpoint between pK_1 and pK_2 is particularly significant in the case of amino acids, which are discussed in Section IV.

Tricarboxylic acids that have nonequivalent carboxyl groups have 8 microscopic forms and 12 microscopic dissociation constants, designated a through l in Eq. (5-38) (83).

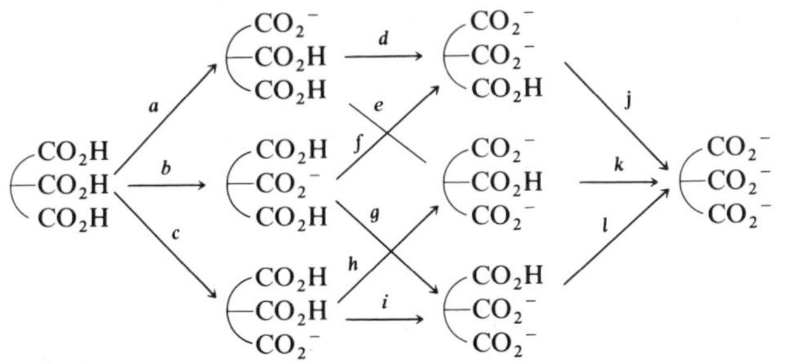

Pearce and Creamer (84) and, earlier, Martin (85) made a complete analysis of the ionization of citric acid, a tricarboxylic acid in which two of the carboxyl groups are equivalent and hence a simpler system than that shown in Eq. (5-38). In this case there are only six microscopic species and seven microscopic dissociation constants [Eq. (5-39)].

$$
\begin{array}{ccccc}
 & & \begin{array}{c} CH_2CO_2H \\ | \\ HOCCO_2^- \\ | \\ CH_2CO_2H \end{array} & \xrightarrow{\ c\ } & \begin{array}{c} CH_2CO_2^- \\ | \\ HOCCO_2^- \\ | \\ CH_2CO_2H \end{array} \\[2em]
\begin{array}{c} CH_2CO_2H \\ | \\ HOCCO_2H \\ | \\ CH_2CO_2H \end{array} & \overset{a}{\underset{b}{\rightrightarrows}} & & & \\[2em]
 & & \begin{array}{c} CH_2CO_2^- \\ | \\ HOCCO_2H \\ | \\ CH_2CO_2H \end{array} & \xrightarrow{\ e\ } & \begin{array}{c} CH_2CO_2^- \\ | \\ HOCCO_2H \\ | \\ CH_2CO_2^- \end{array} \\
\end{array}
\qquad
\begin{array}{c} CH_2CO_2^- \\ | \\ HOCCO_2^- \\ | \\ CH_2CO_2^- \end{array}
$$

(with pathways d, f, g connecting the central species)

(5-39)

The three macroscopic constants of citric acid are accurately known from potentiometric measurements, and only two of the seven microscopic values need be obtained from data on the appropriate methyl esters for the entire scheme to be delineated. The pK values for the macroscopic equilibrium constants K_1, K_2, and K_3 are 3.13, 4.76, and 6.40, respectively; the pK values determined by Pearce and Creamer for the seven microscopic equilibrium constants indicated in Eq. (5-39) by the symbols a through g are, respectively, 3.23, 4.10, 5.01, 4.14, 4.76, 6.05, and 5.42. These values reveal that the central carboxyl group is a stronger acid than a terminal carboxyl group in the neutral compound and in both the monoanion and the dianion. As a result about 80% of the monoanion and 10% of the dianion consist of the symmetrical species. These values can be derived from the appropriate microequilibrium constants as described in Section IV.

Although two independently determined microscopic constants are the minimum number for a complete solution of the citric acid problem, Pearce and Creamer obtained ionization constants for four different methyl citrates. This enables one to make checks that show that the principle of using a carbomethoxyl group as a surrogate for a carboxyl group is somewhat imperfect. Nonetheless, the values for a through g listed earlier are consistent with the thermodynamic requirement that the conversion of citric acid to citrate trianion be independent of the pathway taken, that is, that Eq. (5-40)

$$acf = bdf = beg = K_1 K_2 K_3 \qquad (5\text{-}40)$$

hold. The macroscopic and microscopic constants whose pK values are given earlier satisfy Eq. (5-40) to a high degree of precision.

In some polyprotic acids it is possible to measure microscopic dissociation constants directly. This occurs when the spectral change accompanying dissociation is sufficiently distinct for it to be followed directly. The microscopic ionization of a phenolic hydroxyl group or a sulfhydryl group can usually be determined in this way since the ultraviolet spectra of both groups undergo marked changes on ionization. In addition, ^{1}H-, ^{13}C-, and ^{31}P-nmr spectra can sometimes be used to measure directly the concentrations of isomeric ions in an acid–base equilibrium, such species being indistin-guishable from potentiometric measurements (86).

As the number of carboxyl groups increases, the number of microscopic structures and equilibrium constants increases rapidly. For a hexacarboxylic acid in which none of the groups are equivalent there are 6 macroscopic constants, 64 microscopic structures, and 192 microscopic equilibrium constants (77). The carboxyl groups in many polycarboxylic acids are equivalent, however, which greatly reduces the number of possibilities. Mellitic acid (benzenehexacarboxylic acid), for example, has only 12 different ionic forms, instead of the 63 it would have if there were no degeneracy. The number of microscopic equilibrium constants is 20 instead of 192. The number of macroscopic constants, of course, remains 6; their values are 0.68, 2.21, 3.52, 5.09, 6.32, and 7.49 (90).

Polyhydric phenols seldom provide a puzzle as to which hydroxyl group is the most acidic. The Hammett equation can usually be used to estimate the microscopic dissociation constant of a particular phenolic unit to a fairly high degree of precision. Thus, pyrogallol (1,2,3-benzenetriol) can be calculated to have microscopic dissociation constants a and b of 5.76×10^{-10} and 1.82×10^{-10} for exterior and interior hydroxyl groups, respectively [Eq. (5-40a)]

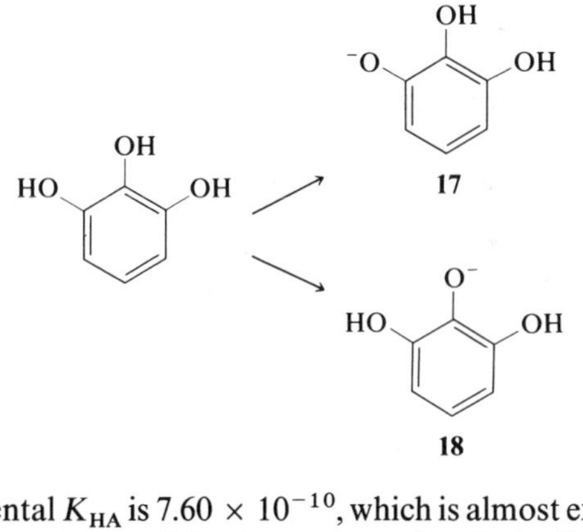

(5-40a)

(94). The experimental K_{HA} is 7.60×10^{-10}, which is almost exactly the sum of the derived microscopic quantities. It corresponds principally to formation of ion 17, although the fairly small gap between the microscopic constants means that the experimental quantity is a mixed constant. Since $a/b = [17]/[18]$ the mixture of monoanions consists of 76% of 17 and 24% of 18, regardless of the pH of the solution.

Phloroglucinol (1,3,5-benzenetriol) undergoes a normal first ionization to give the symmetrical 3,5-dihydroxyphenoxide ion. It, in turn, generates not the diphenoxide ion but rather the isomer 19 as the principal dianionic species, the latter being more highly stabilized by resonance [Eq. (5-41)] (95).

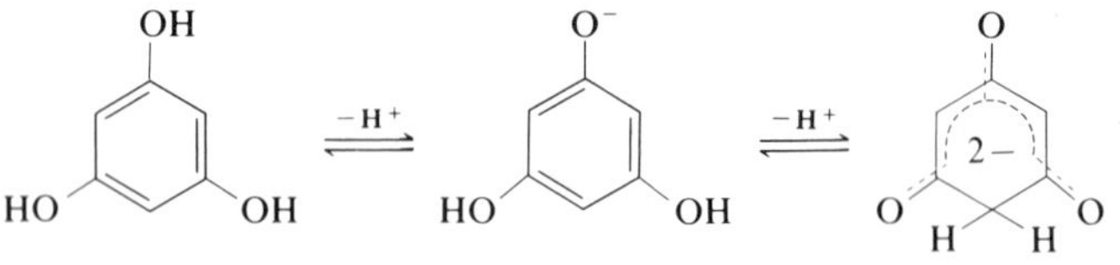

$$
\underset{\text{HO}}{\overset{\text{OH}}{\bigcirc}}\text{OH} \quad \overset{-\text{H}^+}{\rightleftharpoons} \quad \underset{\text{HO}}{\overset{\text{O}^-}{\bigcirc}}\text{OH} \quad \overset{-\text{H}^+}{\rightleftharpoons} \quad \underset{\substack{\text{O} \quad \text{O} \\ \text{H} \quad \text{H}}}{\overset{\text{O}}{\bigcirc}}\;2- \qquad (5\text{-}41)
$$

$$
\mathbf{19}
$$

The unusual nature of the second deprotonation step is signaled by the greater than expected acid strength of the monophenoxide ($pK_2 = 8.88$). It is two to three pK units stronger than the second ionization of analogous dihydroxyarenes that are unable to isomerize in this manner (88). This isomerization was foreshadowed by Baeyer's observation about a century ago that phloroglucinol forms a trioxime [Eq. (5-42)] (96).

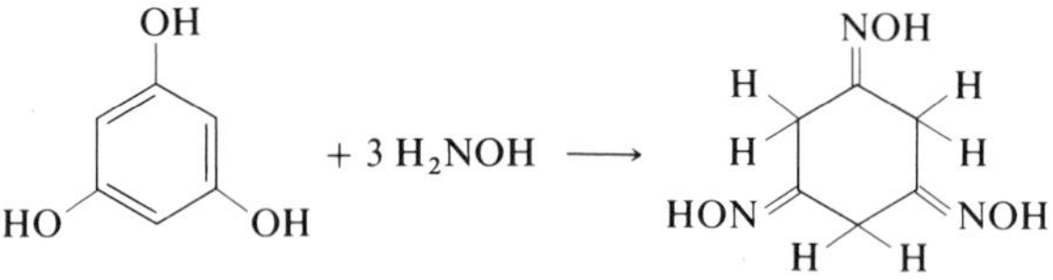

$$
\underset{\text{HO}}{\overset{\text{OH}}{\bigcirc}}\text{OH} \quad + \; 3\,\text{H}_2\text{NOH} \quad \longrightarrow \quad \underset{\substack{\text{HON} \qquad \text{NOH} \\ \text{H} \quad \text{H}}}{\overset{\substack{\text{NOH} \\ \text{H} \qquad \text{H}}}{\bigcirc}} \qquad (5\text{-}42)
$$

IV. Amphiprotic Systems That Form Zwitterions

There are five commonly encountered acidic functional groups in organic chemistry: those present in sulfonic acids, phosphonic acids, carboxylic acids, phenols, and thiols. The four most commonly encountered basic functional groups in organic chemistry are those present in amidines, alkylamines, pyridines, and arylamines. The order of acid strength of typical compounds containing these groups in their acidic forms, that is, the bases in the form of their conjugate acids, is shown here, with the pK_{HA} values in parenthesis:

$$
\text{CH}_3\text{SO}_3\text{H}(-1.9) > \text{CH}_3\text{PO}_3\text{H}_2(2.4) > \text{C}_6\text{H}_5\text{NH}_3{}^+ \;(4.6) \sim \text{CH}_3\text{CO}_2\text{H}\;(4.8) \sim
$$

$$
\text{C}_5\text{H}_5\text{NH}^+ \;(5.2) > \text{C}_6\text{H}_5\text{OH}\;(10.0) \sim \text{CH}_3\text{SH}\;(10.3) \sim
$$

$$
\text{CH}_3\text{NH}_3{}^+ \;(10.6) > \text{CH}_3\text{C(NH}_2)_2{}^+ \;(12.4)
$$

It follows that a sulfonic or phosphonic acid group that is present in any alkyl- or arylamine, pyridine, or amidine should be able to protonate the basic center and form a dipolar ion, or zwitterion. The acidity of a carboxyl group, on the other hand, is comparable to that of the conjugate acids of most arylamines or pyridines, and so compounds containing both kinds of group can be expected to exist only partly as zwitterions; with alkylamines and amidines, however, the zwitterionic form will strongly dominate. Finally, the acidic units in phenols and thiols are able to form a high proportion of zwitterion only with amidines.

The proportion of zwitterion to uncharged molecule, that is, the tautomeric equilibrium constant T, can be calculated with a fair degree of precision by using appropriate model compounds, since the situation in Eq. (5-43), for

$$\overset{+}{H_3}NCH_2CO_2H \underset{b}{\overset{a}{\rightleftarrows}} \begin{array}{c} H_2NCH_2CO_2H \\ \Big| T \\ \overset{+}{H_3}NCH_2CO_2{}^- \end{array} \overset{c}{\underset{d}{\rightleftarrows}} H_2NCH_2CO_2{}^- \qquad (5\text{-}43)$$

example, is entirely analogous to that encountered in the previous section with dicarboxylic acids. The microequilibrium constant a can be approximated by the K_{HA} of the protonated methyl ester $\overset{+}{H_3}NCH_2CO_2CH_3$ (2.57×10^{-8}) (32), and b can be approximated by the K_{HA} of the quaternary compound $(CH_3)_3\overset{+}{N}CH_2CO_2H$ (1.48×10^{-2}) (97).

Provided that the two macroequilibrium constants K_1 and K_2 are known, only one microequilibrium constant need be obtained in order to determine all the other constants, and having surrogates for two of the species in Eq. (5-43) makes it possible for a check to be made on the principle employed.

Taking a in Eq. (5-43) equal to 2.57×10^{-8} and combining it with the experimental values of K_1 and K_2 of the glycine cation (4.47×10^{-3} and 1.66×10^{-10}, respectively) and using Eqs. (5-25), (5-29), and (5-30) gives the following values for the microequilibrium constants b through d and the tautomeric constant T:

$$b = K_1 - a = 4.47 \times 10^{-3} - 2.57 \times 10^{-8} = 4.47 \times 10^{-3}$$

$$c = K_2(1 + b/a) = 1.66 \times 10^{-10}\{1 + [(4.47 \times 10^{-3})/(2.57 \times 10^{-8})]\}$$
$$= 2.88 \times 10^{-5}$$

$$d = ac/b = (2.57 \times 10^{-8} \times 2.88 \times 10^{-5})/4.47 \times 10^{-3} = 1.66 \times 10^{-10}$$

$$T = b/a = c/d = 1.74 \times 10^5$$

The value of T is independent of pH but varies with solvent in the expected manner; that is, the proportion of zwitterion decreases as organic solvents are added to the aqueous solution (98). In aqueous solution the concentration of zwitterion is very much larger than that of its uncharged tautomer. It can be seen from the figures just given that there is a threefold discrepancy between the calculated value of b (4.47×10^{-3}) and the dissociation constant of the trimethylammonio model compound (14.8×10^{-3}). This discrepancy casts only a small shadow on the use of the carbomethoxyl surrogate since we know from examining Hammett substituent constants that the groups $-CO_2CH_3$ and $-CO_2H$ have very similar electronic effects in other systems, whereas the groups $-\overset{+}{N}H_3$ and $-\overset{+}{N}(CH_3)_3$ sometimes differ appreciably in this respect. It is significant that the $-\overset{+}{N}H_3$ group, which can disperse its charge by hydrogen bonding, tends to be less electron withdrawing than the $-\overset{+}{N}(CH_3)_3$ group, in agreement with the direction of the discrepancy just referred to (28, 99, 100). Most α-amino acids containing one amino group and one carboxyl group have tautomeric equilibrium constants that favor the zwitterionic form by factors of 10^5 to 10^6 in water.

The pH at which the concentrations of the zwitterion and its minor tautomer, the uncharged neutral form, are highest is called the isoelectric point. The general expression for this quantity is given by Eq. (5-44), where K_1 and K_2 are the first and second (macroscopic) dissociation constants of the protonated amino acid.

$$[H^+] = (K_1 K_2)^{1/2} \qquad (5\text{-}44)$$

A closely related parameter, which, however, is concentration dependent, is called the isoionic point. This quantity can be defined operationally as the pH that would be obtained if the pure neutral amino acid were to be dissolved in pure water, that is, when no ions other than those from the substrate or water are present. The isoionic point is sometimes defined in such a way that it is identical with the isoelectric point; in the following discussion we use the operational definition, which makes a useful distinction between these quantities.

Equation (5-45) gives the hydrogen ion concentration at the isoionic point

$$[H^+] = \left(\frac{K_1 K_2 C + K_1 K_w}{C + K_1} \right)^{1/2} \qquad (5\text{-}45)$$

in terms of C, the total concentration of the substrate, K_1, K_2, and K_w, the ion product of water. It can be seen that at vanishingly low concentrations of substrate the pH of the isoionic point will approach 7.00 and that at high concentrations of substrate it will approach the isoelectric point (101) [compare Eqs. (5-44) and (5-45)].

In the case of glycine the pH at the isoelectric point is 6.07, a value midway between pK_1 (2.35) and pK_2 (9.78), in agreement with Eq. (5-44). The isoionic point(s) of glycine can be calculated from Eq. (5-45) as a function of the glycine concentration, giving the values 6.43 ($C = 10^{-3}$ M), 6.14 ($C = 10^{-2}$ M), and 6.08 ($C = 10^{-1}$ M). The acidity thus represented is a result of dissociation of the ammonio group [Eq. (5-46)], which occurs to only a very slight extent. In

$$H_3\overset{+}{N}CH_2CO_2^- \qquad H^+ + H_2NCH_2CO_2^- \qquad (5\text{-}46)$$

order to repress this dissociation and maximize the concentration of zwitterion and reach the isoelectric point, a minute amount of mineral or other strong acid must be added to the solution.

There are a number of α-amino acids that contain more than one acidic or basic group. Perhaps the most thoroughly studied of these is cysteine, which contains one basic group (amino) and two acidic groups (carboxyl and thiol); the much greater strength of carboxyl than thiol ensures the zwitterion having the structure **19a**.

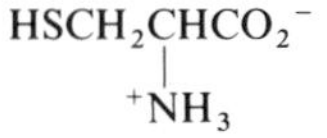

$$HSCH_2\underset{\overset{|}{^+NH_3}}{CHCO_2^-}$$

19a

The hydrochloride or other salt of cysteine in which the carboxylate unit is protonated is a triprotic acid containing three different kinds of acidic group. As we have seen conversion of such species to the basic form in which all three functional groups have lost a proton produces six intermediates, which are linked to the reactant and product through 12 microscopic dissociation processes. The values of all the microscopic equilibrium constants can be determined if four of them can be independently obtained, provided that the three macroscopic dissociation constants are known. In the case of cysteine the use of model compounds shows that the strength of the carboxyl group is so much greater than that of either the thio or ammonio groups that it can be taken as being equal to the first macroscopic dissociation constant. The strengths of the other two acid groups are quite close, and there was considerable doubt as to their relative strengths until the classic work of Benesch and Benesch, who took advantage of the ultraviolet spectral change that accompanies thiol ionization to show that the thiol group in the zwitterion is slightly stronger than the ammonio group (102). Subsequent work using various spectroscopic and kinetic techniques confirmed a ratio of about $2:1$ for their dissociation constants (103–106).

The microscopic constants for cysteine, drawn from the work of Kallen (105), are given in Eq. (5-47) and Table 5-2. Note that the derived microscopic

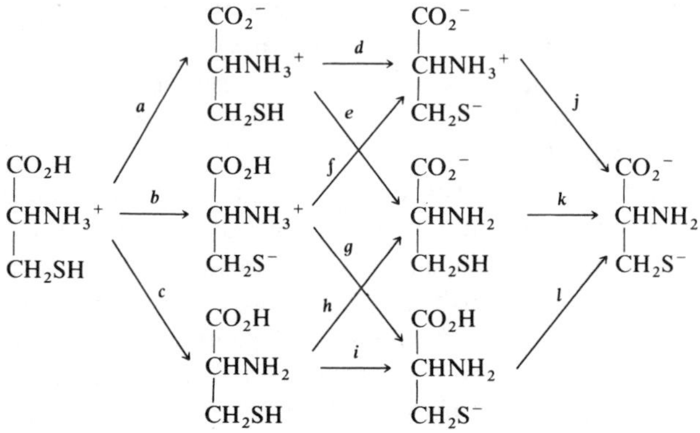

$$(5\text{-}47)$$

Table 5-2

MACROSCOPIC[a] and MICROSCOPIC[b] pK VALUES OF
PROTONATED CYSTEINE IN WATER AT 25°C, IONIC
STRENGTH 1.0[c]

First, second, and third pK values		
First dissociation	Second dissociation	Third dissociation
$K_1 = 2.00$	$K_2 = 8.35$	$K_3 = 10.39$
$a = 2.00$	$d = 8.55$	$j = 10.20$
$b = 7.44$	$e = 8.88$	$k = 9.87$
$c = 6.88$	$f = 3.10$	$l = 4.99$
	$g = 8.32$	
	$h = 4.00$	
	$i = 8.88$	

pK according to functional group		
$-CO_2H$	$-SH$	$-\overset{+}{N}H_3$
$a = 2.00$	$b = 7.44$	$c = 6.88$
$f = 3.10$	$d = 8.55$	$e = 8.88$
$h = 4.00$	$i = 8.88$	$g = 8.32$
$l = 4.99$	$k = 9.87$	$j = 10.20$

[a] Average of values tabulated in refs. 104 and 105.
[b] Values of Kallen (105).
[c] See Eq. (5-43).

constants for protonated cysteine indicate that the ammonio group is here stronger than the thiol group, which is the same order found in the ethyl ester of cysteine. It is clear that the carboxylate unit in the zwitterion decreases the acidity of the adjacent ammonio group more than it does that of the more distant thiol group, accounting for the change in the order of acidity of these groups in going from the cation to the zwitterion.

The 12 microscopic constants a through l correspond to four dissociations for each of the three functional groups in cysteine. The second part of Table 5-2 lists these values.

Given a complete knowledge of the microscopic equilibrium constants for a triprotic system such as is found in cysteine, glutamic acid, or tyrosine (107), the concentration of any intermediate species can be determined at a given pH. The general expressions for such calculations are given by Kallen (108). The general relationships between the number of protic centers, the number of microscopic intermediates, and the number of independent microscopic equilibrium constants is given by King (109).

The amino acid histidine has two basic and two acidic centers, which means that its diprotonated form (20) is a tetraprotic acid. The sequential deprotonation of the latter species is shown in Eq. (5-48).

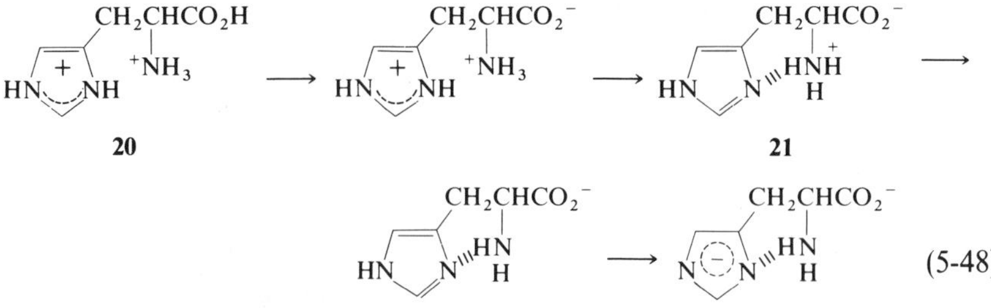

(5-48)

Ruterjans et al. (110) and Roberts et al. (111) obtained evidence from ^{15}N-nmr spectra that there is internal hydrogen bonding in 21 and 22 [Eq. (5-48)], and so it seems reasonable to expect that 23 would behave similarly. The pK values for the successive deprotonations of the histidine dication, 1.8, 6.0, 9.1, and ~ 14, are well separated (111, 112). The tautomeric forms of 21 and 22, in which hydrogen bonding could take place between the carboxylate unit and the NH proton of the imidazole ring might have been expected to be the stable forms of these species, but the nmr evidence seems unequivocal that 21 and 22 are the preferred forms, even in the solid state (113).

A. Aminobenzoic Acids

Aminobenzoic acids differ from aliphatic amino acids in not being overwhelmingly in the zwitterionic form. The case of 4-aminobenzoic acid has been carefully examined by Wepster et al. (114). [See Eq. (5-49), where, as

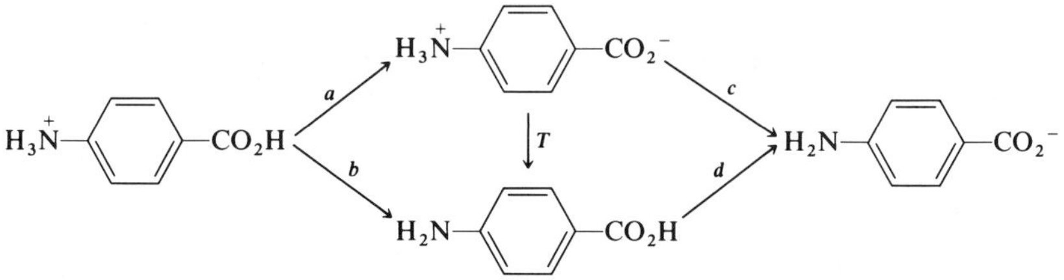

(5-49)

before, the letters a through d are microscopic equilibrium constants, T is the tautomeric constant, and single arrows are used to represent equilibria.] Earlier attempts had been made to determine the equilibrium constant b by

Table 5-3

MACROSCOPIC AND MICROSCOPIC
EQUILIBRIUM CONSTANTS (AS pK VALUES)
FOR 4-AMINO- AND
4-DIMETHYLAMINOBENZOIC ACIDS IN WATER
AT 25°C[a]

Equilibrium constant	4-NH$_2$	4-(CH$_3$)$_2$N
K_1	2.42	2.57
K_2	4.88	5.00
a	3.40	3.28
b	2.47	2.66
c	3.90	4.28
d	4.83	4.90
T	0.93	0.62
Zwitterion (%)	10.5	19.4

[a] Reference *114*. See Eq. (5-49).

using the methyl ester (*115*). This technique had been used with success in other cases, but it is not very satisfactory here, the reason being that b is too close to K_1, the first macroscopic constant, with the result that small errors in b produce large errors in a, c, and T.

Wepster used 4-(trimethylammonio)benzoic acid to estimate the value of a. We saw earlier in the case of glycine that such an approach was less satisfactory than the one that uses the ester. Here, however, the well-studied effect of charged substituents in aromatic systems (*116*) allows a reasonable estimate to be made of a for both the amino and dimethylamino compounds. When these are combined with the appropriate macroscopic equilibrium constants, the values given in Table 5-3 result.

Both 2- and 4-aminobenzoic acids, and also picolinic acid, undergo fairly ready decarboxylation, and the role of minor protolytic forms in this process has been extensively studied (*117*, *118*).

B. Summary

There are a number of general methods for determining microequilibrium constants and the concentration of intermediate species in complex equilibria.

First, when only macroscopic data from glass electrode measurements are available, surrogates may be used, carbomethoxyl for carboxyl or, less reliably, trimethylammonio for ammonio, in order to simplify the protolytic equilibria and enable one to estimate the values of one or more microscopic equilibrium constants. Depending on the number of protolytic sites this may be enough to determine the constants for all the equilibria in the system. A single microscopic constant and the two macroscopic constants are sufficient to solve the case of an unsymmetrical diprotic acid. When a macroscopic and a microscopic equilibrium constant almost coincide, any errors made in estimating the value of the latter will be reflected in large deviations in some of the derived equilibrium values.

Second, direct measurement of a single microscopic protonation or deprotonation step can sometimes be made by spectroscopic or other means. Ultraviolet spectroscopy is particularly useful for thiols and phenols; in certain other cases Raman and nmr spectroscopy can be used. Rates of reaction of a microscopic species with some external reagent may also provide the information required.

Third, substituent effects derived from other equilibrium or kinetic measurements can sometimes be used to determine microequilibrium constants. Aromatic systems give particularly reliable estimates in this regard.

References

1. M. Liler, *Adv. Phys. Org. Chem.* **11**, 267 (1975).
2. R. B. Homer and C. D. Johnson, "Chemistry of Amides" (J. Zabicky, ed.), Chap. 3. Interscience, London, 1970.
3. See, for example, C. R. Smith and K. Yates, *Can. J. Chem.* **50**, 771 (1972).
4. See, for example, C. L. Perrin, C. P. Lollo, and E. R. Johnston, *J. Am. Chem. Soc.* **106**, 2749 (1984).
5. G. Fraenkel and C. Franconi, *J. Am. Chem. Soc.* **82**, 4478 (1960).
6. A. Berger, A. Loewenstein, and S. Meiboom, *J. Am. Chem. Soc.* **81**, 62 (1959).
7. R. J. Gillespie and T. Birchall, *Can. J. Chem.* **41**, 148 (1963).
8. T. Birchall and R. J. Gillespie, *Can. J. Chem.* **41**, 2642 (1963).
9. R. Stewart and L. J. Muenster, *Can. J. Chem.* **39**, 401 (1961).
10. R. Stewart and L. J. Muenster, *Chem..Ind.* p. 1906 (1961); see also ref. *11*.
11. M. J. Janssen, *Spektrochim. Acta* **17**, 475 (1961).
12. C. L. Perrin and E. R. Johnston, *J. Am. Chem. Soc.* **103**, 4697 (1981).
13. C. L. Perrin, E. R. Johnston, C. P. Lollo, and P. A. Kobrin, *J. Am. Chem. Soc.* **103**, 4691 (1981).
14. M. Liler, *J. Chem. Soc., Perkin Trans. 2* p. 334 (1971).
15. M. Liler, *J. Chem. Soc., Perkin Trans. 2* p. 71 (1974).
16. M. Liler and D. Markovic, *J. Chem. Soc., Perkin Trans. 2* p. 551 (1982).
17. M. Liler and C. M. M. Thwaites, *J. Chem. Soc., Perkin Trans. 2* p. 201 (1983).
18. A. R. Fersht, *J. Am. Chem. Soc.* **93**, 3504 (1971).
19. H. Pracejus, M. Kehlen, H. Kehlen, and H. Matschiner, *Tetrahedron* **21**, 2257 (1965).
20. P. Haake and T. Koizumi, *Tetrahedron Lett.* p. 4849 (1970).
21. R. Stewart and K. Yates, *J. Am. Chem. Soc.* **82**, 4059 (1960).
22. M. J. Scanlan, I. H. Hillier, and A. A. MacDowell, *J. Am. Chem. Soc.* **105**, 3568 (1983).
23. O. S. Tee and M. Paventi, *J. Am. Chem. Soc.* **104**, 4142 (1982).
24. P. Beak, F. S. Fry, J. Lee, and F. Steele, *J. Am. Chem. Soc.* **98**, 171 (1976) and refs. therein.
25. H. H. Jaffé, *J. Am. Chem. Soc.* **77**, 4445 (1955).
26. J. N. Gardner and A. R. Katritzky, *J. Chem. Soc.* p. 4375 (1957).

27. P. Beak and J. M. White, *J. Am. Chem. Soc.* **104**, 7073 (1982).

28. D. D. Perrin, B. Dempsey, and E. P. Serjeant, "pK_a Prediction for Organic Acids and Bases." Chapman and Hall, London, 1981.

29. J. Elguero, C. Marzin, A. R. Katritzky, and P. Linda, "Tautomerism of Heterocycles," p. 25. Academic Press, New York, 1976.

30. P. Tomasik and C. D. Johnson, *Adv. Heterocycl. Chem.* **20**, 1 (1976).

31. R. G. Bates and H. B. Hetzer, *J. Res. Natl. Bur. Stand., Sect. A* **64A**, 427 (1960).

32. D. D. Perrin, "Dissociation Constants of Organic Bases in Aqueous Solution." Butterworths, London, 1965.

33. A. Albert, "Heterocyclic Chemistry," 2nd ed., Chap. 4. Athlone Press, London, 1968.

34. A. Albert, R. Goldacre, and J. Phillips, *J. Chem. Soc.* p. 2240 (1948); pK_{BH^+} corrected to 25°C as described in Chapter 3, Section III,A.

35. N. Whittaker, *J. Chem. Soc.* p. 1565 (1951).

36. A. H. Cain, G. R. Sullivan, and J. D. Roberts, *J. Am. Chem. Soc.* **99**, 6423 (1977).

37. F. Jordan and Y. H. Mariam, *J. Am. Chem. Soc.* **100**, 2534 (1978).

38. J.-M. E. H. Chahine and J.-E. Dubois, *J. Am. Chem. Soc.* **105**, 2335 (1983).

39. J. W. Bunting, *Adv. Heterocycl. Chem.* **25**, 1 (1979).

40. J. S. Kwiatkowski and B. Pullman, *Adv. Heterocycl. Chem.* **18**, 200 (1975).

41. J. Elguero and R. Jacquier, *Tetrahedron Lett.* p. 1175 (1965).

42. J.-L. Aubagnac, J. Elguero, and R. Jacquier, *Bull. Soc. Chim. Fr.* p. 3516 (1967).

43. J. Elguero, R. Jacquier, and D. Tizane, *Tetrahedron* **27**, 123 (1971).

44. M. Nardelli and G. Fava, *Acta Crystallogr.* **15**, 214 (1962).

45. M. R. Ellenberger, D. A. Dixon, and W. E. Farneth, *J. Am. Chem. Soc.* **103**, 5377 (1981).

46. M. J. Cook, A. R. Katritzky, P. Linda, and R. D. Tack, *J. Chem. Soc., Perkin Trans. 2* p. 1080 (1973).

47. A. J. Kresge, D. S. Sagatys, and H. L. Chen, *J. Am. Chem. Soc.* **99**, 7228 (1977) and refs. therein.

48. N. C. Gonnella and J. D. Roberts, *J. Am. Chem. Soc.* **104**, 3162 (1982).

48a. See also refs. *91–93*.

49. A. Albert and D. J. Brown, *J. Chem. Soc.* p. 2060 (1954).

50. R. W. Alder, M. R. Bryce, and N. C. Goode, *J. Chem. Soc., Perkin Trans. 2* p. 477 (1982).

51. See Refs. *52* and *53* and refs. therein for examples.

52. Y. K. Lau, K. Nishiwaza, A. Tse, R. S. Brown, and P. Kebarle, *J. Am. Chem. Soc.* **103**, 6291 (1981).

53. J. M. Buschek, F. S. Jorgensen, and R. S. Brown, *J. Am. Chem. Soc.* **104**, 5019 (1982).

54. R. F. Evans and W. Kynaston, *J. Chem. Soc.* p. 3151 (1963).

55. A. Fischer, D. A. R. Harper, and J. Vaughan, *J. Chem. Soc.* p. 4060 (1964).

56. N. Yui, *Bull. Inst. Phys. Chem. Res (Tokyo)* **20**, 256 (1941); *Chem. Abstr.* **35**, 4660 (1941).

57. O. Exner, *in* "Correlation Analysis in Chemistry" (N. B. Chapman and J. Shorter, eds.), p. 450. Plenum, London, 1978.

58. R. L. Hinman, *J. Org. Chem.* **23**, 1587 (1958).

59. P. J. Krueger, "The Chemistry of the Hydrazo, Azo, and Azoxy Groups" (S. Patai, ed.), Chap. 7. Wiley, London, 1975.

60. F. E. Condon, *J. Am. Chem. Soc.* **87**, 4491 (1965).

61. J. M. Carpentier and P. Dominique, *J. Chem. Res. (S)* p. 92 (1981); see also refs. *62–65*.

62. S. Yamamoto, Y. Tenno, and N. Nishimura, *Aust. J. Chem.* **32**, 41 (1979).

63. S.-J. Yeh and H. H. Jaffé, *J. Am. Chem. Soc.* **81**, 3283 (1959).

64. G. E. Lewis, *Tetrahedron* **10**, 129 (1960).

65. Y. Tanizaki, T. Kobayashi, and T. Hoshi, *Bull. Chem. Soc. Jpn.* **39**, 558 (1966).

65a. M. T. Nguyen and A. F. Hegarty, *J. Am. Chem. Soc.* **106**, 1552 (1984).

66. R. Stewart and M. G. Harris, *J. Org. Chem.* **43,** 3123 (1978).

67. R. Wagner and W. von Phillipsborn, *Helv. Chim. Acta* **53,** 299 (1970).

68. D. V. Griffiths and S. P. Swetnam, *Chem. Commun.* p. 1224 (1981).

69. R. F. Cookson and G. W. H. Cheeseman, *J. Chem. Soc., Perkin Trans 2* p. 392 (1972).

70. R. Stewart, S. J. Gumbley, and R. Srinivasan, *Can. J. Chem.* **57,** 2783 (1979).

71. B. Roth and J. Z. Strelitz, *J. Org. Chem.* **35,** 2696 (1970).

72. R. C. Hirt and R. G. Schmitt, *Spectrochim. Acta* **12,** 127 (1958).

73. Ref. *1*, p. 318, 323, or 325.

74. A. Albert, *J. Chem. Soc.* p. 4653 (1965).

75. Ref. *29*, p. 111, 113, 123, or 524.

76. See, however, ref. *1*, p. 324.

77. E. J. King, "Acid–Base Equilibria," Chap. 9. Pergamon, Oxford, 1965,

78. H. le Moal, *Bull. Soc. Chim. Fr.* p. 418 (1956).

79. L. Ebert, *Z. Phys. Chem.* **121,** 385 (1926).

80. D. H. R. Barton, *Nature (London)* **160,** 752 (1947).

81. H. le Moal and F. Salmon-Legagneur, *C. R. Hebd. Seances Acad. Sci.* **241,** 706 (1955).

82. G. E. Dunn, P. Leggate, and I. E. Scheffler, *Can. J. Chem.* **43,** 3080 (1965).

83. R. B. Martin, J. T. Edsall, D. B. Wetlaufer, and B. R. Hollingsworth, *J. Biol. Chem.* **233,** 1429 (1958).

84. K. N. Pearce and L. K. Creamer, *Aust. J. Chem.* **28,** 2409 (1975).

85. R. B. Martin, *J. Phys. Chem.* **45,** 2053 (1961).

86. See, for example, refs. *87–89.*

87. P. H. C. Heubel and A. I. Popov, *J. Solution Chem.* **8,** 615 (1979).

88. R. J. Highet and T. J. Batterham, *J. Org. Chem.* **29,** 475 (1964).

89. R. Matusch, *Z. Naturforsch., B* **32B,** 562 (1977).

90. N. Purdie, M. B. Tomson, and N. Riemann, *J. Solution Chem.* **1,** 465 (1972).

91. G. W. Buchanan and M. J. Bell, *Can. J. Chem.* **61,** 2445 (1983).

92. R. M. Izatt, J. J. Christensen, and J. H. Rytting, *Chem. Rev.* **71,** 439 (1971).

93. R. L. Benoit and M. Fréchette, *Can. J. Chem.* **62,** 995 (1984).

94. Ref. *28*, p. 48.

95. R. J. Highet and F. E. Chou, *J. Am. Chem. Soc.* **99,** 3538 (1977) and refs. therein.

96. A. Baeyer, *Chem. Ber.* **19,** 159 (1886).

97. See Table 2-5.

98. G. Wada, E. Tamura, and M. Okina, *Bull. Chem. Soc. Jpn.* **55,** 3064 (1982).

99. See ref. *57.*

100. A. V. Willi, *Z. Phys. Chem. (Wiesbaden)* **27,** 233 (1961).

101. W. P. Bryan, *Biochem. Educ.* **6,** 14 (1978).

102. R. E. Benesch and R. Benesch, *J. Am. Chem. Soc.* **77,** 5877 (1955).

103. D. M. E. Reuben and T. C. Bruice, *J. Am. Chem. Soc.* **98,** 114 (1976).

104. S. D. Lewis, D. C. Misra, and J. A. Shafer, *Biochemistry* **19,** 6130 (1980).

105. R. G. Kallen, *J. Am. Chem. Soc.* **93,** 6227 (1971).

106. E. L. Elson and J. T. Edsall, *Biochemistry* **1,** 1 (1962).

107. R. B. Martin, J. T. Edsall, D. B. Wetlaufer, and B. R. Hollingsworth, *J. Biol. Chem.* **233,** 1429 (1958).

108. R. G. Kallen, *J. Am. Chem. Soc.* **93,** 6236 (1971).

109. Ref. *77*, p. 223.

110. F. Blomberg, W. Maurer, and H. Ruterjans, *J. Am. Chem. Soc.* **99,** 8149 (1977).

111. J. D. Roberts, C. Yu, C. Flanagan, and T. R. Birdseye, *J. Am. Chem. Soc.* **104,** 3945 (1982).

112. D. D. Perrin, "Dissociation of Organic Bases in Solution," p. 190 and p. 385. Butterworths, London, 1965.

113. M. Munowitz, W. W. Bachovchin, J. Herzfeld, C. M. Dobson, and R. G. Griffin, *J. Am. Chem. Soc.* **104,** 1192 (1982).

114. B. van de Graaf, A. J. Hoefnagel, and B. M. Wepster, *J. Org. Chem.* **46,** 653 (1981).

115. R. A. Robinson and A. J. Biggs, *Aust. J. Chem.* **10,** 128 (1957).

116. A. J. Hoefnagel, M. A. Hoefnagel, and B. M. Wepster, *J. Org. Chem.* **43,** 4720 (1978).

117. G. E. Dunn, H. F. Thimm, and R. K. Mohanty, *Can. J. Chem.* **57,** 1098 (1979).

118. Ref. *82* and refs. therein.

6

Acidity and Basicity of Unstable and Metastable Organic Species

I. Introduction

The acid and base strengths of most of the functional groups of organic chemistry have been considered in previous chapters. Although protonation of very weak bases produces high-energy cations and deprotonation of very weak acids produces high-energy anions, the high-energy species considered in this chapter are of a different type. They are, for the most part, reactive molecules with rather short lifetimes—for example, free radicals, electronically excited states, and tetrahedral reaction intermediates. We shall see that they often have acid and base strengths that are quite different from those of their nearest stable structural analogues.

II. Organic Free Radicals

Many organic radicals can be generated as transient species in aqueous solution by treating organic molecules with reactive entities such as hydroxyl or the solvated electron, which are themselves generated by radiolysis of water (1). Measurement of the acidic and basic pK values of such radicals is most often accomplished by means of ultraviolet spectrophotometry, although esr spectroscopy and kinetic and electrochemical techniques are also used (2).

A. Radical-Anions

The pK_{HA} of hydroxyl is 11.9 (3, 4) [Eq. (6-1)]; this makes it almost four pK units stronger than water, whose pK_{HA} as a solute in water is 15.7. The hydroperoxyl radical is even stronger, having a pK_{HA} near 4.7 (5, 6) [Eq. (6-2)],

$$HO\cdot \;\rightleftharpoons\; H^+ + O^{\overline{\cdot}} \tag{6-1}$$

$$HOO\cdot \;\rightleftharpoons\; H^+ + O_2^{\overline{\cdot}} \tag{6-2}$$

although its very rapid disproportionation to oxygen and hydrogen peroxide

$(k_2 = 10^6 M^{-1} \text{ sec}^{-1}$ at room temperature) means that it can have only a transient existence as a solute in water (7). Its anion, the superoxide radical O_2^{-}, may function as a radical, an electron transfer agent, a nucleophile, or a base (7a, b).

Hydrogen atom abstraction from methanol does not produce the analogous methoxyl radical but rather gives rise to the hydroxymethyl radical. Its pK_{HA} of 10.7 (8, 9) [Eq. (6-3)] shows it to be almost five units stronger than methanol, whose pK_{HA} is 15.5.

$$\cdot CH_2OH \rightleftharpoons H^+ + \cdot CH_2O^- \qquad (6\text{-}3)$$

Removing a hydrogen atom from C-1 of an alcohol invariably increases the acidity of the alcoholic proton. The species so produced, which are usually called ketyls, benefit in their anionic forms from resonance of the sort shown in Eq. (6-4), particularly if the groups Z are able to aid in dispersing the negative

$$\overset{Z}{\underset{Z}{\diagdown}}\dot{C}\text{—OH} \xrightleftharpoons{-H^+} \overset{Z}{\underset{Z}{\diagdown}}\dot{C}\text{—O}^- \longleftrightarrow \overset{Z}{\underset{Z}{\diagdown}}C\text{—O}\cdot \qquad (6\text{-}4)$$

charge. The radical produced by removing a hydrogen atom from acetoin, for example, has a pK_{HA} in water of 4.4 [Eq. (6-5)], making it stronger than acetic

$$H_3C\text{—}\overset{OH}{\underset{}{\underset{|}{CH}}}\text{—}\overset{O}{\underset{}{\overset{||}{C}}}\text{—}CH_3 \xrightarrow{-H\cdot} H_3C\text{—}\overset{OH}{\underset{}{\underset{|}{\dot{C}}}}\text{—}\overset{O}{\underset{}{\overset{||}{C}}}\text{—}CH_3 \xrightleftharpoons{-H^+} H_3C\text{—}\overset{O^-}{\underset{}{\underset{|}{\dot{C}}}}\text{—}\overset{O}{\underset{}{\overset{||}{C}}}\text{—}CH_3 \longleftrightarrow$$

$$H_3C\text{—}\overset{O\cdot}{\underset{}{\underset{|}{C}}}\text{=}\overset{O^-}{\underset{}{\underset{|}{C}}}\text{—}CH_3 \longleftrightarrow \text{ etc.} \qquad (6\text{-}5)$$

acid. See Table 6-1 and ref. 2 for additional examples of the acid strengths of ketyls.

Removing a hydrogen atom from a carbon atom more distant from the hydroxyl group (C-2 and up) has little effect on the acid strength of an alcohol.

Carboxylic acids can generate acidic radicals of various kinds, depending on the structure of the acid and on the means of radical generation. Most aliphatic carboxylic acids react with hydroxyl, or other reactive radicals, at the α-carbon atom to generate radicals of the type shown in Eq. (6-6). Aliphatic carboxylic acids lacking an α-hydrogen atom, such as pivalic acid, are less reactive but are usually able to generate radicals at more remote sites [Eq. (6-7)].

$$HO\cdot + CH_3CO_2H \longrightarrow H_2O + \cdot CH_2CO_2H \qquad (6\text{-}6)$$

$$HO\cdot + (CH_3)_3CCO_2H \longrightarrow H_2O + \cdot CH_2C(CH_3)_2CO_2H \qquad (6\text{-}7)$$

Table 6-1

ACID STRENGTHS OF KETYL RADICALS
AND RELATED COMPOUNDS[a]

Radical	pK_{HA}	Ref.
$\cdot CH_2OH$	10.7	2, 8, 9
$CH_3\dot{C}HOH$	11.6	2, 8, 9
$(CH_3)_2\dot{C}OH$	12.2	2, 8, 9
$(HOCH_2)_3C\dot{C}HOH$	10.4	2
$H_2C{=}CH\dot{C}HOH$	9.6	77
$C_6H_5\dot{C}HOH$	10.5	78
$(C_6H_5)_2\dot{C}OH$	9.2	79–81
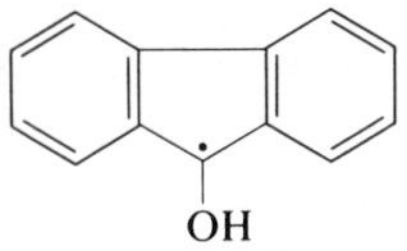 fluorenyl-9-ol radical (9-hydroxyfluorenyl)	6.3	79
$H_5C_6{-}\overset{O}{\overset{\|}{C}}{-}\overset{OH}{\overset{\|}{\underset{\cdot}{C}}}{-}C_6H_5$	5.9	78
$H_5C_6{-}\overset{OH}{\overset{\|}{\underset{\cdot}{C}}}{-}OH$	5.3	13
$CH_3CH{=}CH{-}\overset{OH}{\overset{\|}{\underset{\cdot}{C}}}{-}OH$	7.5	2, 82
$H_5C_6{-}\overset{OH}{\overset{\|}{\underset{\cdot}{C}}}{-}OCH_3$	5.5	13
$H_5C_6{-}\overset{OH}{\overset{\|}{\underset{\cdot}{C}}}{-}NH_2$	7.7	79
$H_5C_6{-}\overset{OH}{\overset{\|}{\underset{\cdot}{N^+}}}{-}O^-$	3.2	83

[a] Additional data can be found in ref. 2.

Radicals of a different sort are produced when the aqueous solvated electron reacts with an aromatic carboxylic acid [Eq. (6-8)]. [Most aliphatic carboxylic acids are not sufficiently reactive to compete effectively with the disappearance of the solvated electron (10).]

$$e^-(aq) + C_6H_5CO_2H \longrightarrow C_6H_5CO_2H\dot{^-} \qquad (6\text{-}8)$$

The pK_{HA} values of the three kinds of radical shown in Eqs. (6-6) to (6-8) are not greatly different from those of typical carboxylic acids. This can be seen in

Eqs. (6-9) through (6-11), where the neutral forms of the radicals are compared with their nonradical precursors.

$$CH_3CO_2H \rightleftharpoons H^+ + CH_3CO_2^- \qquad pK_{HA} = 4.76$$

$$\cdot CH_2CO_2H \rightleftharpoons H^+ + \cdot CH_2CO_2^- \qquad pK_{HA} = 4.9 \ (11)$$

(6-9)

$$(CH_3)_3CCO_2H \rightleftharpoons H^+ + (CH_3)_3CCO_2^- \qquad pK_{HA} = 5.03$$

$$\cdot CH_2C(CH_3)_2CO_2H \rightleftharpoons H^+ + \cdot CH_2C(CH_3)_2CO_2^- \qquad pK_{HA} = 4.8 \ (12)$$

(6-10)

$$C_6H_5CO_2H \rightleftharpoons H^+ + C_6H_5CO_2^- \qquad pK_{HA} = 4.20$$

$$C_6H_5\dot{C}(OH)_2 \rightleftharpoons H^+ + C_6H_5CO_2H^{\cdot} \qquad pK_{HA} = 5.3 \ (13)$$

(6-11)

The benzoic acid radical-anion shown in Eqs. (6-8) and (6-11) retains an ionizable proton, which is removed by aqueous hydroxide [Eq. (6-12)]. It can be seen that adding an electron to benzoic acid reduces the acidity of the carboxylic hydrogen by almost eight pK units.

$$C_6H_5CO_2H^{\cdot} \rightleftharpoons H^+ + C_6H_5CO_2^{\cdot 2-} \qquad pK_{HA} = 12.0 \ (13) \qquad (6-12)$$

α-Carboxylic acid radicals are stabilized by resonance of the sort shown in Eq. (6-13). Judging from the pK values shown in Eqs. (6-9) and (6-10) radical-

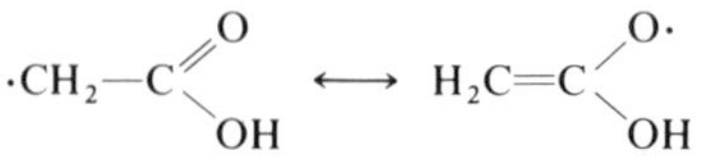

(6-13)

anions appear to possess little extra in the way of resonance stabilization. Neutral radicals of the type shown in Eq. (6-11) are greatly stabilized by resonance involving the aromatic ring [Eq. (6-14)]. They can be regarded as

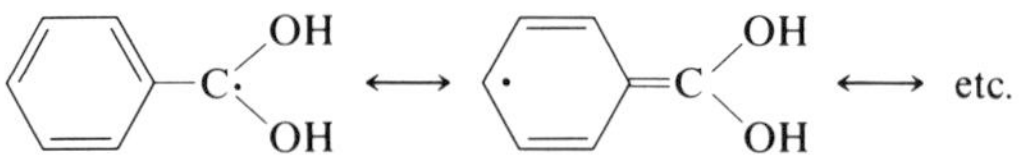

(6-14)

ketyl radicals in which an organic group is replaced by hydroxyl. The pK_{HA} of $(C_6H_5)_2\dot{C}OH$ is 9.2 (Table 6-1), making this molecule some four pK units weaker than $C_6H_5\dot{C}(OH)_2$. This indicates that the acid strength of ketyl radicals is strongly dependent on the polar effect of the groups attached to the ketyl carbon atom, since hydroxyl has a larger electron-withdrawing inductive effect than phenyl [e.g., σ^* is 1.34 for hydroxyl and 0.75 for phenyl (14)] but less capacity to stabilize radicals as such. [Electron spin resonance studies have shown that the spin density in the diphenyl ketyl anion $(C_6H_5)_2CO^{\cdot}$ is largely concentrated on the carbonyl carbon atom (15, 16).]

Semiquinone radical-anions can be generated from quinones by electron transfer from such species as $\cdot CH_2O^-$ or the solvated electron (2, 17, 18). Their protonated forms are much stronger acids than their dihydroxyl counterparts [Eqs. (6-15a) and (6-15b)].

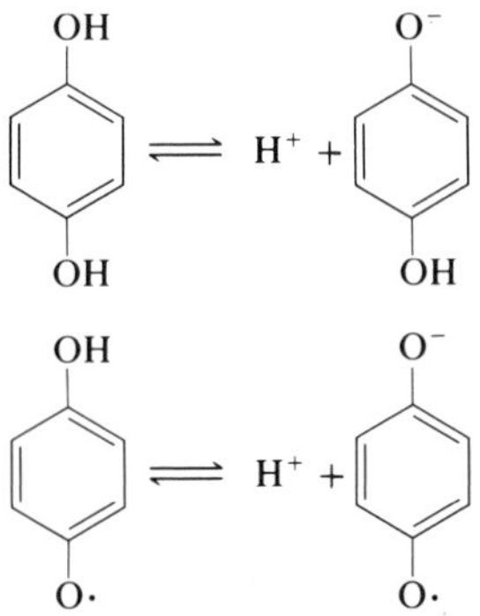

$$pK_{HA} = 9.9$$

$$(6\text{-}15a)$$

$$pK_{HA} = 4.1 \ (17, 19)$$

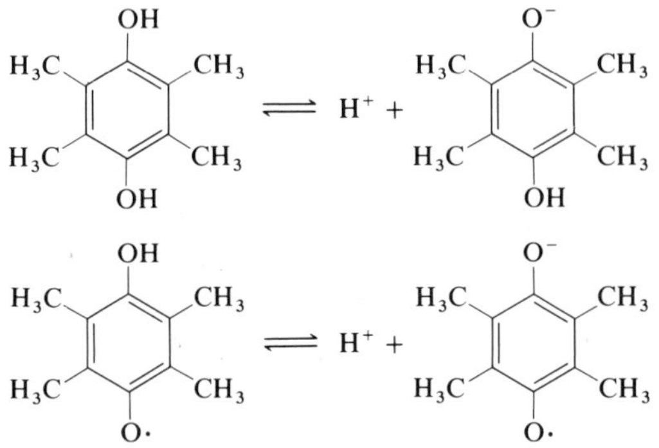

$$pK_{HA} = 11.2$$

$$(6\text{-}15b)$$

$$pK_{HA} = 5.1 \ (17, 20)$$

The symmetry and consequent resonance stabilization in the semiquinone anions shown here are obvious at a glance and account for the strengths of the conjugate acids and the relatively long life that many of the radical-anions possess.

B. Radical-Cations

When alkylammonium salts react with hydroxyl radicals, they generate radical-cations of the sort shown in Eq. (6-16). The pK_{BH^+} of this radical

$$(CH_3)_2\overset{+}{N}H_2 \xrightarrow{\ HO\cdot\ } (CH_3)_2\overset{+\cdot}{N}H \qquad (6\text{-}16)$$

is markedly lower than that of its precursor, dimethylamine [Eqs. (6-17) and (6-18)].

$$(CH_3)_2\overset{+\cdot}{N}H \rightleftharpoons H^+ + (CH_3)_2\overset{\cdot}{N}: \qquad pK_{BH^+} = 7.0 \ (21) \qquad (6\text{-}17)$$

$$(CH_3)_2\overset{+}{N}H_2 \rightleftharpoons H^+ + (CH_3)_2NH: \qquad pK_{BH^+} = 10.73 \qquad (6\text{-}18)$$

The pK_{BH^+} values of a number of other amino compounds and their radical derivatives are compared in Eqs. (6-19) through (6-22).

$$NH_4{}^+ \rightleftharpoons H^+ + NH_3 \qquad pK_{BH^+} = 9.2$$
$$NH_3{}^{\cdot\,+} \rightleftharpoons H^+ + \cdot NH_2 \qquad pK_{BH^+} = 6.7\ (24) \tag{6-19}$$

$$\overset{+}{NH_2}NH_3 \rightleftharpoons H^+ + NH_2NH_2 \qquad pK_{BH^+} = 8.07\ (22)$$
$$NH_2NH_2{}^{\cdot\,+} \rightleftharpoons H^+ + NH_2NH\cdot \qquad pK_{BH^+} = 7.1\ (23) \tag{6-20}$$

$$\overset{+}{H_3}NOH \rightleftharpoons H^+ + H_2NOH \qquad pK_{BH^+} = 6.0$$
$$\overset{+\cdot}{H_2}NOH \rightleftharpoons H^+ + H\dot{N}OH \qquad pK_{BH^+} = 4.2\ (23,\ 24) \tag{6-21}$$

$$\overset{+}{H_3}NOCH_3 \rightleftharpoons H^+ + H_2NOCH_3 \qquad pK_{BH^+} = 4.6$$
$$\overset{+\cdot}{H_2}NOCH_3 \rightleftharpoons H^+ + H\dot{N}OCH_3 \qquad pK_{BH^+} = 2.9\ (24) \tag{6-22}$$

In all these cases the amines are more basic than the radicals to which they give rise (23a), although it may be significant that the difference is least, approximately one pK unit, in the case of hydrazine and hydrazyl [Eq. (6-20)] (24a). The protonated form of hydrazyl is planar with all hydrogen atoms and both nitrogen atoms equivalent (25, 26). The double-quartet theory of Linnett (27, 28), which is so useful in describing the bonding of many odd-electron species, finds ready application in this system, the structure of the neutral radical and that of the radical-cation each being written with a three-electron bond (1 and 2).

$$\overset{\frac{1}{2}+}{H_2N}\!\!-\!\!\overset{\frac{1}{2}-}{NH} \qquad\qquad \overset{\frac{1}{2}+}{H_2N}\!\!-\!\!\overset{\frac{1}{2}+}{NH_2}$$

$$\textbf{1} \qquad\qquad\qquad \textbf{2}$$

The partial negative charge on one of the nitrogen atoms of 1 doubtless makes the molecule more basic than it would otherwise be and contributes to the difference between the basicity of hydrazine and hydrazyl being as small as it is. Similar structures can be drawn for hydroxylamine and its derivatives, but in these cases the partial positive charge appears on oxygen, which is a somewhat less satisfactory arrangement.

Perhaps the best known radical-cations are those called Würster's salts (29, 30). They are intermediates in the oxidation of compounds such as p-phenylenediamine by bromine and other oxidants (31) and, indeed, are stable in slightly acidic aqueous solution (between pH 3.5 and 6.0 for 3). Neither the dication 4, which is formed in more acidic solution, nor the neutral molecule 5, which is formed in basic solution, has the same potential for resonance, and these species, accordingly, decompose rapidly.

$$\cdot \overset{+}{N}H_2 \qquad \cdot \overset{+}{N}H_2 \qquad \cdot NH$$

$$NH_2 \qquad {}^+NH_3 \qquad NH_2$$

$$\mathbf{3} \qquad\qquad \mathbf{4} \qquad\qquad \mathbf{5}$$

The radical-cations formed from phenols are highly acidic as might be expected. The pK_{HA} of $C_6H_5\overset{+\cdot}{OH}$ is approximately -2.0 (*32, 33*). The acid strengths of radical-cations of phenols and arenes have been reviewed by Nicholas and Arnold (*34*).

C. Amino Acid Radicals

α-Amino acids such as glycine readily generate free radicals when subjected to high-energy radiation. At very low temperatures a variety of radical types have been identified, but at room temperature the predominant radical in the solid state has the zwitterionic structure **6** (*35–37*). In aqueous solution, however, the predominant neutral radical is the isomeric form **7** (*2, 38, 39*).

$$H_3\overset{+}{N}-\overset{\cdot}{C}H-CO_2^- \qquad\qquad H_2N-\overset{\cdot}{C}H-CO_2H$$

$$\mathbf{6} \qquad\qquad\qquad \mathbf{7}$$

Presumably, the orientation of the radical in the crystal lattice, which is composed mainly of ordinary zwitterionic glycine units, is such as to favor **6**, since it can form strong hydrogen bonds to its neighbors (*40, 41*). In solution this advantage is lost, but it is not obvious at first glance why the aqueous neutral radical should prefer to have its amino and carboxyl groups un-ionized. Indeed, the neutral glycine radical does not undergo protonation in solution even at pH 1 (*2, 39*), which means that there is a difference of more than seven pK units in the basicity of the amino groups in the neutral radical and its nonradical analogue [Eqs. (6-23) and (6-24)].

$$H_3\overset{+}{N}CH_2CO_2H \rightleftharpoons H^+ + H_2NCH_2CO_2H \qquad pK_{BH^+} \approx 7.6\,(42) \qquad (6\text{-}23)$$

$$H_3\overset{+}{N}\overset{\cdot}{C}HCO_2H \rightleftharpoons H^+ + H_2N\overset{\cdot}{C}HCO_2H \qquad pK_{BH^+} < 1.0\,(39) \qquad (6\text{-}24)$$

There is clearly a special sort of stabilization present in the neutral glycine radical, the most attractive explanation for which can be found in the concept of *merostabilization*. This is a term introduced by Katritzky *et al.* (*45*) to refer to the additional resonance that is present when radical centers serve as the link between electron donors and acceptors (*46*). [Others have used different terms for this phenomenon (*47–49*).] In the case of ordinary glycine the amino and the carboxyl units cannot interact by resonance because they are separated by the saturated methylene unit, and so they are forced to interact by means of proton transfer. In the radical, however, extensive resonance involving both groups becomes possible [Eq. (6-25)].

$$H_2N-\overset{\cdot}{C}H-C\overset{O}{\underset{OH}{\big\langle}} \longleftrightarrow H_2\overset{+}{N}=CH-\overset{\cdot}{C}\overset{O^-}{\underset{OH}{\big\langle}} \longleftrightarrow$$

$$H_2\overset{+\cdot}{N}-CH=C\overset{O^-}{\underset{OH}{\big\langle}} \longleftrightarrow H_2N-CH=C\overset{O\cdot}{\underset{OH}{\big\langle}} \quad (6\text{-}25)$$

A much smaller degree of resonance interaction is possible in the protonated form of the radical [Eq. (6-26)], thus providing an explanation for the very low basicity of the amino group.

$$H_3\overset{+}{N}-CH-C\overset{O}{\underset{OH}{\big\langle}} \longrightarrow H_3\overset{+}{N}-CH=C\overset{O\cdot}{\underset{OH}{\big\langle}} \quad (6\text{-}26)$$

How is the acidity of the carboxyl group in the glycine affected by radical formation? Its acidity is lower than that of the carboxyl group in the un-ionized form of glycine [Eqs. (6-27) and (6-28)], but the effect is not as

$$H_2NCH_2CO_2H \rightleftharpoons H^+ + H_2NCH_2CO_2^- \quad pK_{HA} \approx 4.5\,(42) \qquad (6\text{-}27)$$

$$H_2N\overset{\cdot}{C}HCO_2H \rightleftharpoons H^+ + H_2N\overset{\cdot}{C}HCO_2^- \quad pK_{HA} = 6.6\,(11,\,12) \qquad (6\text{-}28)$$

dramatic as in the case of the amino group, being only two pK units instead of seven or more.

The reason for the relatively modest reduction in the acidity of the carboxyl group that accompanies radical formation is presumed to be the following. Although the number of major resonance forms is the same in the anionic and neutral radicals, the accumulation of negative charge in two of the structures will diminish the contribution they make to the stability of the radical-anion [Eq. (6-29)]. The net result is a limited amount of merostabilization and a

$$H_2N-\overset{\cdot}{C}H-CO_2^- \longleftrightarrow H_2\overset{+}{N}=CH-\overset{\cdot}{C}\overset{O^-}{\underset{O^-}{\big\langle}} \longleftrightarrow$$

$$H_2\overset{+\cdot}{N}-CH=C\overset{O^-}{\underset{O^-}{\big\langle}} \longleftrightarrow H_2N-CH=C\overset{O\cdot}{\underset{O^-}{\big\langle}} \quad (6\text{-}29)$$

modest decline in the acidity of the carboxyl group.

The carboxyl group in α-hydroxycarboxylic acids shows smaller reductions in acidity on conversion to radicals [Eqs. (6-30) and (6-31)]. This is doubtless

$$HOCH_2CO_2H \rightleftharpoons H^+ + HOCH_2CO_2^- \quad pK_{HA} = 3.83 \qquad (6\text{-}30)$$

$$HO\overset{\cdot}{C}HCO_2H \rightleftharpoons H^+ + HO\overset{\cdot}{C}HCO_2^- \quad pK_{HA} = 4.6\,(50) \qquad (6\text{-}31)$$

because the hydroxyl group is less able than amino to bear a positive charge.

The glycolic acid radical, of course, contains two acidic units, the second being the ketyl hydroxyl group; its pK_{HA} has been measured as 8.8 (50).

D. Some Free Radicals of Biological Importance

Many molecules involved in biological processes are subject to 1-equivalent oxidation to give radicals whose subsequent modes of decay are a matter of great interest (51–53). We saw earlier that hydrogen atom abstraction from an organic molecule often produces a radical that is a stronger acid than its precursor; monohydric phenols that lose their ionizable proton and nonamino carboxylic acids are the most prominent exceptions to this rule of thumb. Simple electron loss from any molecule, of course, always causes an increase in acid strength. The acid strengths of a number of radicals formed from molecules of biological importance are considered in the following paragraphs.

Vitamin C (L-ascorbic acid) has the structure **8**. It is an acid ($pK_{HA} = 4.1$) as a consequence of the enolic hydroxyl group at C-3 being conjugated to the carbonyl group at C-1 [Eq. (6-32)].

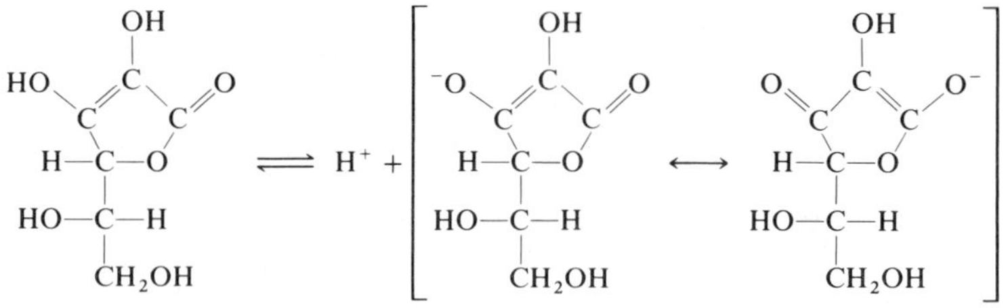

$$(6\text{-}32)$$

8

One-equivalent oxidation of ascorbic acid produces a relatively long lived radical-anion of structure **9** (54–57) whose ease of formation and stability result from the delocalization of both negative charge and electron spin. Protonation of this radical-anion [almost certainly at the C-2 oxygen to give the radical analogue of the ion shown in Eq. (6-32)] does not begin to occur in aqueous solution until the solution is made quite acidic. Both **9** and the radical-anion obtained from the model compound **10** are half-protonated in

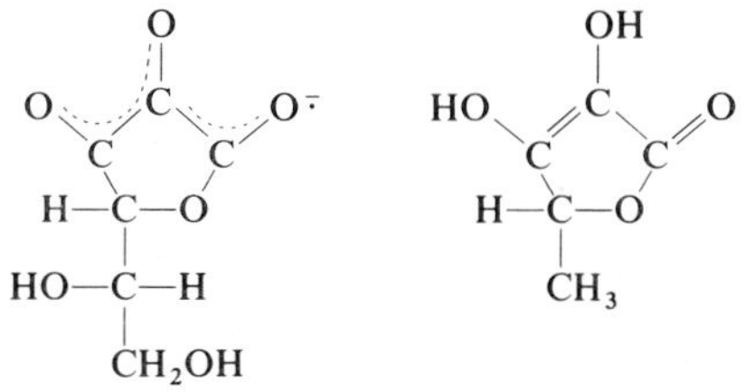

9 **10**

$\sim 2\ M$ perchloric acid $(58, 59)$. The H_- value of such a solution is $-0.9\ (60)$, and this can be taken as an approximate value of the pK_{HA} of the neutral radical species.

A number of important naturally occurring quinones give rise to semiquinone radical-anions as a result of electron addition (Section II,A). The pK_{HA} values of the conjugate acids of several of these are shown in **11** to **13**, with the

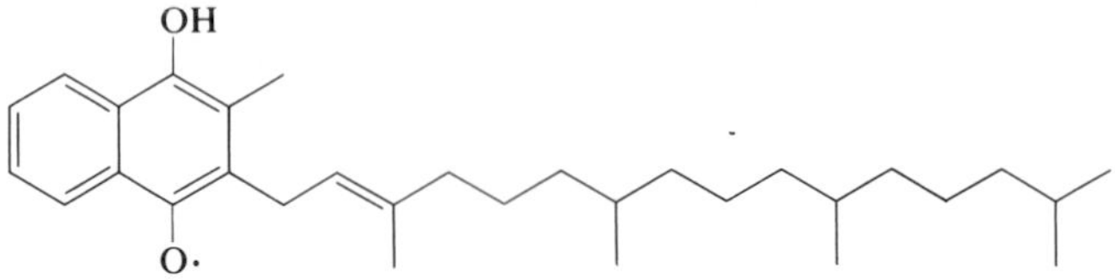

11

$pK_{HA} = 5.5\ (61)$

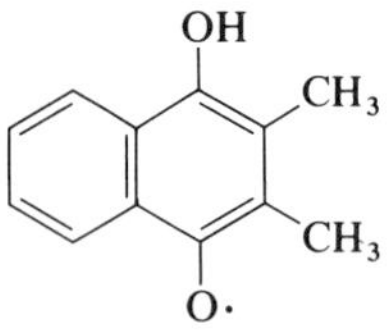

12

$pK_{HA} = 4.25\ (61)$

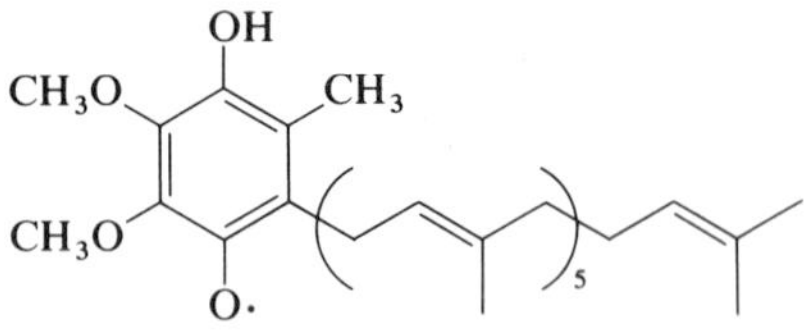

13

$pK_{HA} = 6.5\ (62)$

oxygen atom bearing the proton being arbitrarily assigned. The quinone precursors of **11**, **12**, and **13** are, respectively, vitamin K_1, 2,3-dimethyl-1,4-naphthoquinone (a vitamin K analogue), and one of the ubiquinone oxidative coenzymes, coenzyme Q_6.

Semiquinone radicals are important participants in the 1-equivalent oxidation–reduction of flavin coenzymes (Chapter 4, Section IV,B). The acid strengths of numerous flavin-type radicals have been measured $(63–67)$.

Two nucleotide bases, thymine and uracil, contain the pyrimid-2,4-dione ring system shown in **14**. The parent molecule reacts with hydrated electrons

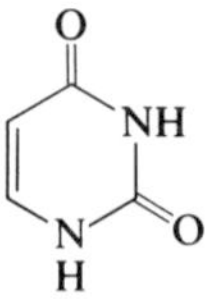

14

in aqueous solution to give a radical-anion that is half-protonated very near pH 7. The favored site of protonation is believed to be that shown in Eq. (6-33), since it permits allylic resonance in the neutral radical; extensive delocalization of the unpaired electron in the radical-anion almost certainly occurs $(68–70)$.

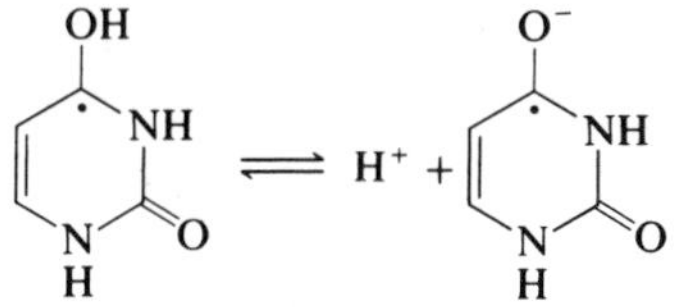

$$pK_{HA} = 7.3 \qquad (6\text{-}33)$$

Nucleotide bases can also form radicals by means of hydroxyl addition or, if activated methyl groups are present, by hydrogen atom abstraction. The acid strengths of a number of such radicals have been measured (*68*).

E. Father–Son Protonations

There are a number of examples in the literature of what has been called "father–son" protonation. This involves transfer of a proton between the initial product of a reaction and a molecule of starting material; that is, the offspring reacts with its immediate progenitor. A process in which the product of an initial reaction becomes the reactant in a second process is commonly called a "mother–daughter" reaction; in a father–son reaction as defined by Elving the product reacts specifically with its precursor (*71, 72*).

Most reactions of this sort involve the generation of anions by electrochemical means or with the help of alkali metals. The anions so produced are often strong bases that readily abstract a proton from starting materials. Several such reactions are shown in Eqs. (6-34) through (6-36) (*71–76*).

$$CH_3CONHCl \xrightarrow{2e^-} CH_3CONH^- + Cl^-$$

$$CH_3CONH^- + CH_3CONHCl \longrightarrow CH_3CONH_2 + CH_3CONCl^-$$

$$(6\text{-}34)$$

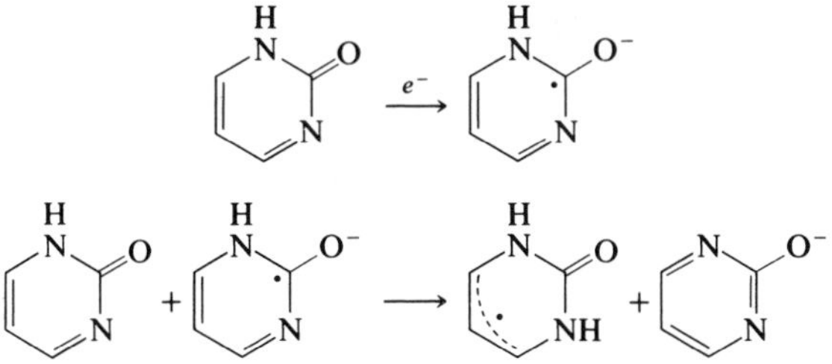

$$(6\text{-}35)$$

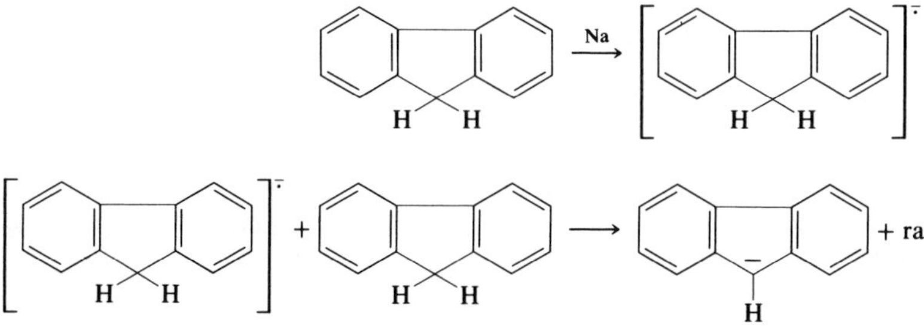

$$(6\text{-}36)$$

III. Electronically Excited States

Profound changes in the acidity or basicity of an organic molecule often accompany the absorption of a quantum of visible or ultraviolet light. Although the lifetimes of most of the electronically excited states so produced are extremely short, the lowest excited singlet (S_1) and triplet (T_1) states are often sufficiently long-lived to establish protolytic equilibria with their surroundings. Indeed, acid- or base-catalyzed processes involving such species sometimes take place.

Protonation or deprotonation of most organic molecules causes changes in their ultraviolet (or visible) absorption spectra, and it is a simple matter to show in such cases that the excited state produced by light absorption must have a different pK than the ground state. Thus, if HA has its principal absorption band at, say, 350 nm and A^- has the corresponding band red-shifted to 400 nm (a bathochromic shift), it takes a smaller quantity of energy to convert A^- to A^-* than to convert HA to HA*. Consequently, HA* must be a stronger acid than HA. Such a situation is illustrated in Fig. 6-1, where it is clear that there is a smaller energy gap between HA* and A^-* than between HA and A^- and that consequently HA* is a stronger acid than HA. When ionization produces a *blue* shift (a hypsochromic shift), HA* is a *weaker* acid than HA. This argument is due to T. Förster, and the general arrangement shown in Fig. 6-1 is called the Förster cycle (84–87).

Since only single energy levels are shown in Fig. 6-1 for each species, the entropic components of the equilibrium are missing; thus, one can calculate only an approximate value for the pK difference between ground and excited states by this means. The formulas that relate ΔpK to the shift in frequency or wavelength of absorption are given in Eqs. (6-37) and (6-38) for a temperature

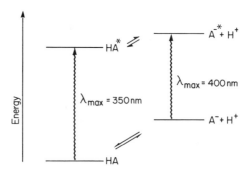

Fig. 6-1 Schematic diagram of the ionization in the ground state (HA) and excited state (HA*) for a weak acid whose ionization in the ground state is accompanied by a shift in its absorption maximum from 350 to 400 nm.

of 25°C. Because of neglect of entropy they are written as approximations. The derivation is as follows:

$$Nhv_{HA} + \Delta H_{HA*} = Nhv_{A-} + \Delta H_{HA}$$

where N is Avogadro's number and ΔH_{HA*} and ΔH_{HA} are the respective enthalpies of ionization;

$$\Delta H_{HA*} - \Delta H_{HA} = \Delta\Delta H \approx \Delta\Delta G$$

since

$$\Delta G = 2.303RT \, pK$$

$$\Delta\Delta G = 2.303RT \, \Delta pK$$

$$\Delta pK \approx \frac{Nh(v_{A-} - v_{HA})}{2.303RT}$$

Taking T as 298 K and inserting the other appropriate numerical values produces Eqs. (6-37) and (6-38). Since wavelength (λ) and energy are not linearly related there is no simple expression for ΔpK in terms of $\Delta\lambda_{max}$.

$$\Delta pK \approx 0.0021 \, \Delta\tilde{v} \, (cm^{-1}) \qquad (6\text{-}37)$$

$$\Delta pK \approx 2.1 \times 10^4 \left[\frac{1}{\lambda_1 \, (nm)} - \frac{1}{\lambda_2 \, (nm)} \right] \qquad (6\text{-}38)$$

Using for purposes of illustration the 50-nm shift in λ_{max} shown in Fig. 6-1, we can calculate a negative pK change (the acid becomes stronger) of 7.5 units; a change in absorption from 200 to 250 nm, which is also a 50-nm red shift, corresponds to a negative pK change of 21 units.

Apart from neglect of entropic factors in the derivation of Eqs. (6-37) and (6-38) there are other, rather serious drawbacks to simply using ultraviolet absorption values for an acid and its conjugate base to calculate pK^* values. First, there is the question of the relevance of the excited states that correspond to the absorption bands being considered. Since the lowest excited singlet (S_1) and triplet (T_1) states are those that are involved in the vast majority of photochemical reactions, it will not usually be helpful to know the pK values of higher levels whose lifetimes are so short as to preclude the establishment of protolytic equilibria. In other words, the absorption being considered for both the acid and its conjugate base should correspond to excitation to S_1, the first excited singlet state, an assignment that often cannot be made with assurance. Furthermore, most organic molecules in solution exhibit broad absorption bands and, even when these can be confidently assigned to transitions to S_1 of the acid and its conjugate base, it is clear that a variety of vibrational levels of the S_1 state are involved, not to mention the problems caused by varying degrees of molecular solvation. A common practice is to use the first absorption peak of each species, but it should be noted that a misreading of just 1 nm at 275 nm, for example, corresponds to a deviation in the calculated value of pK^* of 0.3 unit.

A preferable approach, but one that cannot always be applied, is to use the fluorescence spectra of both species. (Most organic acids lacking an aromatic ring do not show measurable fluorescence; this, in turn, indicates a very short S_1 lifetime in which protolytic equilibria may not have time to become established.) Fluorescence virtually always involves decay from S_1 to S_0 (Fig. 6-2). It too, though, usually gives a broad band, this time because of higher vibrational states of S_0 being formed. Again, $\tilde{v}_{max}$ or λ_{max} and Eq. (6-37) or (6-38) are used to calculate ΔpK.

Do absorption and fluorescence data generally give pK^* values in agreement? Fluorescence bands (corresponding to $S_1 \rightarrow S_0$) appear at longer wavelength than the $S_0 \rightarrow S_1$ absorption bands because upper vibrational states are formed in both cases. If comparable shifts in the frequencies of absorption and emission accompany acid dissociation, then the two methods give comparable results. Frequently, however, the results are widely discrepant as, for example, in the case of phenol. The lowest energy absorption maxima for phenol and phenolate ion are the so-called secondary bands that appear in water at 37,000 cm^{-1} (270 nm) and 34,950 cm^{-1} (286 nm), respectively. Using these values and Eq. (6-37) or (6-38) gives a pK^*_{HA} of 5.7, compared with the ground state pK_{HA} of 9.99. Using the same equations and the fluorescence emission maxima for phenol (33,450 cm^{-1}, or 299 nm) and

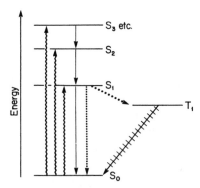

Fig. 6-2 Schematic diagram showing singlet and triplet electronic states and transitions between them. Light absorption, $\rightsquigarrow$; internal conversion, ———; intersystem crossing, ·····; fluorescence, ----; phosphorescence, +++.

phenolate ion (29,750 cm^{-1}, or 336 nm) gives a pK^*_{HA} of 2.3 (88). It has become common practice, following the suggestion of Weller (89), to average the frequencies of the absorption and emission bands for the acid, and similarly for the conjugate base, and then to use these values in Eq. (6-37) to calculate ΔpK. This procedure gives a value of 4.0 for p$K_{HA}(S_1)$ of phenol.

It is often possible to measure the intensity of fluorescence of HA* or A⁻* (or B* or BH⁺*) as a function of the acidity of the medium. Provided that certain assumptions are made regarding the lifetimes of HA* and A⁻* and their rates of proton transfer, p$K_{HA}(S_1)$ is obtained from the midpoint of such fluorescence titration curves (106, 110–114).

Triplet state acidity constants can be obtained using the Förster cycle, provided that phosphorescence can be observed for both HA* and A⁻* (or B* and BH⁺*). The $\tilde{v}_{max}$ of phosphorescence of both species is measured and Eq. (6-37) applied. The variation in the intensity of phosphorescence as a function of pH can yield phosphorescence titration curves but, as with fluorescence, some knowledge of the excited state lifetimes is required to obtain pK^*. The flash photolysis method, introduced by Porter, is a powerful tool for measuring energies of triplet states, particularly as tunable lasers have become available. Here, the T_1 state, achieved by the usual route of $S_0 \rightarrow S_x \rightarrow S_1 \rightarrow T_1$ (Fig. 6-2), absorbs some of the massive input of radiation and forms higher triplet states. Provided that either the acid or its conjugate base undergoes triplet–triplet absorption, measurement of the optical density at an appropriate wavelength as a function of pH usually yields a simple titration curve whose midpoint is p$K_{HA}(T_1)$ in the case of neutral acids and p$K_{BH^+}(T_1)$ in the case of neutral bases. The λ_{max} of triplet–triplet absorption is invariably shifted well to the red, which essentially eliminates interference by ground state absorption.

A. Variation of pK^* with Molecular Structure

The two excited states of importance in acid–base chemistry are S_1 and T_1. A glance at the p$K(S_1)$, p$K(T_1)$, and ground state pK values assembled in Tables 6-2 and 6-3 will show that p$K(T_1)$ lies between the ground state pK and p$K(S_1)$ in virtually all cases and, more often than not, the ground and triplet state pK values are fairly close. These conclusions find support in the much more extensive listings of excited state pK values in the review article of Ireland and Wyatt (85, 116), although in a few cases p$K(T_1)$ may be found to lie outside the ground state pK–p$K(S_1)$ range. Since many molecules do not undergo phosphorescence and since relatively few experiments using the flash photolysis triplet–triplet absorption method have been undertaken, there are great gaps in our knowledge of triplet state acidities. Furthermore, virtually all

Table 6-2

pK_{HA}^* VALUES OF EXCITED STATES OF SOME NEUTRAL
ORGANIC ACIDS[a]

| Acid | pK_{HA} | pK_{HA}^* | | Ref. |
		$pK_{HA}(S_1)$	$pK_{HA}(T_1)$	
Phenol	9.99	4.0	8.5	88
3-Bromophenol	9.03	2.8	7.8	88
4-Bromophenol	9.36	2.9^b	7.7	88
3-Methoxyphenol	9.65	4.6	8.4	88
4-Methoxyphenol	10.21	5.6	8.6	88
2-Nitrophenol	9.1^c	-0.6^c	8.0^c	90
3-Nitrophenol	10.5^c	0^c	9.2^c	90
4-Nitrophenol	9.0^c	-4.0^c	7.3^c	90
2-Naphthol	9.46	3.1	7.9	91
Benzoic acid	4.20	~6	—	92
1-Naphthoic acid	3.70	7.7^d	3.8	92
2-Naphthoic acid	4.20	6.6^d	4.0	92
1-Anthroic acid	3.7	6.9	5.6	92
2-Anthroic acid	4.2	6.6	6.0	92
9-Anthroic acid	3.0	6.5	4.2	92
Pyrene-1-carboxylic acid	4.0	8.1	5.2	95
Purine	8.93	—	5.1	96
6-Aminochrysene	≫14	12.6	—	109
3,5-Diphenylpyrazole	12.9	~8.6	—	113

[a] Where Förster cycle calculations and fluorescence intensity measurements are
available, average values for $pK_{HA}(S_1)$ have been given in most cases, and similarly
for $pK_{HA}(T_1)$ with regard to Förster cycle calculations and flash photolysis
measurements; aqueous or nearly aqueous conditions apply unless otherwise
indicated.
[b] Recalculated from spectral data in ref. 88.
[c] In absolute ethanol.
[d] See also refs. 93 and 94.

compounds for which $pK(T_1)$ values have been measured contain aromatic
rings. On the other hand, $pK(S_1)$ values can be assigned to most organic
compounds by using the Förster cycle and the shift in absorption spectrum
that usually accompanies proton loss or proton gain. As we have seen, such
values may be in error by several pK units, judging from the results of
fluorescence intensity measurements; the latter experiments, of course, cannot
be made in all cases. Furthermore, aliphatic compounds, in particular, have S_1
lifetimes that are so fleeting as to render any $pK(S_1)$ value of theoretical
interest only, since protolytic equilibrium will not have time to become
established.

Table 6-3

pK^*_{BH} VALUES OF EXCITED STATES OF SOME NEUTRAL ORGANIC BASES[a]

Base	pK_{BH^+}	$pK^*_{BH^+}$		Ref.
		$pK_{BH^+}(S_1)$	$pK_{BH^+}(T_1)$	
2-Nitroaniline	-0.29	-5.0	—	98
4-Methoxyaniline	5.34	-0.8	—	99
2,5-Dinitroaniline	-2.5	-7.9	—	98
Nitrobenzene	-12.1	1.5^b	Approx. -10	90
Azobenzene	-2.9	13.7	—	100
2-Naphthylamine	4.1	-2.0^c	3.2	91
N,N-Dimethyl-1-				
naphthylamine	4.9	—	2.8	91
1-Aminoanthracene	3.3	-5.5	2.8	101
1-Aminopyrene	2.8	-5.8	3.5	95
6-Aminochrysene	3.2	-3.3	—	109
1-(4-Aminophenyl)pyrene	4.05	3.5	—	102
8-Methylquinoline	4.80	9.7	5.9	103
6-Nitroquinoline	2.10^d	4.4^d	2.1^d	90
Acridine	5.45	10.2	5.6	103, 108
Phenazine	1.21	6.0	4.0	104
3-Phenylpyridine	4.85	12.2	—	105
Purine	2.39	$—^e$	5.1	96, 97
3,5-Diphenylpyrazole	1.43	$2.7^f, 6.1^g$	—	113

[a] See footnote a of Table 6-2.
[b] Recalculated on the basis of ground state pK_{BH^+} being -12.1 (Chapter 3).
[c] See also ref. 102.
[d] In absolute ethanol.
[e] The $S_0 \rightarrow S_1$ transition is $n-\pi^*$ (107).
[f] From fluorescence data.
[g] From absorbance data.

The direction of the shift in $pK(S_1)$ for many types of aromatic molecules can be predicted on the basis of simple resonance structures, for example, those shown in Eqs. (6-39) to (6-41).

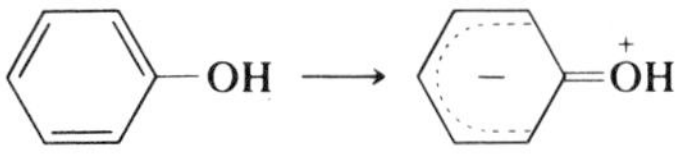

$$\tag{6-39}$$

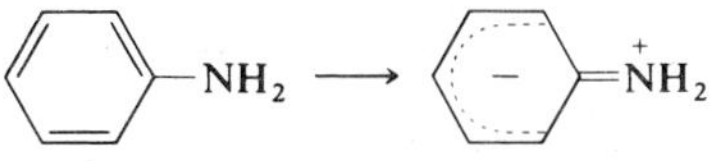

$$\tag{6-40}$$

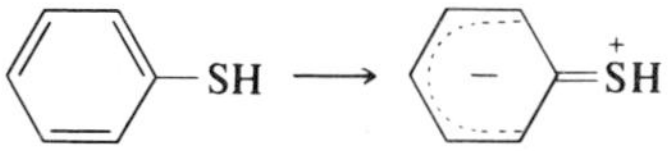

$$\tag{6-41}$$

Thus, phenols, arylamines, and aryl thiols all appear to be stronger acids in S_1 than in S_0 (86), and all show red shifts on being converted to their conjugate bases. The change in polarity that accompanies light absorption is such as to suggest that the S_1 state has a large contribution from the dipolar structures shown in Eqs. (6-39) to (6-41) than does the ground state and, if this is so, a greater acid strength for the excited state is only to be expected. The capacity of HO, NH_2, and SR groups to donate negative charge to the ring is indicated by the large difference between their σ and σ^+ values (117) and, in the case of aniline, by a reversal in sign of the charge on the nitrogen atom on excitation to S_1 (118).

In the case of arylcarboxylic acids there is no direct conjugation between the acidic hydroxyl group and the ring; there are indications, however, that structures such as 15 are of some importance in the S_1 state (86, 91, 92), which would have the effect of lowering the acidity of the carboxyl group, as a result of electron transfer from the ring [Eq. (6-42)]. Carboxylic acids are, indeed,

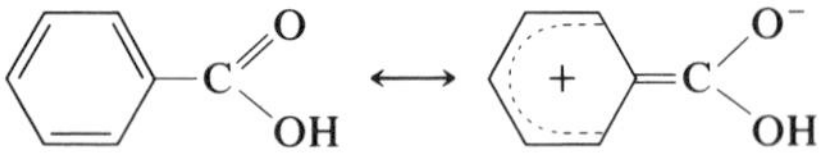

$$(6-42)$$

15

somewhat weaker acids in S_1 than in S_0, although ΔpK is usually only two units or so.

It can be seen in Table 6-2 that the acid strengths of the triplet states of phenols and carboxylic acids are fairly near those of the ground state. In the case of carboxylic acids this has been rationalized on the basis that the triplet state is not dipolar but rather diradicaloid to a high degree, as shown in 16 (86,

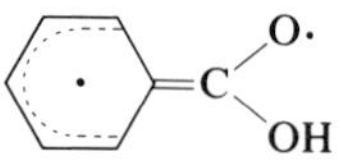

16

91, 92, 119). Such diradical structures do not involve a substantial shift of charge either toward or away from the acidic functional group, and hence they would be expected to have only a small effect on acidity.

Carbon acids that have accessible absorption bands may also undergo dissociation upon irradiation. Thus, flash photolysis of 2,4,6-trinitrotoluene generates the carbanion, even in acidic aqueous solution (120).

With regard to base strength, almost all of the functional groups that have been examined have been attached to or been part of aromatic rings, and in most of these cases the S_1 state is more basic than S_0. The principal exception, and a very important one, is the amino group. As might be expected on the basis of the singlet excited state having a high degree of dipolar character

[Eq. (6-40)], arylamines (and presumably aryl thiols and phenols) become much less basic on excitation to S_1 (Table 6-3). On the other hand, when the basic nitrogen center is part of a heterocyclic ring, the S_1 state is usually considerably more basic than the ground state (*121*). For example, *trans*-3-styrylpyridine (17) has a ground state pK_{BH^+} of 4.76 and a $pK_{BH^+}(S_1)$ of 12.5

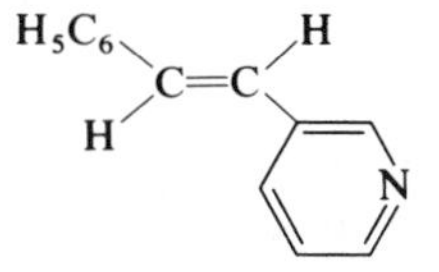

17

(*106*). The latter is the mean of the values obtained using the Förster cycle and the average maxima for absorption and fluorescence (12.4) and that obtained by fluorescence titration (12.6).

The red shift of the lowest energy band in the spectrum of **17** that accompanies either protonation or a change to a more polar solvent indicates a $\pi-\pi^*$ transition. Curiously, the lowest absorption band in pyridine itself is an $n-\pi^*$ transition in which negative charge migrates from nitrogen to the ring (*122, 123*), thus reducing the attraction of the nitrogen atom for protons. In 2-aminopyridine, however, and in most other pyridine derivatives containing electron-donating groups, the S_1 state is π,π^*, which is a more polar and hence a more basic state than S_0. Indeed, most heterocyclic amines that have been examined follow this pattern (*85*). Quinoline (**18**) for example, resembles

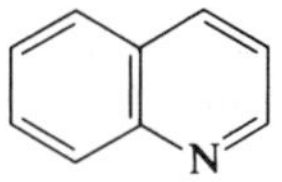

18

naphthalene in showing only $\pi-\pi^*$ transitions in solution, with no indication of the presence of a low-lying n,π^* state (*103*). Actually, the energies of n,π^* and π,π^* states of azanaphthalenes are often rather close, and it has not always been easy to characterize S_1 using such terms. The nature of excited states in heterocyclic molecules has been examined and reviewed by Ross (*123, 123a*) and by Schulman (*124*), who has also considered the acidity and basicity of such states.

As was found to be the case with neutral organic acids, the triplet state of most organic bases is much closer to that of the ground state than to that of the lowest singlet state (Table 6-3).

The excited state of benzoic acid is much more basic than the ground state, as would be expected if S_1 were a π,π^* state having a large contribution from the dipolar structure **15**. The absorption and fluorescence spectra of arenecarboxylic acids invariably show a red shift on protonation, and application of the Förster cycle to these maxima suggests that the S_1 states are possibly 7–8 pK units more basic than the ground states (*92*). This would put the p$K_{BH^+}(S_1)$ of benzoic acid somewhere near 3.0. Nitroarenes are similar in having greater polarity in the excited state (*125*), which means a larger negative charge on the oxygen atoms of the nitro group. They are consequently expected to be more basic in S_1 than in S_0, a conclusion that is supported by the red shift that accompanies protonation. Application of the Förster cycle to the latter leads to a ΔpK of more than 13 units for nitrobenzene (Table 6-3). As is the case with benzoic acid the triplet and ground states of nitrobenzene have rather similar basicities.

The lowest singlet state of simple aldehydes and ketones is n,π^*. Since the dipole moment of the carbonyl group is smaller in this state than in the ground state the basicity of the molecule would be expected to be decreased by excitation. In fact, little is known about the base strength of the S_1 states of aliphatic aldehydes and ketones; in any case they seldom exhibit fluorescence, and so it is unlikely that their S_1 lifetimes are long enough for them to enter into protolytic equilibria (*126*).

Aromatic ketones have been much more thoroughly studied. The first singlet excited state of benzophenone is predominantly n,π^*, although it is believed that it has some π,π^* character as a result of mixing with nearby π,π^* states (*127*). Its n,π^* designation derives from its relatively low intensity ($e \approx 100$) and from the blue shift that accompanies an increase in solvent polarity ($\lambda_{max} \sim 350$ nm in cyclohexane and ~ 335 in ethanol). By way of contrast, the intense $\pi-\pi^*$ transition that occurs at 253 nm in cyclohexane undergoes a red shift to 270 nm in ethanol. Following earlier arguments it might be expected that photoexcitation of benzophenone would lower its basicity, since S_1 (n,π^*) has a lower dipole moment than S_0 and consequently a lower electron density on the carbonyl oxygen atom (*128, 129*). However, the protonated state should also be considered. Benzophenone is half-protonated in 74% aqueous sulfuric acid (Table 3-12); the lowest energy absorption of the conjugate acid thus formed appears to be the intense $\pi-\pi^*$ band whose maximum is just below 350 nm and which extends into the visible. Thus, while the S_1 state of $(C_6H_5)_2C{=}O$ is n,π^*, the S_1 state of $(C_6H_5)_2C{=}OH^+$ is π,π^*. Most other aromatic carbonyl compounds also undergo red shifts of their lowest energy absorption and fluorescent bands on protonation, and as a result they, like benzophenone, are stronger bases in S_1 than in S_0 (*85, 110, 111*). They are frequently stronger still in T_1; the resulting order, p$K_{BH^+}(S_0)$ < p$K_{BH^+}(S_1)$ < p$K_{BH^+}(T_1)$, is unusual since with most other functional groups the acid and base strengths of the ground and triplet states are fairly close, with the lowest singlet state occupying an extreme position.

With regard to determining absolute values of $pK_{BH^+}(S_1)$ and $pK_{BH^+}(T_1)$ of carbonyl compounds by the Förster cycle, the uncertainty implicit in the method is compounded by uncertainty regarding the absolute values of $pK_{BH^+}(S_0)$ (Chapter 3). Nonetheless, it is clear that the excited states of some aromatic ketones, xanthone, for example, are protonated in the dilute aqueous region (110, 112). More recently, the triplet spectra of duroquinone (tetramethyl-p-benzoquinone) and its conjugate acid have been obtained by flash photolysis and the compound shown by a plot of optical density against pH to have a $pK_{BH^+}(T_1)$ of -0.1 (130). The same value can be obtained from a study of the quenching of duroquinone triplets by various substrates, and so the pK^* value seems firm. Our knowledge of the ground state pK_{BH^+} values of quinones is even less certain than that of ordinary ketones; it is clear, however, that the basicity of the triplet state of quinones is very much greater than that of their ground states.

B. Phototautomerism

The most common form of tautomerism in ground state molecules is prototropy, or proton migration. Photoexcitation can cause new prototropic equilibria to become established, which those without an aversion to jingles (131) might wish to call the photoprototropo phenomenon.

We have seen that light absorption can cause profound changes in the order of acid or base strengths of various functional groups. It is not surprising, therefore, that excited states may sometimes relocate their protons, always provided that the lifetime of the state is sufficiently long to permit such a process to occur. In the case of adjacent groups propinquity may provide a ready intramolecular pathway. Four cases of excited state tautomerism are considered here:

1. An early example of an intramolecular proton transfer was provided by Weller, who showed that salicylic acid forms the zwitterion in the excited state [Eq. (6-43)] (132). This is caused by the large increases in the acidity of the

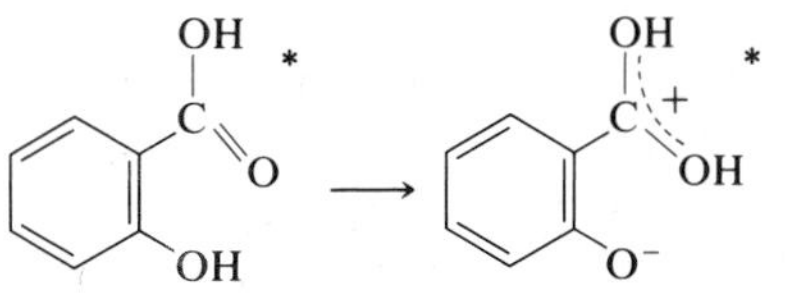

$$(6\text{-}43)$$

phenolic group and in the basicity of the carboxyl group that occur when the first excited singlet state is formed.

The methyl ester of salicylic acid behaves the same way as the parent compound (133), as does the amide (134), but the methyl ether, of course, does not. Nor do 3- and 4-hydroxybenzoic acids exhibit phototautomerism, possibly because the intermolecular route required in these cases is not sufficiently fast to allow equilibrium to take place during the short lifetimes of the S_1 states.

2. The tautomeric equilibrium in the ground state of azaphenols is more nearly balanced than is the case with the carboxyphenols already discussed. In 6-hydroxyquinoline, for example, the neutral form **18a** in Eq. (6-44) is favored over the zwitterion **19** by a factor of 10^2; in the S_1 state, however, it is the

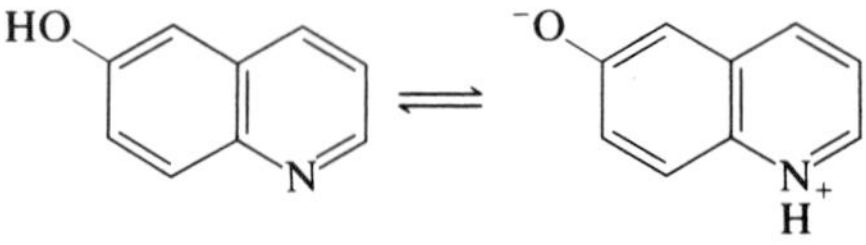

$$ \text{(6-44)} $$

18a   **19**

zwitterion that becomes favored, and by a very large factor (10^{12}) (*135*). This is the expected direction for the equilibrium to shift, in view of the increase in acidity of phenolic groups and the increase in basicity of quinolinic nitrogen atoms that are known to be brought about by excitation.

3. 3-Hydroxyflavone (**20**) sometimes called flavonol, has a bright green fluorescence. Indeed, its appearance after enzymatic hydrolysis of the nonfluorescent diphosphate ester has been used as a method of assay of phosphatase enzymes (*136, 137*). Wolfbeis *et al.* showed that in neutral aqueous or organic solvent the green fluorescence emanates from the tautomeric zwitterion **21** formed by an intramolecular proton transfer between the adjacent hydroxyl and carbonyl groups [Eq. (6-45)]. Again the

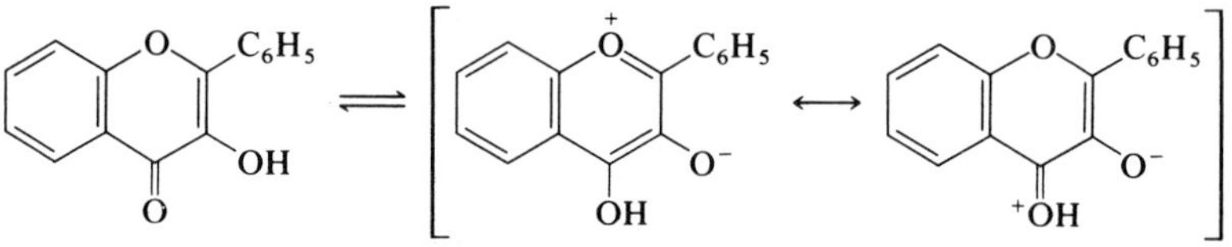

20   **21**

$$ \text{(6-45)} $$

strength of the acidic group and the strength of the basic group are increased by photoexcitation, allowing the reaction to take place.

4. 5-Aminoindazole (**22**) provides an interesting case of tautomerism in both cation and anion (*138*). In 0.1 M sulfuric acid almost all the substrate is present in the ammonio form **23** [Eq. (6-46)]. After absorbing light, however, the dominant species becomes the tautomeric cation **24**.

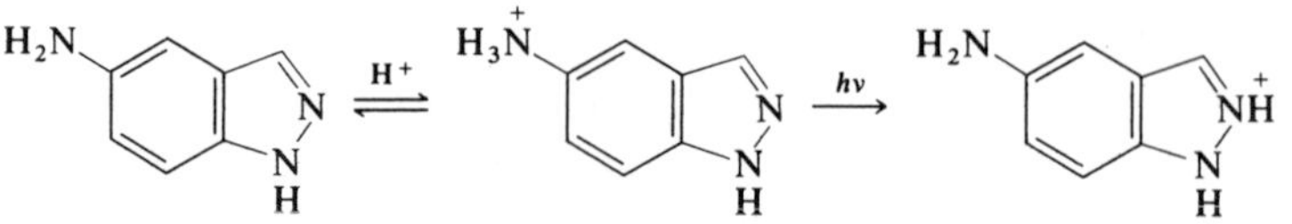

$$ \text{(6-46)} $$

22   **23**   **24**

In strongly basic aqueous solution the anion **25** is the dominant form, whereas in the exicted state under the same conditions the most stable monoanion is the tautomer **26** [Eq. (6-47)].

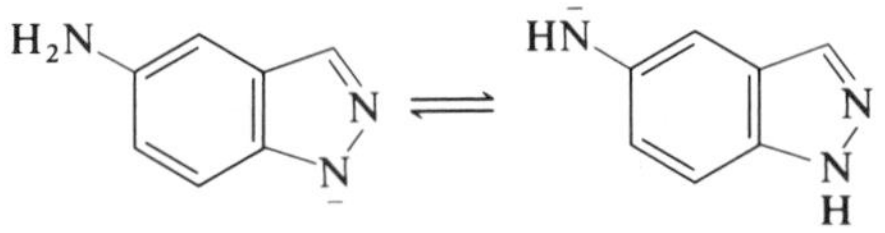

$$(6-47)$$

25   **26**

The $pK_{HA}(S_1)$ of **22** is 11.9, corresponding to proton loss from the amino group to form **26**. The ground state pK for the same reaction must be well above 20, since aniline has a pK_{HA} of 27.7 and since the five-membered diaza ring is so constructed as to be unable to offer direct resonance stabilization to the anionic nitrogen atom at position 5.

Other examples of ground and excited states having different sites of protonation have been given by Schulman (*124, 124a*).

C. Other Acid–Base Reactions of Excited States

It is possible to take advantage of a compound's having different pK values in ground and excited states to bring about abrupt changes in the pH of aqueous solutions. Thus, irradiation of 2-naphthol and its derivatives with an intense laser pulse can cause an almost instantaneous drop in pH of three to five units for an aqueous solution that is near neutrality (*142, 142a*). The maximum proton concentration is built up during the laser pulse and decays to the initial level within a few microseconds. Similarly, laser irradiation of certain triaryl carbinols can bring about comparable *increases* in pH (*143*). These reactions [Eq. (6-48) and (6-48a)] are convenient means of perturbing

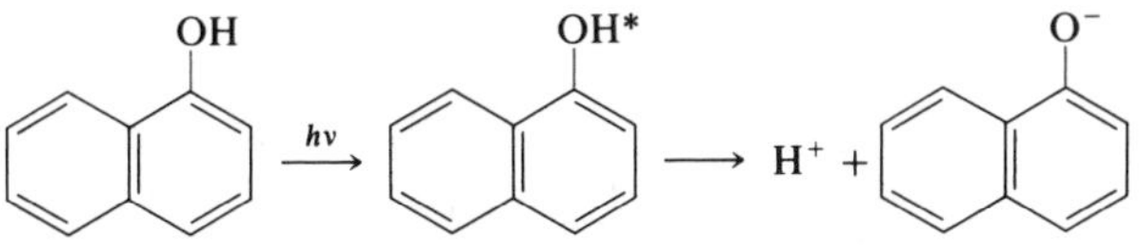

$$(6-48)$$

$$Ar_3COH \xrightarrow{h\nu} Ar_3COH^* \longrightarrow HO^- + Ar_3C^+ \qquad (6\text{-}48a)$$

chemical and biochemical reactions via brief, intense pulses of protons or hydroxide ions. Laser pulses have also been used to generate unstable species (e.g., enols), whose acid strengths can then be measured (Chapter 2, Section II,F) (*143a*).

The polarity of carbon–carbon double and triple bonds generally rises on photoexcitation, with the result that proton addition to one or the other carbon atom is facilitated. Little is known about the pK_{BH^+} values of most alkenes and alkynes but, judging from some of the catalytic effects that are observed, there must be a very large increase in basicity on forming the S_1 states. Furthermore, in those cases where proton addition to alkenes gives rise to detectable amounts of carbonium ion, large red shifts are observed, indicating a substantial increase in basicity of the alkene in the excited state.

Many of the early acid-catalyzed photochemical additions to alkenes were carried out using photosensitizers and hence involved the triplet state. More recently direct irradiation of alkenes and alkynes has been shown to produce excited singlet states that readily undergo acid-catalyzed addition reactions. In some cases of acid catalysis the singlet and triplet states lead to the same intermediates and products, whereas in others (e.g., aliphatic acyclic alkenes) triplet sensitization is ineffective and only direct irradiation leads to product. Furthermore, those reactions that are photosensitized may lead to side reactions as a result of radical intermediates being formed. Full accounts of the development of the field of photochemical acid catalysis and its application to organic chemistry can be found in the articles of Marshall (*144*), Wubbels (*145*), Kropp (*146*), and Yates (*147*).

The acid-catalyzed photoaddition of water or alcohols to six-, seven-, or eight-membered cycloalkenes appears to involve the production of a highly strained *trans*-cycloalkene from either the triplet or singlet excited state. Subsequent rapid protonation gives the carbonium ion and then the alcohol, ether, or alkene. The evidence is strong that *trans*-alkenes are, indeed, intermediates in these processes (*148–150*), which are illustrated in Eq. (6-49) using the reaction of 1-methylcyclohexene in aqueous methanol as an example.

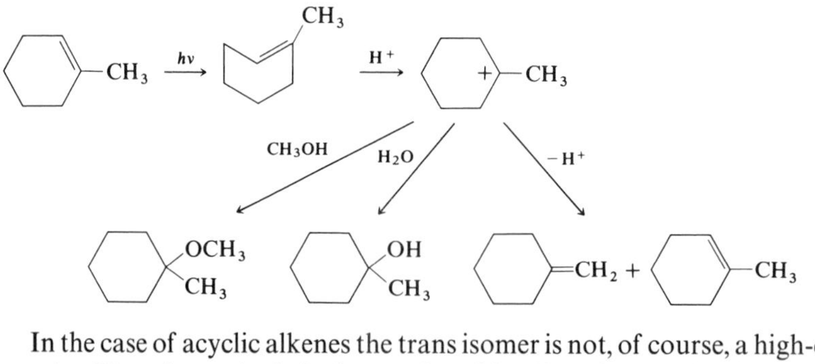

In the case of acyclic alkenes the trans isomer is not, of course, a high-energy species. Here the excited singlet produced by direct irradiation is believed by some to be a dipolar Rydberg state, which is able to trap nucleophiles (*146*). When the double bond is in a ring having fewer than six carbon atoms or when it is part of a rigid polycyclic system, the *trans*-alkene cannot form. In these cases, too, photosensitization is ineffective, and again a dipolar excited singlet state appears to be involved (*151*).

Yates *et al.* examined the effect of pH (or H_0) on the rate of the photohydration of a number of arylalkenes (*152, 153*). They found that irradiating styrene and most of its derivatives in water in the dilute acid region (pH 1–7) resulted in rather inefficient conversion to the corresponding 1-arylethanol. Adding sulfuric or other strong acid to the system to produce solutions more acidic than $H_0 \approx 0$ causes a sharp increase in quantum yield and an efficient conversion to alcohol. The observation that general acid catalysis is operative (Chapter 7) suggests a mechanism involving rate-limiting protonation of the S_1 state of the alkene [Eq. (6-50)].

$$ArHC{=}CH_2 \xrightarrow{h\nu} ArHC{=}CH_2{*} \xrightarrow{H^+} Ar\overset{+}{C}HCH_3 \xrightarrow{fast}$$

$$ArCHOHCH_3 (+ArHC{=}CH_2) \quad (6\text{-}50)$$

The 3- and 4-nitro derivatives of styrene are unusual in forming anti-Markownikoff photohydration products and in showing no indication of acid catalysis. It is believed that the reactive intermediate here is the T_1 triplet state, which has a lower proton affinity than the S_1 singlet.

Simple arylallenes also undergo acid-catalyzed photohydration via S_1 states and yield the corresponding cinnamyl alcohols (*154*). On the other hand, 3- and 4-nitro derivatives give nitrophenylacetones and appear to react via T_1 (*154, 154a*).

Many arylalkynes are subject to facile acid-catalyzed photohydration via a route that is analogous to that of arylalkenes (*147*). The products are substituted acetophenones [Eq. (6-51)] formed via a bent, highly polar S_1 state

$$ArC{\equiv}CH \xrightarrow{h\nu} ArC{\equiv}CH{*} \xrightarrow{H^+} Ar\overset{+}{C}{=}CH_2 \xrightarrow{fast} ArCOH{=}CH_2 \longrightarrow ArCOCH_3$$

$$(6\text{-}51)$$

(**27**). Again the 3- and 4-nitro derivatives are anomalous in giving the anti-Markownikoff products, the arylacetaldehydes. Unlike the nitroarylalkene case referred to earlier these reactions were found to be subject to acid catalysis. Here T_1, which sensitization and quenching experiments show to be the reactive state, has enough polarity (in the reverse sense to that of **27**) to

$$\underset{\textbf{27}}{\overset{\displaystyle Ar\diagdown}{\underset{\displaystyle H}{\overset{\delta+\;\;\;\delta-}{C{:\!:}C}\diagup}}}$$

react with H_3O^+ to form the hydration products, although it is believed that in these cases discrete vinyl cations are not formed.

2-Hydroxyphenylacetylene and 2-hydroxystyrene are interesting in that they readily undergo photohydration in neutral solution, giving the corresponding acetophenone and 1-arylethanol. In these cases it is clear from the evidence that there is an intramolecular proton transfer from the o-hydroxyl group to the β-carbon atom in the S_1 states of these molecules (155).

Other photochemical processes that involve acid catalysis or proton transfer in the excited state include the conversion of 4-nitrobenzyl alcohol to 4-nitrosobenzaldehyde (156), the conversion of nitroarenes and hydrochloric acid to trichloroanilines (157), photoinduced electron transfer (158), photochemical additions to benzene (159), the photochemical Wallach rearrangement (160), protium–deuterium exchange in arenes (161), cyclizations (162–165), and various other processes (166).

IV. Tetrahedral Intermediates

Acyl substitution reactions of carbonyl compounds generally take place via tetrahedral intermediates formed by addition of nucleophile to the carbonyl group as, for example, in the case of ester interchange [Eq. (6-52)].

$$
\text{R—CO}_2\text{CH}_3 + \text{C}_2\text{H}_5\text{OH} \;\rightleftharpoons\; \text{R}\!-\!\overset{\displaystyle \text{OH}}{\underset{\displaystyle \text{OC}_2\text{H}_5}{\overset{|}{\underset{|}{\text{C}}}}}\!-\!\text{OCH}_3 \;\rightleftharpoons\; \text{RCO}_2\text{C}_2\text{H}_5 + \text{CH}_3\text{OH} \quad (6\text{-}52)
$$

28

Intermediate **28** is a hemiortho ester, and there has been much interest in the properties of this and analogous species formed in other carbonyl addition reactions (167–169). They are rarely isolable, and so their acid strengths must be determined indirectly. Useful models in this respect are aldehyde and ketone carbonyl hydrates, some acidities of which are listed in Table 2-15; most of these compounds contain strongly electron-withdrawing groups, which stabilize the tetrahedral form relative to the carbonyl form.

It is sometimes possible to determine the acidity constant of a tetrahedral species by relating it to the equilibria that link the carbonyl form and its alkoxide and alcohol adducts. This approach, which has been used by Hine (170) and by McClelland (141, 141a), is most readily applied when adduct formation is intramolecular, as in Eq. (6-53).

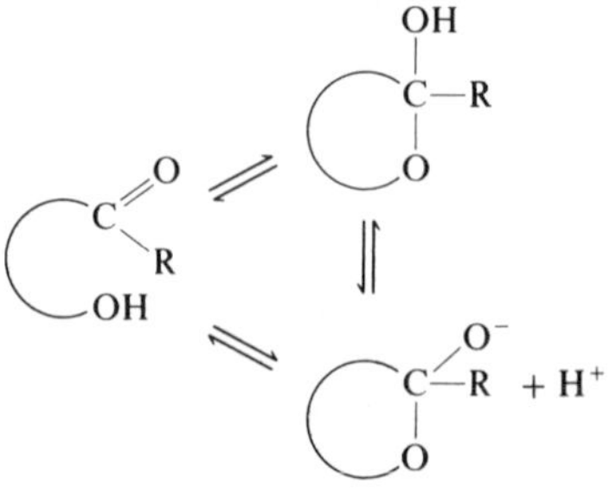

$$(6\text{-}53)$$

Guthrie used calorimetric values to determine equilibrium constants for tetrahedral intermediate formation and estimated the relevant pK_{HA} values using substituent effects (139, 171, 179). Jencks also used substituent effects to estimate the changes in acidity and basicity that occur at reactive sites during acid- and base-catalyzed reactions of carbonyl compounds (169, 172–175).

Table 6-4 lists pK_{HA} and pK_{BH^+} values that have been estimated by various means for a number of unstable tetrahedral species.

Table 6-4

ACIDITY AND BASICITY CONSTANTS OF
TETRAHEDRAL INTERMEDIATES[a,b]

Molecule	pK_{HA} or pK_{BH^+}	Ref.
$\begin{array}{c} \text{OH} \\ \vert \\ H_3C-C-N(CH_3)_2 \\ \vert \\ {}^+OH_2 \end{array}$	-2.5	139
$\begin{array}{c} \text{O}\underline{\text{H}} \\ \vert \\ H_3C-C-N(CH_3)_2 \\ \vert \\ {}^+OH_2 \end{array}$	9.5	139
$\begin{array}{c} \text{OH} \\ \vert \\ H_3C-C-OCH_3 \\ \vert \\ {}^+OH_2 \end{array}$	-3.5	139
$\begin{array}{c} \text{O}\underline{\text{H}} \\ \vert \\ H_3C-C-OCH_3 \\ \vert \\ {}^+OH_2 \end{array}$	7.9	139
$\begin{array}{c} \text{OH} \\ \vert \\ H_3C-C-SC_2H_5 \\ \vert \\ {}^+OH_2 \end{array}$	-3.3	139
$\begin{array}{c} \text{O}\underline{\text{H}} \\ \vert \\ H_3C-C-SC_2H_5 \\ \vert \\ {}^+OH_2 \end{array}$	8.3	139
$\begin{array}{c} \text{O}\underline{\text{H}} \\ \vert \\ H_3C-C-N(CH_3)_2 \\ \vert \\ \text{O}\underline{\text{H}} \end{array}$	13.4	139

(cont.)

Table 6-4 (*cont.*)

Molecule	pK_{HA} or pK_{BH^+}	Ref.
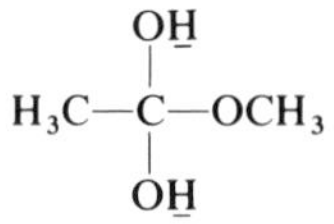 $H_3C-\overset{\underset{\displaystyle OH}{\displaystyle OH}}{\underset{}{C}}-OCH_3$	11.8	*139*
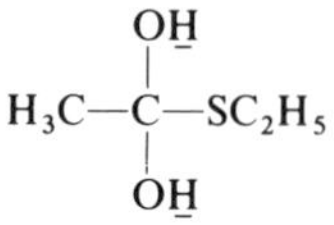 $H_3C-\overset{\underset{\displaystyle OH}{\displaystyle OH}}{\underset{}{C}}-SC_2H_5$	12.1	*139*
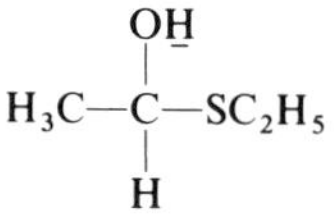 $H_3C-\overset{\underset{\displaystyle H}{\displaystyle OH}}{\underset{}{C}}-SC_2H_5$	14.3	*139*
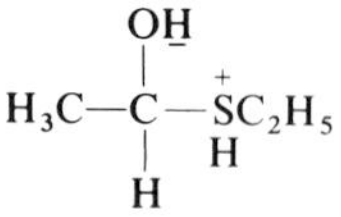 $H_3C-\overset{\underset{\displaystyle H}{\displaystyle OH}}{\underset{}{C}}-\overset{+}{S}C_2H_5$ (H on S)	6.6	*139*
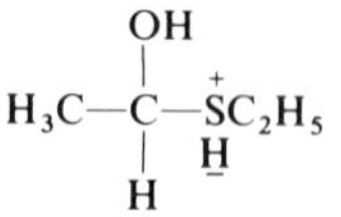 $H_3C-\overset{\underset{\displaystyle H}{\displaystyle OH}}{\underset{}{C}}-\overset{+}{S}C_2H_5$ (H)	−7.9	*139*
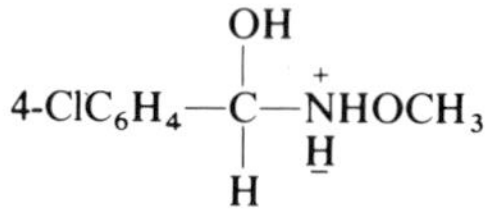 $4\text{-ClC}_6H_4-\overset{\underset{\displaystyle H}{\displaystyle OH}}{\underset{}{C}}-\overset{+}{N}HOCH_3$ (H)	2.0	*140*
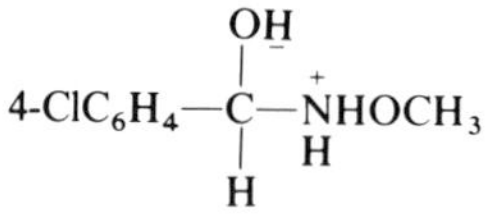 $4\text{-ClC}_6H_4-\overset{\underset{\displaystyle H}{\displaystyle OH}}{\underset{}{C}}-\overset{+}{N}HOCH_3$ (H)	8.5	*140*
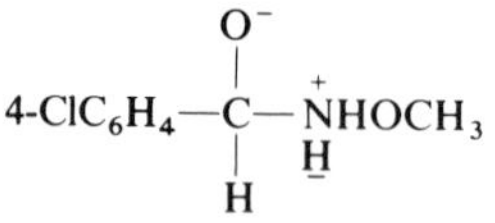 $4\text{-ClC}_6H_4-\overset{\underset{\displaystyle H}{\displaystyle O^-}}{\underset{}{C}}-\overset{+}{N}HOCH_3$ (H)	7.3	*140*
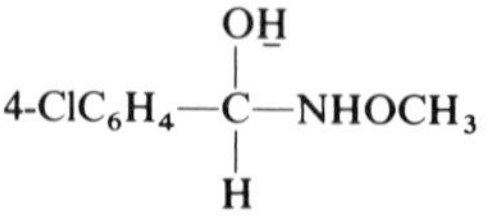 $4\text{-ClC}_6H_4-\overset{\underset{\displaystyle H}{\displaystyle OH}}{\underset{}{C}}-NHOCH_3$	13.8	*140*

Table 6-4 (*cont.*)

Molecule	pK_{HA} or pK_{BH^+}	Ref.
	11.1	*141*
	10.4	*141a*
	11.3	*178*
	7.5	*178*
	11.8	*178*
	11.5	*178*
	12.1	*178*

[a] Dissociating proton is underlined.
[b] Carbonyl hydrates are found in Table 2-15.

References

1. R. W. Fessenden and R. H. Schuler, *in* "Advances in Radiation Chemistry" (M. Burton and J. L. Magee, eds.), p. 1. Wiley (Interscience), New York, 1970.

2. E. Hayon and M. Simic, *Acc. Chem. Res.* **7,** 114 (1974).

3. J. Rabani and M. S. Matheson, *J. Phys. Chem.* **70**, 761 (1966).

4. G. V. Buxton, *Trans, Faraday Soc.* **65**, 2150 (1969).

5. M. S. Matheson and J. Rabani, *Science* **146**, 427 (1964).

6. See also ref. *2, 6a,* and *7b.*

6a. G. Eberlein and T. C. Bruice, *J. Am. Chem. Soc.* **105**, 6685 (1983).

7. B. H. J. Bielski, *Photochem. Photobiol.* **28**, 645 (1978).

7a. M. Sugawara, M. M. Baizer, W. T. Monte, R. D. Little, and U. Hess, *Acta Chem. Scand., Ser. B* **37B**, 509 (1983).

7b. P. M. Allen, U. Hess, C. S. Foote, and M. M. Baizer, *Synth. Commun.* **12**, 123 (1982).

8. K. D. Asmus, A. Henglein, A. Wigger, and G. Beck, *Ber. Bunsenges Phys. Chem.* **70**, 756 (1966).

9. G. P. Laroff and R. W. Fessenden, *J. Phys. Chem.* **77**, 1283 (1973).

10. E. J. Hart and M. Anbar "The Solvated Electron." Wiley (Interscience), New York, 1970.

11. M. Z. Hoffman and E. Hayon, *J. Phys. Chem.* **77**, 990 (1973).

12. P. Neta, M. Simic, and E. Hayon, *J. Phys. Chem.* **73**, 4207 (1969).

13. M. Simic and M. Z. Hoffman, *J. Phys. Chem.* **76**, 1398 (1972).

14. D. D. Perrin, B. Dempsey, and E. P. Serjeant, "pK_{HA} Prediction for Organic Acids and Bases." Chapman and Hall, London, 1981.

15. P. B. Ayscough and R. Wilson, *J. Chem. Soc.* p. 5412 (1963).

16. N. Hirota, *in* "Radical Ions" (E. T. Kaiser and L. Kevan, eds.), Chap. 2. Wiley (Interscience), New York, 1968.

17. R. L. Willson, *Chem. Commun.* p. 1249 (1971).

18. E. J. Land and A. J. Swallow, *J. Biol. Chem.* **245**, 1890 (1970).

19. G. E. Adams, *Trans. Farady Soc.* **63**, 1175 (1967).

20. P. S. Rao and E. Hayon, *J. Phys. Chem.* **77**, 2274 (1973).

21. R. W. Fessenden and P. Neta, *J. Phys. Chem.* **76**, 2857 (1972).

22. R. H. Hinman, *J. Org. Chem.* **23**, 1587 (1958).

23. E. Hayon and M. Simic, *J. Am. Chem. Soc.* **94**, 42 (1972).

23a. See, however, S. F. Nelson, W. P. Parmelee, M. Göbl, K.-O. Hiller, D. Veltwisch, and K.-D. Asmus, *J. Am. Chem. Soc.* **102**, 5606 (1980) for an example of the opposite effect.

24. M. Simic and E. Hayon, *J. Am. Chem. Soc.* **93**, 5982 (1971).

24a. See also S. F. Nelson, J. M. Buschek, M. Göbl, and K.-D. Asmus, *J. Chem. Soc., Perkin Trans. 2* p. 11 (1984).

25. J. Q. Adams and J. R. Thomas, *J. Chem. Phys.* **39**, 1904 (1963).

26. H. R. Falle, *Can. J. Chem.* **46**, 1703 (1968).

27. J. W. Linnett, "The Electronic Structure of Molecules." Methuen, London, 1964.

28. W. B. Jensen, *Can. J. Chem.* **59**, 807 (1981) and refs. therein.

29. C. Würster and R. Sendtner, *Chem. Ber.* **12**, 1803 (1879).

30. L. Michaelis, *Chem. Rev.* **16**, 243 (1935).

31. R. Stewart, "Oxidation Mechanisms: Applications to Organic Chemistry." Benjamin, New York, 1964.

32. W. T. Dixon and D. Murphy, *J. Chem. Soc., Faraday Trans. 2* **74**, 432 (1978).

33. D. M. Holton and D. Murphy, *J. Chem. Soc., Faraday Trans. 2* **75**, 1637 (1979).

34. A. M. de P. Nicholas and D. R. Arnold, *Can. J. Chem.* **60**, 2165 (1982); see also C. J. Schlesener, C. Amatore, and J. K. Kochi, *J. Am. Chem. Soc.* **106**, 7472 (1984).

35. J. Sinclair, *J. Chem. Phys.* **55**, 245 (1971).

36. C. Iacona, J. P. Michaut, and J. Roucin, *J. Chem. Phys.* **67**, 5658 (1977).

37. C. J. Smith, C. P. Poole, and H. A. Farach, *J. Chem. Phys.* **74**, 993 (1981).

38. P. Smith, W. M. Fox, D. J. McGinty, and R. D. Stevens, *Can. J. Chem.* **48**, 480 (1970).

39. P. Neta, M. Simic, and E. Hayon, *J. Phys. Chem.* **76**, 3507 (1972).

40. A. Hedberg and A. Ehrenberg, *J. Chem. Phys.* **48**, 4822 (1968).

41. H. Taniguchi, K. Fukui, D. Ohnishi, H. Hatano, H. Hasegawa, and T. Maruyama, *J. Chem. Phys.* **72**, 1926 (1968).

42. Calculated using estimates of the glycine tautomeric constant given in refs. *43* and *44*. See also Chapter 5, Section IV.

43. P. Haberfield, *J. Chem. Educ.* **57**, 346 (1980).

44. J. T. Edsall and M. H. Blanchard, *J. Am. Chem. Soc.* **55**, 2337 (1933).

45. R. W. Baldock, P. Hudson, A. R. Katritzky, and F. Soti, *J. Chem. Soc., Perkin Trans. 1* p. 1422 (1974).

46. See W. J. Leigh, D. R. Arnold, R. W. R. Humphreys, and P. C. Wong, *Can. J. Chem.* **58**, 2537 (1980) and refs. therein.

47. A. T. Balaban, *Rev. Roum. Chim.* **16**, 725 (1971).

48. A. T. Balaban, M. T. Caproiu, N. Negoita, and R. Baican, *Tetrahedron* **33**, 2249 (1977).

49. L. Stella, Z. Janousek, R. Merenyi, and H. G. Viehe, *Angew. Chem. Int. Ed. Engl.* **17**, 691 (1978).

50. M. Simic, P. Neta, and E. Hayon, *J. Phys. Chem.* **73**, 4214 (1969).

51. "Oxygen and Oxy-Radicals in Chemistry and Biology" (M. A. J. Rodgers and E. L. Powers, eds.), Academic Press, New York, 1981.

52. W. A. Pryor, "Free Radicals in Biology," Vols. I–V. Academic Press, New York, 1976–1982.

53. L. K. Obukhova and N. M. Emmanuel, *Russ. Chem. Rev.* **52**, 201 (1983).

54. M. A. Schuler, K. Bhatia, and R. H. Schuler, *J. Phys. Chem.* **78**, 1063 (1974).

55. D. T. Sawyer, G. Chiericato, and T. Tsuchiya, *J. Am. Chem. Soc.* **104**, 6273 (1982).

56. D. Weir, D. A. Hutchinson, J. Russell, and J. K. S. Wan, *Can. J. Chem.* **60**, 703 (1982).

57. H. M. Swartz and N. J. F. Dodd in ref. *51*, p. 161.

58. G. P. Laroff, R. W. Fessenden, and R. H. Schuler, *J. Am. Chem. Soc.* **94**, 9062 (1972).

59. Y. Kirino and R. H. Schuler, *J. Am. Chem. Soc.* **95**, 6926 (1973).

60. R. H. Boyd, *J. Am. Chem. Soc.* **83**, 4288 (1961).

61. K. B. Patel and R. L. Willson, *J. Chem. Soc., Faraday Trans. 1* **69**, 814 (1973).

62. E. J. Land and A. J. Swallow, *J. Biol. Chem.* **245**, 1890 (1970).

63. E. J. Land and A. J. Swallow, *Biochemistry* **8**, 2117 (1969).

64. R. D. Draper and L. L. Ingraham, *Arch. Biochem. Biophys.* **125**, 802 (1968).

65. G. Eberlein and T. C. Bruice, *J. Am. Chem. Soc.* **105**, 6679 (1983).

66. W. M. Clark, "Oxidation Reduction Potentials of Organic Systems." R. E. Creiger, Huntington, New York, 1972.

67. P. Hemmerich, *Prog. Chem. Org. Nat. Prod.* **33**, 29 (1976).

68. P. S. Rao and E. Hayon, *J. Am. Chem. Soc.* **96**, 1295 (1974).

69. E. Hayon, *J. Chem. Phys.* **51**, 4881 (1969).

70. A. Grimison and M. Eberhardt, *J. Phys. Chem.* **77**, 1673 (1973).

71. P. J. Elving, *Can. J. Chem.* **55**, 3392 (1977).

72. T. E. Cummings and P. J. Elving, *J. Electroanal. Chem.* **94**, 123 (1978).

73. D. Bérubé and J. Lessard, *Can. J. Chem.* **60**, 1127 (1982).

74. E. G. Janzen and J. L. Gerlock, *J. Organomet. Chem.* **8**, 354 (1967).

75. J. J. Eisch and W. C. Kaska, *Chem. Ind.* p. 470 (1961).

76. J. Y. Pape and J. Simonet, *Electrochem. Acta* **23**, 445 (1978).

77. J. Lilie and A. Henglein, *Ber. Bunsenges. Phys. Chem.* **73**, 170 (1969).

78. A. Beckett, A. D. Osborne, and G. Porter, *Trans, Faraday Soc.* **60**, 873 (1964).

79. E. Hayon, T. Ibata, N. N. Lichtin, and M. Simic, *J. Phys. Chem.* **76**, 2072 (1972).

80. A. Beckett and G. Porter, *Trans. Faraday Soc.* **50**, 2038 (1963).

81. G. E. Adams and R. L. Willson, *J. Chem. Soc. Faraday Trans. 1* **69**, 719 (1973).

82. E. Hayon, N. N. Lichtin, and V. Madhavan, *J. Am. Chem. Soc.* **95**, 4762 (1973).

83. K. D. Asmus, A. Wigger, and A. Henglein, *Ber. Bunsenges Phys. Chem.* **70,** 862 (1966).

84. T. Förster, *Naturwissenschaften* **36,** 186 (1949).

85. J. F. Ireland and P. A. H. Wyatt, *in* "Advances in Physical Organic Chemistry" (V. Gold and D. Bethell, eds.), Vol. 12, p. 132. Academic Press London, 1976.

86. A. Weller, *Prog. React. Kinet.* **1,** 189 (1961).

87. E. Vander Donckt, *Prog. React. Kinet.* **5,** 273 (1970).

88. E. L. Wehry and L. B. Rogers, *J. Am. Chem. Soc.* **87,** 4234 (1965).

89. A. Weller, *Z. Elektrochem.* **56,** 662 (1952).

90. S. G. Schulman, L. B. Sanders, and J. D. Winefordner, *Photochem. Photobiol.* **13,** 381 (1971).

91. G. Jackson and G. Porter, *Proc. R. Soc. London, Ser. A* **260A,** 13 (1961).

92. E. Vander Donckt and G. Porter, *Trans. Faraday Soc.* **64,** 3215, 3218 (1968).

93. E. L. Wehry and L. B. Rodgers, *J. Am. Chem. Soc.* **88,** 351 (1966).

94. P. J. Kovi and S. G. Schulman, *Anal. Chim. Acta* **63,** 39 (1973).

95. E. Vander Donckt, R. Draimaix, J. Nasielski, and C. Vogels, *Trans. Faraday Soc.* **65,** 3258 (1969).

96. J. J. Aaron and J. D. Winefordner, *Photochem. Photobiol.* **18,** 97 (1973).

97. Chapter 2, Table 20.

98. J. P. Idoux and C. K. Hancock, *J. Org. Chem.* **33,** 3498 (1968).

99. J. W. Bridges, D. S. Davies, and R. T. Williams, *Biochem. J.* **98,** 451 (1966).

100. R. H. Ellerhorst and H. H. Jaffé, *J. Org. Chem.* **33,** 4115 (1968).

101. R. Rotkiewicz and Z. R. Grabowski, *Trans. Faraday Soc.* **65,** 3263 (1969).

102. S. Hagopian and L. A. Singer, *J. Am. Chem. Soc.* **105,** 6760 (1983).

103. S. G. Schulman and A. C. Capomacchia, *J. Am. Chem. Soc.* **95,** 2763 (1973).

104. A Grabowska and B. Pakula, *Photochem. Photobiol.* **9,** 339 (1969).

105. E. Bouwhuis and M. J. Janssen, *Tetrahedron Lett.* p. 233 (1972).

106. G. Favoro, U. Mazzucato, and F. Masetti, *J. Phys. Chem.* **77,** 601 (1973).

107. M. J. Robey and I. G. Ross, *Photochem. Photobiol.* **21,** 363 (1975).

108. Y. Nishida, K. Kikuchi, and H. Kokobun, *J. Photochem* **13,** 75 (1980).

109. A. K. Mishra and S. K. Dogra, *J. Photochem.* **23,** 163 (1983).

110. J. F. Ireland and P. A. H. Wyatt, *J. Chem. Soc., Faraday Trans. 1* **69,** 161 (1973).

111. A. C. Hopkinson and P. A. H. Wyatt, *J. Chem. Soc. B* p. 1333 (1967).

112. J. F. Ireland and P. A. H. Wyatt, *J. Chem. Soc. Faraday Trans. 1* **68,** 1053 (1972).

113. M. Swaminathan and S. K. Dogra, *J. Photochem.* **21,** 245 (1983).

114. Ref. *85,* p. 140; see also ref. *110.*

115. A. C. Capomacchia and S. G. Schulman, *J. Phys. Chem.* **79,** 1337 (1975).

116. See also R. Bonneau, T. Pereyre, and J. Joussot-Dubien, *Mol. Photochem.* **6,** 245 (1974).

117. O. Exner, *in* "Correlation Analysis in Chemistry" (N. B. Chapman and J. Shorter, eds.), p. 451. Plenum, New York, 1978.

118. Ref. *85,* p. 207.

119. N. J. Turro, "Modern Molecular Photochemistry," p. 219 and p. 229. Benjamin/Cummings, Menlo Park, 1978.

120. N. E. Burlinson, M. E. Sitzman, L. A. Kaplan, and E. Kayser, *J. Org. Chem.* **44,** 3695 (1979).

121. See for example, ref. *105.*

122. R. M. Hochstrasser and J. W. Michaluk, *J. Chem. Phys.* **55,** 4668 (1971).

123. I. G. Ross, *in* "Photochemistry of Heterocyclic Compounds" (O. Buchardt, ed.), Chap. 1. Wiley, New York, 1976.

123a. See also P. S. Mariano, *Tetrahedron* **39,** 3845 (1983).

124. S. G. Schulman, *in* "Physical Methods in Heterocyclic Chemistry" (A. R. Katritzky, ed.), Vol. VI, Chap. 4. Academic Press, New York, 1974.

124a. See also F. T. Vert, I. Z. Sanchez, and A. O. Torrent, *J. Photochem.* **23**, 355 (1983).

125. N. G. Bakhshiev, M. I. Knyazhanskii, V. I. Minkin, O. I. Osipov, and G. V. Saidov, *Russ. Chem. Rev.* **38**, 740 (1969).

126. Ref. *85*, p. 137.

127. Ref. *119*, p. 106.

128. P. H. Gore, J. A. Hoskins, R. J. W. LeFevre, L. Radom, and G. L. D. Ritchie, *J. Chem. Soc. B* p. 741 (1967).

129. R. M. Hochstrasser and J. L. Noe, *J. Mol. Spectrosc.* **38**, 175 (1971).

130. J. C. Scaiano and P. Neta, *J. Am. Chem. Soc.* **102**, 1608 (1980).

131. H. W. Fowler, "Modern English Usage," p. 317. Oxford Univ. Press, London, 1965.

132. A. Weller, *Z. Electrochem.* **60**, 1144 (1956).

133. P. J. Kovi, C. L. Miller, and S. G. Schulman, *Anal. Chim. Acta* **61**, 7 (1972).

134. S. G. Schulman, P. J. Kovi, and J. F. Young, *J. Pharm. Sci.* **62**, 1197 (1973).

135. S. F. Mason, J. Philp, and B. E. Smith, *J. Chem. Soc. A* p. 3051 (1968); see also M. Itoh, T. Adachi, and K. Tokumura, *J. Am. Chem. Soc.* **106**, 850 (1984).

136. D. B. Land and E. Jackim, *Anal. Biochem.* **16**, 481 (1966).

137. O. S. Wolfbeis, A. Knierzinger, and R. Schipfer, *J. Photochem.* **21**, 67 (1983).

138. M. Swaminathan and S. K. Dogra, *J. Am. Chem. Soc.* **105**, 6223 (1983); see also M. Swaminathan and S. K. Dogra, *J. Chem. Soc., Perkin Trans. 2* p. 1641 (1983).

139. J. P. Guthrie, *J. Am. Chem. Soc.* **100**, 5892 (1978).

140. S. Rosenberg, S. M. Silver, J. M. Sayer, and W. P. Jencks, *J. Am. Chem. Soc.* **96**, 7986 (1974).

141. O. S. Tee, M. Trani, R. A. McClelland, and N. E. Seaman, *J. Am. Chem. Soc.* **104**, 7219 (1982).

141a. R. A. McClelland, N. E. Seaman, and D. Cramm, *J. Am. Chem. Soc.* **106**, 4511 (1984).

142. M. Gutman, D. Huppert, and E. Pines, *J. Am. Chem. Soc.* **103**, 3709 (1981).

142a. See also M. J. Politi and J. H. Fendler, *J. Am. Chem. Soc.* **106**, 265 (1984) and refs. therein.

143. M. Irie, *J. Am. Chem. Soc.* **105**, 2078 (1983).

143a. Y. Chiang, A. J. Kresge, Y. S. Tang, and J. Wirz, *J. Am. Chem. Soc.* **106**, 460 (1984).

144. J. A. Marshall, *Acc. Chem. Res.* **2**, 33 (1969).

145. G. G. Wubbels and D. W. Celander, *J. Am. Chem. Soc.* **103**, 7669 (1981); **104**, 2677 (1982) and refs. therein.

146. P. J. Kropp, *Org. Photochem.* **4**, 1 (1979) and refs. therein.

147. P. Wan and K. Yates, *Rev. Chem. Intermed.* **5**, 157 (1984) and refs. therein.

148. W. G. Dauben, H. C. H. A. van Riel, J. D. Robbins, and G. J. Wagner, *J. Am. Chem. Soc.* **101**, 6383 (1979).

149. R. Bonneau, J. Joussot-Dubien, L. Salem, and A. J. Yarwood, *J. Am. Chem. Soc.* **98**, 4329 (1976).

150. P. J. Kropp and H. J. Krause, *J. Am. Chem. Soc.* **89**, 5199 (1967).

151. P. J. Kropp, *J. Am. Chem. Soc.* **95**, 4611 (1973).

152. P. Wan, S. Culshaw, and K. Yates, *J. Am. Chem. Soc.* **104**, 2509 (1982).

153. P. Wan and K. Yates, *J. Org. Chem.* **48**, 869 (1983).

154. K. Rafizadeh and K. Yates, *J. Org. Chem.* **49**, 1500 (1984); see also M. W. Klett and R. P. Johnson, *Tetrahedron Lett.* **24**, 1107 (1983).

154a. See K. Fujita, K. Matsui, and T. Shono, *J. Am. Chem. Soc.* **97**, 6256 (1975) for similar reactions in acetic acid.

155. M. Isaks, K. Yates, and P. Kalenderopoulos, *J. Am. Chem. Soc.* **106**, 2728 (1984).

156. P. Wan and K. Yates, *J. Org. Chem.* **48**, 136 (1983).

157. G. G. Wubbels and R. L. Letsinger, *J. Am. Chem. Soc.* **96**, 6698 (1974).

158. M. J. Thomas and P. J. Wagner, *J. Am. Chem. Soc.* **99**, 3845 (1977).

159. D. Bryce-Smith and A. Gilbert, *Tetrahedron* **33**, 2459 (1977).

160. R. H. Squire and H. H. Jaffé, *J. Am. Chem. Soc.* **95**, 8188 (1973).

161. D. A. de Bie and E. Havinga, *Tetrahedron* **21**, 2359 (1965).

162. K. Ichimura and S. Watanabe, *Bull. Chem. Soc. Jpn.* **49**, 2224 (1976).

163. G. Frater and H. Schmid, *Helv. Chim. Acta* **50**, 255 (1967).

164. P. Mariano and A. A. Leone, *J. Am. Chem. Soc.* **101**, 3607 (1979).

165. R. Lapouyade, R. Koussini, and H. Bouas-Laurent, *J. Am. Chem. Soc.* **99**, 7374 (1977).

166. R. F. Childs, *Rev. Chem. Intermed.* **3**, 285 (1980).

167. R. A. McClelland and L. J. Santry, *Acc. Chem. Res.* **16**, 394 (1983).

168. B. Capon, A. K. Ghosh, D. M. A. Grieve, *Acc. Chem. Res.* **14**, 306 (1981).

169. W. P. Jencks, *Acc. Chem. Res.* **9**, 425 (1976).

170. J. Hine and G. F. Koser, *J. Org. Chem.* **36**, 1348 (1971).

171. J. P. Guthrie and P. A. Cullimore, *Can. J. Chem.* **58**, 1281 (1980).

172. J. P. Fox and W. P. Jencks, *J. Am. Chem. Soc.* **96**, 1436 (1974).

173. N. Gravitz and W. P. Jencks, *J. Am. Chem. Soc.* **96**, 499 (1974).

174. A. C. Satterthwait and W. P. Jencks, *J. Am. Chem. Soc.* **96**, 7018 and 7031 (1974).

175. See also refs. *176* and *177*.

176. T. Okuyama, T. C. Pletcher, D. J. Sahn, and G. L. Schmir, *J. Am. Chem. Soc.* **95**, 1253 (1973).

177. P. Deslongchamps and R. J. Taillefer, *Can. J. Chem.* **53**, 3029 (1975).

178. B. Capon and M. de N. de M. Sanchez, *Tetrahedron* **39**, 4143 (1983).

179. J. P. Guthrie, *J. Am. Chem. Soc.* **102**, 5286 (1980).

7

Activation of Organic Molecules by Acids and Bases

I. Introduction

During the 1920s, studies of homogeneous acid and base catalysis by Brønsted, Lowry, and others set the stage for the burgeoning interest in mechanistic organic chemistry that was to follow in the next several decades. From the utilitarian point of view acids and bases have long been the dominant catalysts for organic processes, and even with the striking advances in the field of homogeneous transition metal catalysis they continue to be of the greatest importance to practicing organic chemists.

Catalysis is generally defined in some such way as the following. A substance brings about catalysis when it increases the rate of a reaction without itself being consumed. It is sometimes used in a looser sense to indicate activation of any sort, even when the agent responsible for the rate increase is consumed in the process. Thus, esters are activated toward hydrolysis by acids, and this is a case of true catalysis, since the activating species, the proton, is regenerated in a later stage of the reaction. Activation can also be brought about by base, but in this case the agent is not regenerated and so catalysis by base is an inappropriate term for the process. The fact that small (i.e., catalytic) amounts of acid are effective here [Eq. (7-1)], whereas small amounts of base are not [Eq. (7-2)], serves as a practical reminder of the definition. Nonetheless, it is

$$RCO_2R' + H_2O \xrightarrow{H^+} RCO_2H + R'OH \qquad (7\text{-}1)$$

$$RCO_2R' + NaOH \longrightarrow RCO_2Na + R'OH \qquad (7\text{-}2)$$

convenient to group such reactions as the base-promoted hydrolysis of esters with those other acid and base processes that are truly catalytic, since they fit into a common mechanistic framework, and it is only the adventitious consumption of the activating agent by one of the products that prevents these agents from qualifying as catalysts.

The term *activation* is useful since it focuses our attention on the relationship between reactants and transition state. Thus, both acids and bases activate esters toward hydrolysis, that is, they lower the free energy of activation, but only acids are regenerated in a subsequent step and hence only they are catalytic. In the reverse reaction, ester formation, the reactants are, of course, activated by acids, since catalysts simply speed the approach of the system to the position of equilibrium. Bases, on the other hand, are ineffective as esterifying agents, and this is a second point of practical importance that derives from the definition of catalysis given earlier.

The mechanism of hydrolysis of a methyl ester brought about by protons or hydroxide ions is given in Eqs. (7-3) and (7-4).

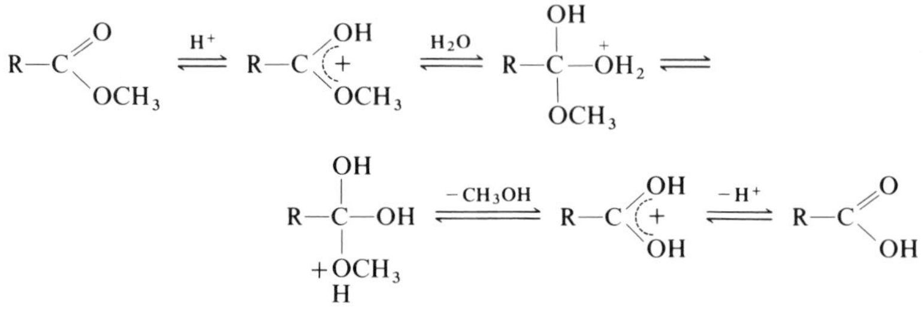

$$
R-C\overset{O}{\underset{OCH_3}{\diagdown}} \xrightleftharpoons{H^+} R-C\overset{OH}{\underset{OCH_3}{(+}} \xrightleftharpoons{H_2O} R-\overset{OH}{\underset{OCH_3}{\overset{|}{C}}}-\overset{+}{O}H_2 \rightleftharpoons
$$

$$
R-\overset{OH}{\underset{\underset{H}{+OCH_3}}{\overset{|}{C}}}-OH \xrightleftharpoons{-CH_3OH} R-C\overset{OH}{\underset{OH}{(+}} \xrightleftharpoons{-H^+} R-C\overset{O}{\underset{OH}{\diagdown}} \qquad (7\text{-}3)
$$

$$
R-C\overset{O}{\underset{OCH_3}{\diagdown}} \xrightleftharpoons{HO^-} R-\overset{O^-}{\underset{OCH_3}{\overset{|}{C}}}-OH \xrightleftharpoons{-CH_3O^-}
$$

$$
R-C\overset{O}{\underset{OH}{\diagdown}} \xrightarrow{CH_3O^-} RCO_2^- + CH_3OH \qquad (7\text{-}4)
$$

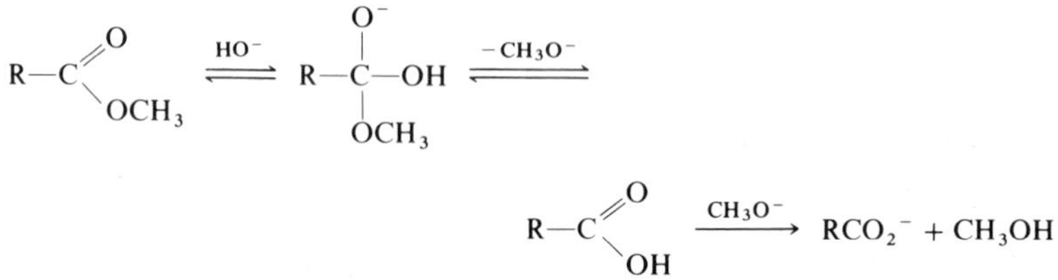

The effect of acid or base on the reaction of a number of common functional groups is given in Table 7-1, with a distinction being drawn between the terms *activation* and *catalysis*. For those reactions that are shown to be subject to activation but not catalysis the acid or base is, in fact, a reagent that must be present in stoichiometric amounts. Strictly speaking, it is molecules that are *activated* and reactions that are *catalyzed*. We shall sometimes find it convenient, however, to link the term *activation* with the overall process, rather than restrict its use to the reaction of activating agent with substrate. Catalysis always includes an activating step, but activation does not necessarily give rise to catalysis, since the activating agent is sometimes not regenerated in a subsequent step.

Why is it that chain reactions are not generally considered to be examples of catalyzed processes? In both cases a reactive species that is recycled and that need be present in only a small amount brings about changes in the chemical composition of the system; in the case of chain reactions, however, the

Table 7-1

EFFECTS OF ACIDS AND BASES ON COMMON ORGANIC REACTIONS

Substrate (reaction)[a]	Activated by		Catalysis by	
	Acid	Base	Acid	Base
Esters (hydrolysis)	+	+	+	−
Amides (hydrolysis)	+	+	−	−
Hemiacetals (cleavage)	+	+	+	+
Acetals (hydrolysis)	+	−	+	−
Ortho esters (hydrolysis)	+	−	+	−
Nitriles (hydration)	+	+	+	+
Cyanohydrins (formation)	−	+	−	+
Alkenes or alkynes (hydration)	+	−	+	−

[a] Including the reverse reaction in the case of catalysis.

recycled species is unstable and subject to decay, which means that it is consumed as the reaction proceeds. One could doubtless find examples of reactions in which a "catalyst" has limited stability, and thus to some extent it is a question of usage; certain reactions, often involving radical intermediates, have the terms *chain reaction* and *initiation* associated with them, while others are recognized as being subject to *catalysis*.

II. Specific Acid and Specific Base Catalysis

Reactions whose rates depend on the equilibrium proton activity of a system are said to be subject to specific acid or specific base catalysis. The presence of additional proton donors or acceptors has no effect on the rate of such reactions, provided that allowances are made for any ionic strength or similar effects that they might have; it is thus only the thermodynamic protonating or deprotonating ability of the system (the pH in dilute aqueous solution) that matters. This means that the substrate is activated in an equilibrium protonation or deprotonation step and that protons are neither added nor removed in the rate-controlling step of the reaction. (Reactions in which the rate-controlling step includes proton transfer are considered under the heading of general acid and general base catalysis, Section III.)

The general form of reactions that are subject to specific acid catalysis are shown in Eq. (7-5), and some examples are listed in Eqs. (7-6) to (7-10). [An extensive listing of reactions of this type is given by Bender (1), who also summarizes the evidence that has been accumulated for the mechanism of acetal hydrolysis given in Eq. (7-6).] Analogously, specific base catalysis is illustrated in Eqs. (7-11) and (7-12).

Specific acid catalysis

General form

$$Z + H^+ \rightleftharpoons ZH^+$$
$$ZH^+ \xrightarrow{\;*\;} \text{[transition state]}$$

(7-5)

where the asterisk (*) indicates no proton transfer.

Examples

Hydrolysis of most acetals (for exceptions see Section III,A)

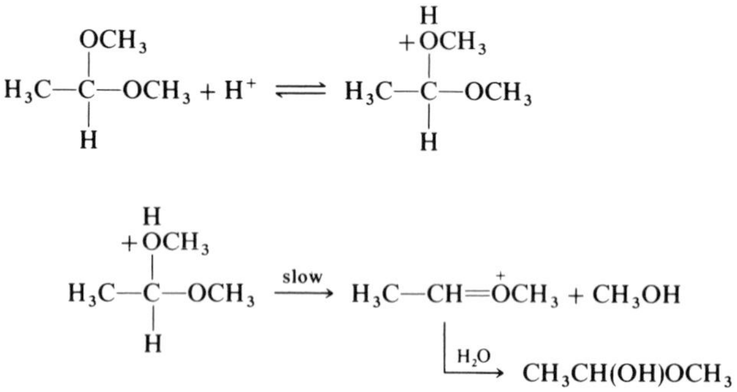

$$\text{H}_3\text{C}-\overset{\overset{\displaystyle \text{OCH}_3}{|}}{\underset{\underset{\displaystyle \text{H}}{|}}{\text{C}}}-\text{OCH}_3 + \text{H}^+ \rightleftharpoons \text{H}_3\text{C}-\overset{\overset{\displaystyle \text{H}}{\overset{\displaystyle +\text{OCH}_3}{|}}}{\underset{\underset{\displaystyle \text{H}}{|}}{\text{C}}}-\text{OCH}_3$$

(7-6)

$$\text{H}_3\text{C}-\overset{\overset{\displaystyle \text{H}}{\overset{\displaystyle +\text{OCH}_3}{|}}}{\underset{\underset{\displaystyle \text{H}}{|}}{\text{C}}}-\text{OCH}_3 \xrightarrow{\text{slow}} \text{H}_3\text{C}-\text{CH}=\overset{+}{\text{O}}\text{CH}_3 + \text{CH}_3\text{OH}$$
$$\xrightarrow[\text{H}_2\text{O}]{} \text{CH}_3\text{CH(OH)OCH}_3$$

tert-*Butyl acetate hydrolysis (2)*

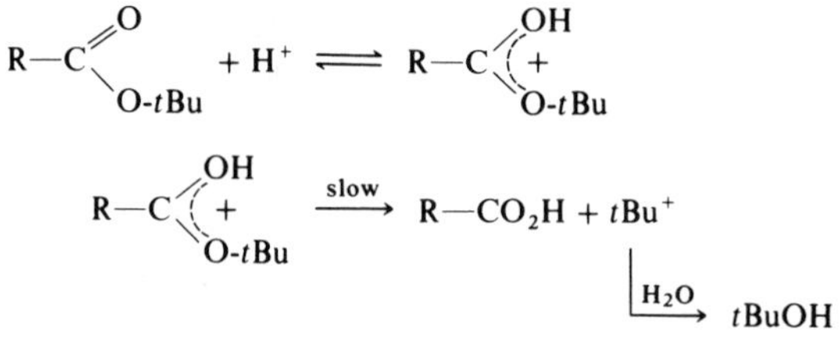

$$\text{R}-\text{C}\overset{\displaystyle O}{\underset{\displaystyle \text{O-}t\text{Bu}}{\diagdown}} + \text{H}^+ \rightleftharpoons \text{R}-\text{C}\overset{\displaystyle \text{OH}}{\underset{\displaystyle \text{O-}t\text{Bu}}{\diagdown}} +$$

$$\text{R}-\text{C}\overset{\displaystyle \text{OH}}{\underset{\displaystyle \text{O-}t\text{Bu}}{\diagdown}} + \xrightarrow{\text{slow}} \text{R}-\text{CO}_2\text{H} + t\text{Bu}^+$$
$$\xrightarrow[\text{H}_2\text{O}]{} t\text{BuOH}$$

(7-7)

Alkoxymethyl ester hydrolysis (3)

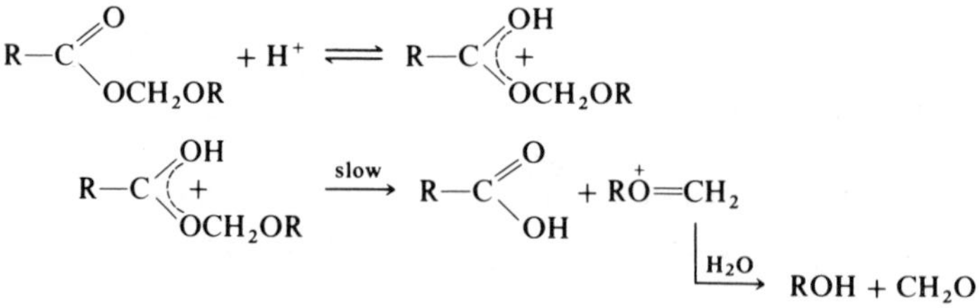

$$\text{R}-\text{C}\overset{\displaystyle O}{\underset{\displaystyle \text{OCH}_2\text{OR}}{\diagdown}} + \text{H}^+ \rightleftharpoons \text{R}-\text{C}\overset{\displaystyle \text{OH}}{\underset{\displaystyle \text{OCH}_2\text{OR}}{\diagdown}} +$$

$$\text{R}-\text{C}\overset{\displaystyle \text{OH}}{\underset{\displaystyle \text{OCH}_2\text{OR}}{\diagdown}} + \xrightarrow{\text{slow}} \text{R}-\text{C}\overset{\displaystyle O}{\underset{\displaystyle \text{OH}}{\diagdown}} + \text{R}\overset{+}{\text{O}}=\text{CH}_2$$
$$\xrightarrow[\text{H}_2\text{O}]{} \text{ROH} + \text{CH}_2\text{O}$$

(7-8)

Triazene decomposition (4)

$$R\!-\!N\!=\!N\!-\!NR_2 + H^+ \rightleftharpoons R\!-\!N\!=\!N\!-\!\overset{+}{\underset{H}{N}}R_2$$

$$R\!-\!N\!=\!N\!-\!\overset{+}{\underset{H}{N}}R_2 \xrightarrow{\text{slow}} RN_2^+ + R_2NH \tag{7-9}$$

Aromatic protodealkylation (5)

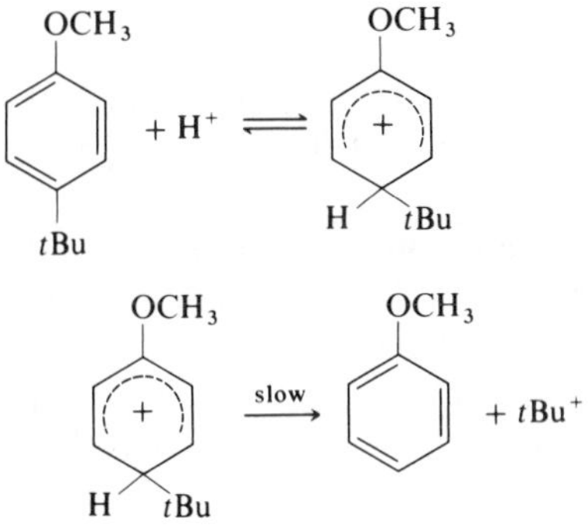

$$\tag{7-10}$$

Specific base catalysis

General form

$$ZH + HO^- \rightleftharpoons Z^- + H_2O$$

$$Z^- \xrightarrow{*} [\text{transition state}] \tag{7-11}$$

where the asterisk (*) indicates no proton transfer.

Example

Condensation of acetaldehyde (at high aldehyde concentrations this reaction resembles many other condensations in being subject to general base catalysis)

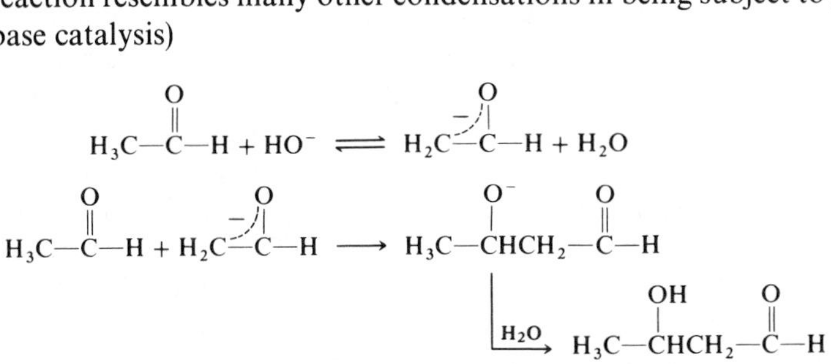

$$\tag{7-12}$$

The rate law for a specific-acid-catalyzed reaction is shown in Eq. (7-13), where K_{BH^+} is the acidity constant of the conjugate acid of the substrate Z.

The analogous expression for specific base catalysis is shown in Eq. (7-14), where K_{HA} is the acidity constant of the substrate ZH.

$$\text{Rate} = k[ZH^+] = \frac{k}{K_{BH^+}}[Z][H^+] \tag{7-13}$$

$$\text{Rate} = k[Z^-] = kK_{HA}[ZH]/[H^+] \tag{7-14}$$

If proton transfer takes place in the rate-controlling step of a reaction, it follows that all acids or bases present in the system will have an opportunity to participate and, as a consequence, they will appear in the rate expression. The reaction will then be general, not specific, acid or base catalyzed, and its rate in water will not be a simple function of the pH of the solution.

A number of reactions that appear to be subject to specific acid catalysis take place at convenient rates only when strongly acidic conditions, say, $> 10\%$ sulfuric acid, are employed. An example would be the hydrolysis of methyl mesitoate (6). Such reactions, along with those shown in Eqs. (7-6) to (7-10), are believed to proceed by way of unimolecular decomposition of the conjugate acid of the substrate, the symbol A-1 often being used to designate the mechanism. As might be expected, the rates of these reactions are found to rise more rapidly as the concentration of sulfuric or other acid is increased than do the rates of those reactions in which the conjugate acid of the substrate reacts with water or other base in the rate-controlling step. Such reactions, designated A-2 processes, give rise to general acid catalysis and will be discussed later.

Although the symbols A-1 and A-2 are generally associated with reactions in fairly acidic media, it is convenient to use the terms in a general way; thus, A-1 reactions are those that follow the mechanism shown in Eq. (7-5), with the additional requirement that the decomposition of ZH^+ be unimolecular, or spontaneous.

The numerical designations in the various symbols used by organic chemists (e.g., A-1, A-2, S_N1, and S_N2) indicate the molecularity of the process, in particular that of the rate-controlling step, and they are thus inferential quantities. On the other hand, kinetic order, catalytic effects, and other phenomena are experimentally observable (7). Thus, one must resist the temptation to equate, for example, the terms *specific acid catalysis* and *A-1 reaction*, since a reaction in which the conjugate acid of the substrate reacts in a bimolecular rate-controlling step will exhibit specific acid catalysis, provided that no proton transfer takes place in this step. There is, of course, a direct connection between mechanism and catalytic type in an acid–base reaction; the bond to the proton must be intact in the transition state of a specific-acid- (or base-) catalyzed reaction and in the process of being broken in the transition state of a general-acid- (or base-) catalyzed reaction. Some authors exploit this strict connection and use the catalytic designation in a mechanistic, as opposed to a kinetic, sense (8) although this practice will not be followed herein.

Although proton transfer in the rate-controlling step is excluded for reactions that are subject to specific acid or base catalysis, it is not safe to assume that all carbon–hydrogen bonds remain intact in the transition states of such reactions. Many oxidation processes of organic compounds involve initial activation of the substrate by base [or occasionally by acid (9)] in an equilibrium step, followed by cleavage of a carbon–hydrogen bond by the oxidant. If the latter has no acidic or basic properties, then catalysis will be specific rather than general; that is, the rate in water at constant ionic strength will be a simple function of the pH of the solution. An example of such a reaction is the permanganate oxidation of 2,2,2-trifluoro-1-arylethanols, whose reaction path is shown in Eqs. (7-15) and (7-16).

$$ArCHOHCF_3 + HO^- \rightleftharpoons \overset{\overset{O^-}{|}}{ArCHCF_3} + H_2O \tag{7-15}$$

$$\overset{\overset{O^-}{|}}{ArCHCF_3} \xrightarrow{\text{Mn(VII)}} \overset{\overset{O}{\|}}{ArCCF_3} \tag{7-16}$$

There is a large isotope effect (16:1 at room temperature) for the oxidation of the isotopically substituted compound $ArCDOHCF_3$, and there is no doubt that the bond to hydrogen is partly cleaved in the transition state [probably by hydrogen atom abstraction by Mn(VII) to form a radical-anion and Mn(VI), followed by further oxidation to the ketone (12, 13)]. The reactions give typical pH–rate profiles for specific base catalysis (Fig. 7-1). The

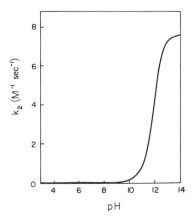

Fig. 7-1 pH–Rate profile for the oxidation of $C_6H_5CHOHCF_3$ by aqueous potassium permanganate. The pK_{HA} of the substrate is 11.9. Data from ref. 12.

degree of activation caused by deprotonation of the alcohol is $\geq 10^4$; this quantity is called the deprotonating factor (dpf) (*14*). There is an analogous quantity, the proton activating factor (paf), which represents the degree of activation caused by protonation of the oxidant; in the case of the reaction shown in Eqs. (7-15) and (7-16) this quantity is $> 10^4$ (*15*).

For reactions that are subject to specific acid or base catalysis the quantity paf or dpf is not always easily obtained since the nonactivated paths may be extremely slow. For reactions that are subject to general acid catalysis we shall see that there are alternative means of determining paf.

III. General Acid and General Base Catalysis

Reactions in which a proton is transferred in the rate-controlling step exhibit general acid or general base catalysis. This means that the rate is not simply a function of the equilibrium acid or base strength of the system but rather depends on the nature and concentration of all the acids HA or bases B present in the system, including, of course, the species H^+ and HO^-. The general form of the rate laws governing both kinds of process for the substrate Z are shown in Eqs. (7-17) and (7-18). Many reactions exhibit both general acid

$$\text{Rate} = [Z](k_a[\text{HA}_a] + k_b[\text{HA}_b] + k_c[\text{HA}_c] + \cdots + k_i[\text{HA}_i]) \quad (7\text{-}17)$$

$$\text{Rate} = [Z](k_a[\text{B}_a] + k_b[\text{B}_b] + k_c[\text{B}_c] + \cdots + k_i[\text{B}_i]) \quad (7\text{-}18)$$

and general base catalysis, and so the rate law in these cases would include all the terms shown in both equations. [The integrated forms of these expressions, which are used in kinetic studies, can differ according to the rate and degree of ionization of the substrate; an analysis of the types of rate equation encountered in studies of acid–base catalysis and a historical account of the subject are given by R. P. Bell (*17*).]

The partial rate constants $k_a \ldots k_i$ in Eqs. (7-17) and (7-18) are the catalytic coefficients for the various acids and bases involved in the reaction. As might be expected, the stronger the equilibrium strength of these species, the larger their catalytic coefficients tend to be (Section III,D). The partial rate contribution depends on the concentration of the catalyzing acid as well as on the magnitude of its catalytic coefficient, and so one can usually find a region of the pH spectrum where the effect of a very weak acid or base is completely dominant and the effect of the proton or hydroxyl ion negligible. At low or high pH, of course, the proton or hydroxide ion becomes the predominant species, and in these circumstances it may not be possible to detect the contributions of weaker components.

In the discussion that follows it will be assumed that the reactions take place in aqueous medium, unless it is otherwise stated.

A. The A-2 and A-S$_E$2 Mechanisms

The two principal routes taken by general-acid-catalyzed processes are those shown in Eqs. (7-19) and (7-20). These reactions are kinetically

A-2 mechanism

$$Z + H^+ \rightleftharpoons ZH^+$$

$$ZH^+ + A^- \xrightarrow{\text{slow}} \text{products}$$

(7-19)

A-S$_E$2 mechanism

$$Z + HA \xrightarrow{\text{slow}} \text{products} \qquad (7\text{-}20)$$

equivalent and give rise to identical rate expressions. In the A-2 mechanism the conjugate acid of the substrate, which is formed in an equilibrium step, reacts with the conjugate base of the catalyzing acid in a rate-controlling step to give products or intermediates that lead to products. If H_3O^+ is the catalyst, the conjugate base is H_2O; if a weak acid HA is the catalyst, the conjugate base is the anion A^- (Eq. 7-19). Some authors restrict the use of the symbol A-2 to reactions in which water is the base (or nucleophile), but we shall find it convenient to use the term in the general sense.

The stronger the acid HA, the weaker is the base strength of the acid anion A^- and the less effective it will be in, for example, removing a proton from ZH^+ in the rate-controlling step. In view of this condition it may seem paradoxical that catalytic effectiveness is proportional to the strength of the acid, but this is indeed so since changes in the rate constant of the rate-controlling step resulting from changes in strength of the base A^- are more than compensated for by changes in the concentration of the latter.

A typical A-2 reaction is acid-catalyzed enolization of carbonyl compounds (*18*), as shown in Eq. (7-21) for the case of acetone and formic acid. The rate law for this reaction takes the form shown in Eq. (7-22), where K_{BH^+} and K_{HA} are the respective ionization constants of acetone and formic acid.

$$\text{H}_3\text{C}-\overset{\text{O}}{\overset{\|}{\text{C}}}-\text{CH}_3 + \text{H}^+ \rightleftharpoons \text{H}_3\text{C}-\overset{\text{+OH}}{\overset{\|}{\text{C}}}-\text{CH}_3$$

$$\text{H}_3\text{C}-\overset{\text{+OH}}{\overset{\|}{\text{C}}}-\text{CH}_3 + \text{HCO}_2^- \longrightarrow \text{H}_3\text{C}-\overset{\text{OH}}{\overset{\|}{\text{C}}}=\text{CH}_2 + \text{HCO}_2\text{H}$$

(7-21)

$$\text{Rate} = k[\text{CH}_3\text{C}(\overset{+}{\text{O}}\text{H})\text{CH}_3][\text{HCO}_2^-]$$

$$= \frac{k}{K_{BH^+}}[\text{CH}_3\text{COCH}_3][\text{H}^+][\text{HCO}_2^-]$$

$$= \frac{kK_{HA}}{K_{BH^+}}[\text{CH}_3\text{COCH}_3][\text{HCO}_2\text{H}] \qquad (7\text{-}22)$$

The second step of an A-2 reaction can also involve nucleophilic addition to the protonated substrate of a molecule of water that undergoes simultaneous deprotonation by the base. The latter case, though not strictly speaking a bimolecular reaction, exhibits the same kinetics as the other A-2 processes. Such nucleophilic reactions are considered in Section III,C.

The major bimolecular alternative to the A-2 route is provided by the A-S_E2 mechanism (5, 19–22). A typical A-S_E2 reaction is isotopic exchange of the ring hydrogens in aromatic compounds [Eq. (7-23)]. Indeed, the symbolism

$$ArH + DA \longrightarrow ArHD^+ + A^-$$

$$ArHD^+ + A^- \xrightarrow{\text{fast}} ArD + HA \tag{7-23}$$

$$Rate = k[ArH][DA]$$

derives from this reaction, the S_E indicating electrophilic substitution. It has become common practice to use this symbol for reactions in which the rate-controlling step is proton transfer from the catalyzing acid to the substrate, for example, the hydrolysis of ketene acetals [Eq. (7-24)] (22), even though the

$$H_2C=C(OR)_2 \xrightarrow{HA} CH_3\overset{+}{C}(OR)_2 \xrightarrow[\text{fast}]{H_2O} CH_3CO_2R + ROH \tag{7-24}$$

overall process may not involve substitution as such (23). Furthermore, it will become clear that both symbols, A-2 and A-S_E2, must have vectorial significance. We thus restrict the use of both symbols directionally and at the same time extend in a sense the meaning of the symbol A-S_E2 in an attempt to bring some coherence to the terminology of the field, as will be made clear later.

The process that we designate A-2 is often called *specific acid–general base catalysis* and that which we designate A-S_E2 is often called *true general acid catalysis*. The problem with such labels is that the former reaction is just as much a case of catalysis by general acids as is the latter. Indeed, the two routes are kinetically equivalent. Moreover, if a reaction follows the A-2 route in one direction, it will of necessity follow the A-S_E2 route in the opposite direction. This follows from the principle of microscopic reversibility, according to which the forward and reverse reactions traverse the same path but in opposite directions. In the transition state of the A-2 reaction, A$^-$ is in the process of removing a proton to form HA. Therefore, in the reverse direction, HA must be in the process of adding a proton in the transition state. Both routes, of necessity, are cases of "true" general acid catalysis. The reverse of enolization [Eq. (7-21)] is ketonization, which is an A-S_E2 reaction [Eq. (7-25)]; ketonization, of course, also exhibits general acid catalysis (24).

$$\underset{\overset{|}{\underset{H_3C-C=CH_2}{}}}{OH} + HA \longrightarrow \underset{\overset{\|}{\underset{H_3C-C-CH_3}{}}}{\overset{+}{O}H} + A^- \tag{7-25}$$

The isotopic exchange shown in Eq. (7-23) is an identity reaction, with the second step and the reversal of the first being chemically identical. It can be seen that they correspond to the second step of an A-2 process, whereas the first step of the forward reaction is that of an A-S_E2 process.

The most important reaction that follows the simple A-2 reaction in the forward direction is enolization, which is, in turn, the rate-limiting process of many acid-catalyzed condensation, racemization, halogenation, oxidation, and isotope exchange reactions of carbonyl and similar compounds. In acid-catalyzed enolization there are two proton transfers, the first a rapid equilibration of the substrate and its conjugate acid and the second the rate-controlling transfer of a different proton to the conjugate base of the catalyzing acid. Evidence for this and a number of other mechanistic assignments will be described in Sections III,F and G. Reactions that take place by nucleophilic attack of water on the conjugate acid of the substrate and thus qualify as A-2 processes include the hydrolysis of most ethers, esters, and amides. The connection between nucleophilic processes and acid catalysis is considered in Section III,C.

In addition to isotopic exchange of ring hydrogens the A-S_E2 mechanism is followed in the hydrolysis of vinyl ethers (25), the cleavage of alkylmercuric iodides (26), the hydrolysis of some but by no means all diazo ketones (27, 28), and the ketonization of enols (24).

Although most acetals are hydrolyzed by way of the A-1 mechanism and exhibit specific acid catalysis, some are subject to general acid catalysis. In these cases protonation of an alkoxyl group of the acetal becomes rate determining, with expulsion of the cation occurring before the protonation process can be reversed; that is, they follow the A-S_E2 mechanism. The structural features that induce the switch from specific acid catalysis and the A-1 mechanism to general acid catalysis and the A-S_E2 mechanism have been delineated by Fife (29). Most ortho esters are hydrolyzed by the A-S_E2 mechanism [Eq. (7-26)] (30), as is the heterolysis of certain alcohols, such as tropyl alcohol, that yield stabilized cations [Eq. (7-27)] (31, 32).

$$RC(OR)_3 + HA \longrightarrow RC(OR)_2^+ + ROH + A^-$$

$$\downarrow \xrightarrow[H_2O]{fast} RCO_2R + ROH \qquad (7\text{-}26)$$

$$(7\text{-}27)$$

B. The B-2 Mechanism

We saw earlier that specific base catalysis is exhibited by reactions in which equilibrium deprotonation is followed by a slower reaction that does not involve proton transfer (Section II). The example used there was aldol condensation, in which the conjugate base of the substrate reacts in a bimolecular slow step with a second molecule of the substrate, the latter acting as an electrophile. It can be formally designated a B-2 reaction, since it is the counterpart of an A-2 reaction in which a conjugate *acid* reacts in a slow step with a *nucleophile*. Specific base catalysis is also exhibited by reactions in which the conjugate base of the substrate is formed in a preequilibrium step and undergoes unimolecular decomposition. The decomposition of diacetone alcohol [Eq. (7-28)] is such a reaction; it is designated a B-1 process and is the

$$
\begin{array}{c}
\underset{\underset{CH_3}{|}}{\overset{\overset{OH}{|}}{H_3C-C-CH_2-}}\overset{\overset{O}{\|}}{C}-CH_3 + HO^- \rightleftharpoons \underset{\underset{CH_3}{|}}{\overset{\overset{O^-}{|}}{H_3C-C-CH_2-}}\overset{\overset{O}{\|}}{C}-CH_3 + H_2O
\end{array}
$$

$$
\underset{\underset{CH_3}{|}}{\overset{\overset{O^-}{|}}{H_3C-C-CH_2-}}\overset{\overset{O}{\|}}{C}-CH_3 \longrightarrow \overset{\overset{O}{\|}}{H_3C-C}-CH_3 + H_2C\overset{\overset{O}{\|}}{\underset{}{-C}}-CH_3 \quad (7\text{-}28)
$$

$$
\overset{\overset{O}{\|}}{H_3C-C}-CH_3 + HO^- \longleftarrow \quad H_2O
$$

counterpart of an A-1 reaction, in which the conjugate acid of the substrate undergoes unimolecular decomposition.

The bimolecular pathway followed by most reactions of interest is rate-controlling proton abstraction from the unactivated substrate, as in Eq. (7-29), where the base is shown as A^-, the anion of a weak acid HA. A second possible mechanism that has a bimolecular slow step and identical kinetics is shown in Eq. (7-30). Both routes lead to the same rate expression, and in both cases there will be contributions from all bases that are present in the system.

$$
ZH + A^- \xrightarrow{\text{slow}} Z^- + HA
$$
$$
\qquad\qquad\quad \longrightarrow \text{products} \qquad\qquad (7\text{-}29)
$$

$$
ZH + HO^- \rightleftharpoons Z^- + H_2O
$$
$$
Z^- + HA \xrightarrow{\text{slow}} \text{products} \qquad\qquad (7\text{-}30)
$$

In Eq. (7-30) the general acid adds a proton to the conjugate base of the substrate, which is formed in a preequilibrium step. This makes it the basic counterpart of an A-2 reaction, in which a general base removes a proton from the conjugate acid of the substrate, which is similarly formed in a pre-equilibrium step. Hence, the designation B-2 is entirely appropriate for this mechanism. In Eq. (7-29) the rate-controlling step is direct proton abstraction by the general base, and so this reaction is the counterpart of the A-S_E2 reaction, in which the slow step is direct protonation of the substrate by the general acid. Although considerations of symmetry would suggest that this form of base catalysis be designated B-S_N2, the reaction is neither nucleophilic nor a substitution (except on hydrogen) and so we shall use the B-2 designation for this route as well. Actually, being able to distinguish alternative routes by means of symbols is less important for general base than for general acid catalysis, since most general-base-catalyzed reactions that are of interest proceed in the direction given by Eq. (7-29); that is, rate-controlling proton abstraction by the general base takes place.

There is little doubt that the base-catalyzed enolization of ketones takes place by the mechanism shown in Eq. (7-31) (Section III,G,1 and ref. *18*). The

$$
\begin{array}{cc}
\underset{\displaystyle R-\overset{\textstyle O}{\overset{\|}{C}}-CH_3 + A^-}{} \longrightarrow & R-\overset{\textstyle O^-}{\overset{|}{C}}CH_2 + HA \\[2mm]
R-\overset{\textstyle O^-}{\overset{|}{C}}CH_2 + H_2O \rightleftharpoons & R-\overset{\textstyle OH}{\overset{|}{C}}=CH_2 + HO^- \\[2mm]
HO^- + HA \rightleftharpoons & A^- + H_2O
\end{array}
\tag{7-31}
$$

reverse reaction, ketonization of enols, is one of the few cases amenable to study that follows the alternative route (*24*). It is apparent that base-catalyzed enolization and ketonization traverse the same path in opposite directions, one corresponding to the route shown in Eq. (7-29) and the other to that shown in Eq. (7-30).

The B-2 designation is frequently used for nucleophilic reactions as well, particularly for reactions such as ester hydrolysis, with the place of bond cleavage being indicated thus: B-Ac2 (acyl–oxygen cleavage) or B-Al2 (alkyl–oxygen cleavage). Catalysis involving nucleophilic addition to carbonyl and similar groups is considered in the next section.

C. Nucleophilic Reactions at Unsaturated Centers

A molecule with an unshared pair of electrons can act either as a base or as a nucleophile, the distinction resting entirely on whether it takes up a proton in the reaction in question. Nucleophilic additions to unsaturated centers take place in many important reactions of organic chemistry, particularly reactions at carbonyl carbon atoms, and these reactions may be subject to both nucleophilic and basic catalysis. Perhaps the most thoroughly studied case involving both types of catalysis is the hydrolysis of esters catalyzed by imidazole (**1**). The latter compound is an effective nucleophile, particularly in neutral solution, for a number of reasons: first, it is a nitrogen base, and such compounds tend to have higher nucleophilicities than their basicities would suggest; second, it is analogous to a tertiary amine, which is the most effective type of amino compound (provided that steric effects are minimized, as they are with imidazole); and third, its pK_{BH^+} of 7 allows it to function effectively at neutral pH. (If the pK_{BH^+} were lower, the nucleophilicity would be less; if it were higher, the molecule would be protonated.)

Esters that have good leaving groups, such as *p*-nitrophenyl acetate, are activated by imidazole and undergo ready hydrolysis in water. Acetylimidazole can be detected as an intermediate in the reaction [Eq. (7-32)],

$$\text{H}_3\text{C}-\overset{\overset{\text{O}}{\|}}{\text{C}}-\text{OC}_6\text{H}_4(\textit{p}\text{-NO}_2) + \text{:N} \underset{\textbf{1}}{\overset{\frown}{}}\text{NH} \longrightarrow \text{H}_3\text{C}-\overset{\overset{\text{O}}{\|}}{\text{C}}-\text{N}\overset{\frown}{}\overset{+}{\text{NH}} + {}^-\text{OC}_6\text{H}_4(\textit{p}\text{-NO}_2)$$

$$(7\text{-}32)$$

$$\text{H}_3\text{C}-\overset{\overset{\text{O}}{\|}}{\text{C}}-\text{N}\overset{\frown}{}\overset{+}{\text{NH}} \xrightarrow{\text{H}_2\text{O}} \text{CH}_3\text{CO}_2\text{H} + \text{N}\overset{\frown}{}\text{NH}$$

its concentration building up and then decaying as the reaction proceeds (*33, 34*). Imidazole is an effective catalyst here because it is a good leaving group, as well as a good nucleophile.

Hydrolysis of esters that have very poor leaving groups, ethyl acetate, for example, shows a much smaller degree of catalysis by imidazole, and effective catalysis requires more strongly basic conditions, in which direct nucleophilic attack by hydroxide ion at carbonyl carbon takes place. Indeed, any imidazole catalysis that can be observed appears to be due to general base catalysis and not to nucleophilic catalysis. This is because the tetrahedral intermediate **2**

$$\text{R}-\overset{\overset{\text{O}^-}{|}}{\underset{\underset{\text{Y}}{|}}{\text{C}}}-\text{N}\overset{\frown}{}\overset{+}{}\text{NH}$$

2

that is formed in reactions such as that shown in Eq. (7-32) will expel imidazole instead of Y if the latter is a very poor leaving group. The role of imidazole is then confined to activation of a water molecule by means of general base catalysis to form the intermediate shown in Eq. (7-33), one that is

$$\text{R}-\overset{\overset{\text{O}}{\diagup}}{\underset{\underset{\text{Y}\ \ \text{H}}{\diagdown}}{\text{C}}}\overset{\diagup}{}\text{O}-\text{H}\ \ \text{:N}\overset{\frown}{}\text{NH} \longrightarrow \text{R}-\overset{\overset{\text{O}^-}{|}}{\underset{\underset{\text{Y}}{|}}{\text{C}}}-\text{OH} + \text{HN}\overset{\frown}{}\overset{+}{}\text{NH} \quad (7\text{-}33)$$

of higher energy and hence not so readily formed but one that is much less averse to expelling the poor leaving group Y (*35, 36*).

What evidence is there for depicting imidazole as a general base, as shown in Eq. (7-33)? Perhaps the best evidence comes from the fact that imidazole catalyzes the hydrolysis of acetylimidazole (38). The catalysis cannot be nucleophilic since a displacement at carbonyl carbon would produce no chemical change. It is therefore likely that the reaction occurs via the mechanism shown in Eq. (7-34), with the imidazole catalyst acting as a general

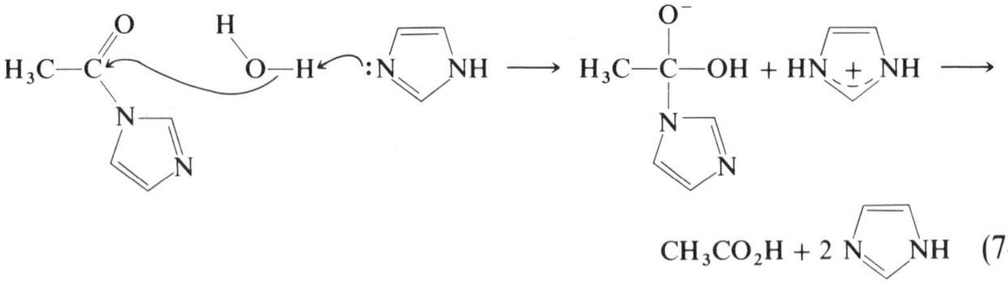

base to activate a water molecule, which acts as the nucleophile (39).

Distinguishing nucleophilic catalysis from general base catalysis can sometimes be done on the basis of adherence, or lack thereof, to the Brønsted catalysis law (Section III,D). Certain molecules or ions, in particular those with polarizable atoms at the nucleophilic center or those with an unshared electron pair on an adjacent atom (the α effect) (42–45), are much more effective as nucleophiles than would be expected from a consideration of their equilibrium base strengths. When these species exhibit enhanced catalytic effects, as expressed through positive deviations from the appropriate Brønsted line, it can be concluded that they are acting as nucleophiles, not as general bases. Furthermore, steric effects tend to be more severe in nucleophilic than in general base reactions (46); indeed, in the latter, bulky groups near the proton transfer site increase the rate in some cases (Section III,D).

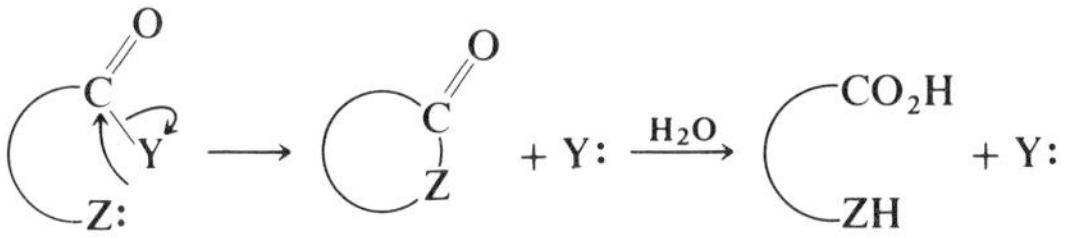

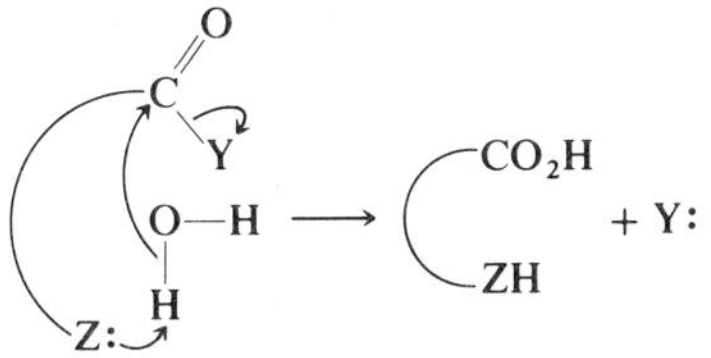

Nucleophilic and general base catalysis can also be distinguished in those reactions that have cyclic analogues using the concept of *effective molarity*. Reactions of the type shown in Eq. (7-35) (nucleophilic catalysis) and Eq. (7-36) (general base catalysis) have an entropic advantage over their acyclic analogues, in which Z, the nucleophile or base, is in a separate molecule. This advantage can be expressed in terms of the ratio of the rate constant of the cyclic reaction to that of an appropriate acyclic analogue. Since the kinetic orders of the reactions differ, the ratio has the dimensions of molarity. The term *effective molarity* (EM) is used to express this ratio. It turns out that most reactions that on other grounds are believed to follow Eq. (7-35), that is, to proceed by way of nucleophilic catalysis, have much larger EM values than do those that are believed to proceed by way of general base catalysis and follow Eq. (7-36).

Kirby suggested that an EM of 80 is a reasonable point of division for the two classes; nucleophilic reactions usually have EM values that are greater than 80 (sometimes very much greater); general base reactions virtually always have EM values that are less than 80 (47). Some examples of the use of EM, taken from Kirby's compilations, are as follows:

General base catalysis

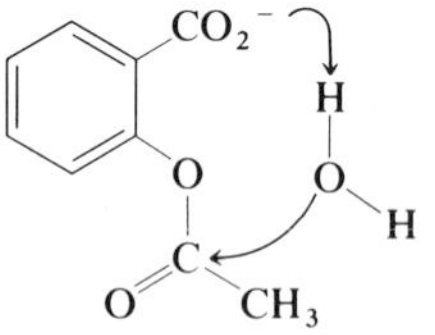

$$EM = 13$$

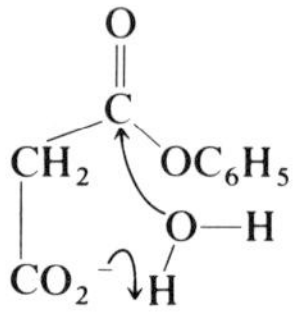

$$EM = 25$$

Nucleophilic catalysis

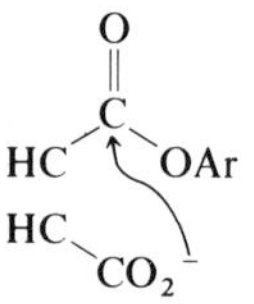

$$EM = 3 \times 10^9$$

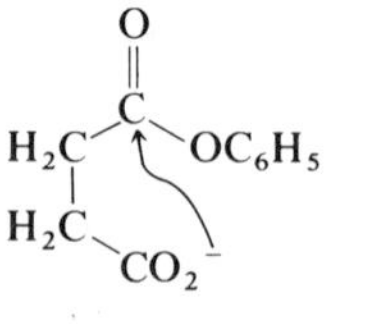

$$EM = 5.1 \times 10^4$$

Hydration of aldehydes and ketones is a reaction that at some stage requires nucleophilic attack at the carbon atom of the carbonyl group. This well-studied reaction (48, 49) is subject to both general acid and general base catalysis. The two possible mechanisms for the reaction catalyzed by general acids are shown in Eqs. (7-37) and (7-38). These reactions take place in one

$$R-C \underset{\overset{|}{H}}{\overset{O \cdot H-A}{\diagdown}} OH_2 \longrightarrow R-\overset{OH}{\underset{\overset{|}{H}}{\overset{|}{C}}}-\overset{+}{O}H_2 + A^- \rightleftharpoons R-\overset{OH}{\underset{\overset{|}{H}}{\overset{|}{C}}}-OH + HA \quad (7\text{-}37)$$

$$R-C \overset{O}{\underset{\diagdown H}{\diagup}} + H^+ \rightleftharpoons R-C \overset{\overset{+}{O}H}{\underset{\diagdown H}{\diagup}} \quad (7\text{-}38a)$$

$$R-C \overset{\overset{+}{O}H}{\underset{\diagdown H}{\diagup}} \underset{H}{\overset{}{\frown}} O-H \overset{}{\frown} A^- \longrightarrow R-\overset{OH}{\underset{\overset{|}{H}}{\overset{|}{C}}}-OH + HA \quad (7\text{-}38b)$$

instance by rate-controlling protonation of the substrate by the general acid and in the other by rate-controlling proton abstraction by the conjugate base of the general acid. They therefore correspond to the $A\text{-}S_E2$ and A-2 routes described earlier, with the difference being that a molecule of water is also involved. In aqueous solution such participation has no effect on the reaction kinetics, and one can consider these mechanisms to be pseudo-$A\text{-}S_E2$ and pseudo-A-2 reactions, respectively. [Although the reactions are formally termolecular, hydrogen bonding between the carbonyl group and the general acid in the one case and between water and the acid anion in the other (50) can be invoked to overcome any objections to the mechanisms that might be raised on this ground.]

It should be noted again that the mechanistic designations must be interchanged for the reverse reactions; that is, the pathway represented by Eq. (7-37) becomes an A-2 mechanism and that of Eq. (7-38) an $A\text{-}S_E2$ mechanism when the reactions are written in the reverse direction.

Considerable effort has been expended in determining which path is followed in which direction, particularly since the same dilemma arises in the addition to the carbonyl group of a number of other weakly basic compounds, such as alcohols and arylamines.

An alternative means of classifying catalyzed additions to carbonyl and similar groups is that of Jencks (51, 52). In his scheme the focus is on the site of catalysis in the alternative pathways, which are called class e or class n reactions, according to the kind of interaction the acid or base catalyst has with the other two components, the electrophile and the nucleophile. (The terms e-s and n-s are sometimes used for the reactions being considered here, the s indicating that the slow step of the reaction is being considered.) In the first of the alternative pathways for general-acid-catalyzed hydration of aldehydes, shown earlier, the acid donates a proton to the carbonyl compound (the electrophile) in the slow step in the forward direction of Eq. (7-37) and receives it back in the opposite direction, again in the slow step. It is thus a class e (or e-s) reaction.

In the alternative route [Eq. (7-38)] the catalyst (in the form of the acid anion) removes a proton from the nucleophile in the slow step and returns it in the slow step of the reverse reaction. There are strong indications that the general-acid-catalyzed hydrations of aldehydes are class e reactions, which follow the mechanism shown in Eq. (7-37); that is, they are pseudo-A-S_E2 reactions in the hydration direction, not pseudo-A-2 reactions. The first indication that Eq. (7-38) is not the operative path came from the observation that the reverse reaction is a model for general-acid-catalyzed acetal hydrolysis. Since the hydrolysis of ordinary aliphatic acetals is not subject to general acid catalysis it seems unlikely that the analogous scission of aldehyde hydrates will occur in this way. [Such analogies are not without flaws; with certain acetals and ortho esters that are structurally similar the acetals show specific acid catalysis and the ortho esters general acid catalysis (*50*, *53*).] It is, of course, obvious that a catalyst must be effective for both forward and reverse reactions, and so if catalysis can be discounted in one direction it can be similarly discounted in the other.

There are a number of other characteristics of the acid-catalyzed aldehyde–aldehyde hydrate interconversion that point to it being a class e reaction, that is, a pseudo-A-S_E2 mechanism for hydration and an A-2 mechanism for dehydration. These include rate–structure correlations of various sorts, the requirement for a faster-than-diffusion step in the principal alternative mechanism, and an analysis of the relative catalytic effectiveness of general acids and general bases (*52*, *54*, *55*).

Carbonyl hydration is also subject to general base catalysis, and here there is no difficulty in assigning the reaction to class n [Eq. (7-39)], the class e alternative being unattractive on a number of grounds. Paths analogous to that shown in Eq. (7-39) are also made use of in other carbonyl reactions (e.g.,

$$R-C\overset{O}{\underset{H}{\diagup}}\cdots O\overset{H}{\underset{H}{\diagup}}\ \overset{\frown}{A^-}\ \longrightarrow\ R-\overset{O^-}{\underset{H}{\overset{|}{C}}}-OH + HA\ \rightleftharpoons\ R-\overset{OH}{\underset{H}{\overset{|}{C}}}-OH + A^-\quad (7\text{-}39)$$

hemiacetal formation), although with strong nucleophiles direct attack at the carbonyl group can occur without a water or similar molecule intervening.

It was at one time thought that carbonyl hydration took place through a cyclic array containing several water molecules (*48*) but, except possibly for the water-catalyzed (or uncatalyzed) reaction, this is now considered unlikely (*52*, *56*).

Finally, it should be pointed out that water and alcohols are much less effective at forming adducts with carbonyl compounds than are compounds with sulfur or nitrogen at the nucleophilic center (49). They or their anionic forms also suffer in comparison with carbon anions (e.g., CN^-, $^-CH_2CHO$) in their capacity to react with aldehydes or ketones to form stable neutral addition compounds; compare aldehyde hydrate and aldehyde cyanohydrin stabilities, for example.

Addition of water or alcohol to carbonium ions is analogous to carbonyl hydration in the sense that a nucleophile combines with an electrophilic center. These reactions, too, are accelerated by the presence of general bases, provided that the alcohol is not too basic nor the carbonium ion too reactive; an example is shown in Eq. (7-40) (57).

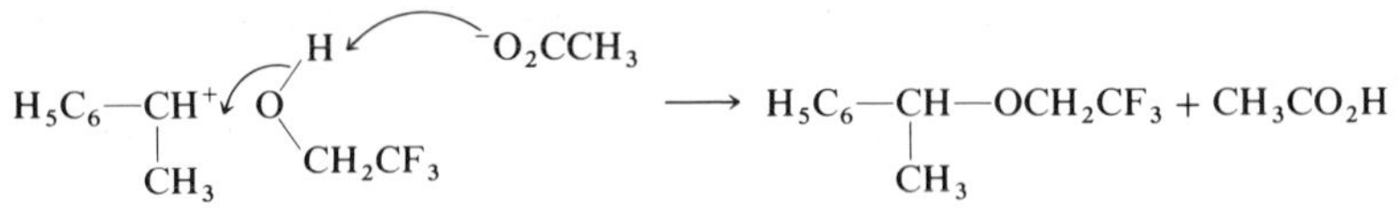

$$H_5C_6-\overset{\underset{|}{CH_3}}{CH^+}\overset{H}{\underset{CH_2CF_3}{O}}\overset{^-O_2CCH_3}{\longrightarrow} H_5C_6-\underset{\underset{CH_3}{|}}{CH}-OCH_2CF_3 + CH_3CO_2H$$

$$(7\text{-}40)$$

Although many nucleophilic additions of the type considered in this section are formally termolecular, preassociation between two of the components, that is, between reactants or between reactant and catalyst, is believed to take place in many cases. The criteria that determine whether such reactions are concerted have been outlined by Jencks (51) and by Guthrie (50). [Even the meaning of the word *concerted* is a matter of debate (58).]

D. The Brønsted Relation

It might be expected that the rate at which a base pulls a proton from an acidic site in a molecule should be a function of the respective base and acid strengths of the components at equilibrium, and by and large that is the case. What has become known as the Brønsted relation was put forward by Brønsted and Pedersen in 1924; it can be expressed in the form of Eq. (7-41)

$$\log k_i = \beta \log K_i + \text{const} \qquad (7\text{-}41)$$

using the rate constants k_i for the reaction of a series of bases with some weak acid (usually a carbon acid) and the equilibrium strengths K_i of the bases. The logarithmic relationship results from it being the free energies of activation and equilibrium that are related.

Provided that the members of the series of weak bases or acids are not too dissimilar in structure and large steric effects are not present, Eq. (7-41) is generally obeyed. The slope β, called the *Brønsted coefficient* or *exponent*, expresses the sensitivity of rate to changes in structure of the base or acid, with the benchmark being the equilibrium values, that is, the ionization constants themselves. The vast majority of reaction series that obey the Brønsted relation have coefficients between 0 and 1.0, with values near 0.5 being fairly common.

Base strengths are usually expressed in terms of the strengths of their conjugate acids, which has the effect of changing the sign of the correlation. Thus, for a series of reactions in which anionic bases A^- remove a proton from a given molecule, Eq. (7-41) takes the form of Eq. (7-42).

$$\log k = \beta \log(K_w/K_{HA}) + \text{const}$$

$$= -\beta \log K_{HA} + \text{const} \tag{7-42}$$

The same general expression applies when a molecule reacts with a series of weak acids HA serving as proton donors, in which case the equilibrium strengths of the latter are used in the correlation and the symbol β is replaced with α, giving Eq. (7-43). Thus, we shall use α to indicate that the variable

$$\log k = \alpha \log K_{HA} + \text{const} \tag{7-43}$$

species is the proton donor and β to indicate that it is the proton acceptor. (As explained at the outset K_{HA} stands for the acid dissociation constant of the species whose formula or name is given, regardless of the species' charge type.)

When polyprotic acids or bases are used as the variable species, whether as substrates or as reagents, statistical corrections can be made to Eqs. (7-42) and (7-43) to reflect this fact. If the acid has p equivalent protons that can dissociate and its conjugate base has q equivalent sites for proton attachment, it can be shown that Eqs. (7-44) and (7-45) should apply, provided that the acidic or basic sites can be considered to operate independently of one another.

$$\log k/q = -\beta \log(q/p)K_{HA} + \text{const} \tag{7-44}$$

$$\log k/p = \alpha \log(q/p)K_{HA} + \text{const} \tag{7-45}$$

Both p and q are sometimes taken as unity for the carboxyl group even though the carboxylate ion has two sites for proton addition, since it can be argued that the two oxygen atoms of the ion constitute a single unit. In the case of ammonium ions the formal values of p and q are, respectively, 4 and 1 for NH_4^+, 3 and 1 for RNH_3^+, 2 and 1 for $R_2NH_2^+$, and 1 and 1 for R_3NH^+, although it is not clear whether the full statistical factor of 4 applies when the catalytic effects of ammonium and trialkylammonium ions are compared. The H_3O^+ ion, which frequently deviates from Brønsted correlations, has formal values of 3 and 1 for p and q, but it has been argued that the two lone-pair lobes of the oxygen atom in the water molecule constitute two basic sites and that q should be taken as 2 for the hydronium ion (59). On this basis, for water acting as an acid, $p = 2$ and $q = 3$.

It turns out that statistical corrections of this type are seldom critical in determining the slope of a Brønsted correlation. Usually, acids or bases of related structure are used in a given study; even when comparison is made between two series containing functional groups whose p and q values differ, it is frequently found that the rate differences between series are greater than could be ascribed to errors in assigning values to p and q.

One of the most extensive studies involving a single substrate and a series of general bases is that of Srinivasan and Stewart. They measured the rate of abstraction of the methylene protons in the quaternary salt of creatinine [3, Eq. (7-46)] by following rates of iodination (60) or isotopic exchange in D_2O

$$\text{B:} + \quad \overset{\text{CH}_3}{\underset{3}{\text{3}}} \quad \longrightarrow \quad \text{BH}^+ + \quad \overset{\text{CH}_3}{\text{3}} \qquad (7\text{-}46)$$

(61), using ~ 75 different bases. (Although the term *catalysis* is used for both the exchange and iodination processes, the base is actually consumed in the latter case. Since the rate-controlling step is identical in the two reactions we follow common practice and gloss over this point and use the convenient term *catalysis* for both.) Figure 7-2 shows the results for all the monoprotic bases in the series, eight of which are pyridines and the rest carboxylate ions.

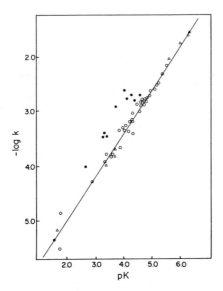

Fig. 7-2 Brønsted plot for reaction of various bases with 3 [Eq. (7-46)] in water at 25°C. ○, Aliphatic carboxylates and meta- and para-substituted benzoates; △, pyridines; ●, ortho-substituted benzoates.

A very good Brønsted correlation is obtained for the aliphatic carboxylates, the meta- and para-substituted benzoates, and the pyridines, the Brønsted coefficient being 0.79, whereas the ortho-substituted benzoates deviate markedly and appear to constitute a separate set entirely. Curiously, the direction of displacement of these hindered bases shows that they are more, not less, effective at removing a proton from the substrate than would be expected from their equilibrium base strengths. On the other hand, we shall see that hindered pyridines usually show Brønsted deviations in the expected direction; that is, they usually react with carbon acids more slowly than would be suggested by the magnitude of their equilibrium base strengths. Furthermore, unhindered carboxylates and pyridines do not always form a single set, as they do in the example shown in Fig. 7-2.

It is obvious that a Brønsted plot such as that shown in Fig. 7-2 will level off when extremely strong bases are employed, simply because of the limit that the rate of diffusion places on a bimolecular process. It turns out that curvature of Brønsted plots for proton removal from carbon acids has been observed in a number of extended series and that leveling off occurs well before the diffusion-controlled limit is reached. With proton transfer from oxygen acids, on the other hand, the diffusion rate is frequently the rate-limiting factor, though slow proton transfer from oxygen is by no means unknown (62, 63). Brønsted plots cover a limited range of acid or base strengths, and curvature is usually concealed by experimental errors or by small and random effects of structural variation in the series of bases that are used.

Bell extended the range of coverage of the Brønsted relation by examining the effect of varying both the substrate and the base (64). He used a selection of esters, monoketones, diketones, and keto esters as weakly acidic substrates and various carboxylate ions, phenolate ions, and pyridines as the bases, omitting, for example, sterically hindered pyridines, which are known sometimes to produce anomalous effects. There is definite curvature to the extended Bell–Brønsted plot, with the slope at the "reactive" end being only about half of that at the "unreactive" end. This is a common effect with carbon acids; the faster the rate of proton loss from a particular compound, the less the effect of altering the strength of the base (see, however, ref. 65). Furthermore, the Brønsted coefficient α or β will tend to be larger for an unreactive series than for a reactive one, particularly if the two series being compared are structurally similar. This can be seen from the results shown in Table 7-2 for the A-S_E2 reaction of acids with methoxybenzenes, followed by tritium exchange [Eq. (7-47)]. The more reactive the substrate, the lower the Brønsted α value.

$$\text{ArH} \xrightarrow{\text{T}^+} \text{ArHT}^+ \xrightarrow{-\text{H}^+} \text{ArT} \tag{7-47}$$

Eigen et al. showed that curvature of a Brønsted plot can be observed even when a single substituent is used, provided that techniques are available to

Table 7-2

BRØNSTED α VALUES AND RELATIVE RATES OF THE
A-S_E2 ISOTOPIC EXCHANGE OF METHOXYBENZENES
IN AQUEOUS PERCHLORIC ACID[a]

Substrate	Brønsted α[b]	Relative rate
Benzene	0.93	1
Methoxybenzene	0.71	10^6
1,3-Dimethoxybenzene	0.64	10^{10}
1,3,5-Trimethoxybenzene	0.55	10^{13}

[a] Data of Kresge et al. (66); see also ref. 67.
[b] Determined by differentiation of a log k/log K plot.

measure reaction rates over a wide span of catalyst activities. They showed that the Brønsted plots for both the general-acid- and general-base-catalyzed enolization of 2,4-pentanedione show curvature (68).

In Table 7-3 are listed the Brønsted α coefficients for three reactions that are subject to general acid catalysis, together with the catalytic constants of a number of acids that are common to the three studies. The reactions are the enolization of acetone (as measured by the rate of iodination), the hydration of 4-nitrobenzaldehyde, and the hydrolysis of dichloroketene dimethyl acetal. The general mechanisms of these three reactions were given earlier: the

Table 7-3

BRØNSTED COEFFICIENTS AND CATALYTIC CONSTANTS[a] FOR THREE
REPRESENTATIVE GENERAL-ACID-CATALYZED REACTIONS

General acid	pK_{HA}	Acetone iodination[b]	4-Nitrobenzaldehyde hydration[c]	Dichloroketene dimethyl acetal hydrolysis[d]
H_3O^+	−1.74	2.8×10^{-5}	92	5.2
$CNCH_2CO_2H$	2.47	—	1.1	0.86
$ClCH_2CO_2H$	2.87	1.3×10^{-6}	0.82	0.96
$HOCH_2CO_2H$	3.83	3.3×10^{-7}	0.4[e]	0.32
CH_3CO_2H	4.76	1.0×10^{-7}	0.15	0.14
$(CH_3)_3CCO_2H$	5.03	7.5×10^{-8}	—	—
α	—	0.55	0.38	0.39

[a] Expressed in units of M^{-1} sec^{-1}, 25°C.
[b] Data of Bell and Lidwell (69).
[c] Data of McClelland and Coe (54); at equilibrium this compound is hydrated to the extent of 15% in water.
[d] Data of Kresge and Straub (22).
[e] Value for methyl ether of acid.

Table 7-4

BRØNSTED COEFFICIENTS AND CATALYTIC CONSTANTS[a] FOR THREE
REPRESENTATIVE GENERAL-BASE-CATALYZED REACTIONS

General base	pK_{HA} of conjugate acid	Acetone iodination[b]	4-Nitrobenzaldehyde hydration[c]	Isotopic exchange in 3[d]
H_2O	-1.74	1.0×10^{-11}	0.001	10^{-8}
$CNCH_2CO_2^-$	2.47	—	0.04	4.9×10^{-5}
$ClCH_2CO_2^-$	2.87	5.0×10^{-9}	0.06	1.1×10^{-4}
$HOCH_2CO_2^-$	3.83	3.9×10^{-8}	0.16^e	3.7×10^{-4}
$CH_3CO_2^-$	4.76	2.4×10^{-7}	0.47	3.0×10^{-3}
$(CH_3)_3CCO_2^-$	5.03	4.1×10^{-7}	—	6.3×10^{-3}
β	—	0.88	0.42	0.79

[a] Expressed in units of $M^{-1} sec^{-1}$.
[b] See footnote b, Table 7-3; $T = 25°C$.
[c] See footnote c, Table 7-3; $T = 25°C$.
[d] Data of Srinivasan and Stewart (61); see Eq. (7-46); $T = 35°C$.
[e] Value for methyl ether of acid anion.

enolization of acetone is an A-2 reaction [Eq. (7-21)]; the hydration of aldehydes we have called a pseudo-A-S_E2 reaction [Eq. (7-37)]; and the hydrolysis of ketene acetals is an A-S_E2 reaction [Eq. (7-24)].

In Table 7-4 are listed the Brønsted β coefficients for three reactions that are subject to general base catalysis, together with the catalytic constants of a number of bases. It will be noted that the latter are the conjugate bases of the general acids that appear in Table 7-3 and that were used as catalysts in the reactions referred to in the previous paragraph. Moreover, two of the reactions, the iodination of acetone and the hydration of 4-nitrobenzaldehyde, are also catalyzed by acids and also appear in Table 7-3. The third reaction, the isotopic exchange of **3**, is subject only to base catalysis, just as the third reaction of Table 7-3, the hydrolysis of the ketene acetal, is subject only to acid catalysis. The mechanisms of the iodination reaction [Eq. (7-29)] and the isotopic exchange [Eq. (7-46)] involve proton abstraction from the substrate. In the case of the aldehyde hydration [Eq. (7-39)] the proton is abstracted in the rate-controlling step from the nucleophile water.

It was noted earlier that the Brønsted coefficient tends to decrease in magnitude as the reaction rate rises, and the examples in Tables 7-3 and 7-4 conform to this notion, though this is certainly not always the case with reactions that are as disparate as those shown there.

What is the microscopic significance of the Brønsted coefficient; that is, what information does it provide about the structure of the transition state of a general-acid- or general-base-catalyzed reaction? Consider the simple proton abstraction process in Eq. (7-48). The stronger the base B and the more exergonic the reaction, the more does the transition state come to resemble reactants rather than products in terms of energy. According to the principle associated with the names of a number of eminent scientists (Bell, Leffler, Polanyi, and others) the resemblance should extend to structure; that is, the transition state in such circumstances should resemble more the reactants and less the products, though it need not be intermediate between them (70). As we have seen, high reactivity is often associated with low values of the Brønsted coefficient, and on this basis one can associate values of β (or α) between, say, 0 and 0.2 with a small degree of proton transfer in the transition state. By the same token a high value of β (between, say, 0.8 and 1.0) implies a high degree of proton transfer in the transition state. This corresponds to the proton being closer to the weaker base in the transition state. (See ref. 67 for an analysis of the significance of the Brønsted coefficient from this point of view.)

The discovery that in certain cases the Brønsted coefficient can lie outside the range of 0 to 1 (see next section) shows that this argument is not without flaws. Nonetheless, there appears to be merit in the general notion that the size of the Brønsted coefficient tells us something about the state of advancement of the transition state along the path between reactants and products. Isotope effects have also been actively studied as transition state probes, but they are subject to a number of ambiguities (Chapter 4, Section III), and many investigators continue to regard the Brønsted coefficients as better, though imperfect, measurements of the degree of proton transfer in the transition state (71). In this connection a number of investigators have examined the problem of rate–equilibrium relationships, of which the Brønsted equation is one, and have analyzed the causes of their not infrequent breakdowns (70, 74–78).

$$Z\text{—}H + B \longrightarrow [Z\cdots H\cdots B]^{\ddagger} \longrightarrow Z^{-} + HB^{+} \qquad (7\text{-}48)$$

Some time ago R. A. Marcus developed an equation that expresses the rate of an electron transfer reaction in terms of a number of simple quantities, and it has subsequently been applied to proton and group transfers. Essentially, the free energy of activation ($\Delta G^{\ddagger}$) for the reaction of X—Y with Z to give X and Y—Z is expressed in terms of the overall free-energy change of the reaction (ΔG^{0}) and the mean of the free energies of activation of the identity reactions X—Y + X and Z—Y + Z; a simplified form of the Marcus equation is given in Eq. (7-49), where $\Delta G^{\ddagger}_{\text{mean}}$ has the meaning just described.

$$\Delta G^{\ddagger} = \Delta G^{\ddagger}_{\text{mean}} \left(1 + \frac{\Delta G^{0}}{4\Delta G^{\ddagger}_{\text{mean}}} \right)^{2} \qquad (7\text{-}49)$$

The Brønsted coefficient expresses the variation of rate with respect to equilibrium (in logarithmic terms), and hence it is the differential of the free energy of activation with respect to free energy of reaction. Differentiating Eq. (7-49) in this way gives Eq. (7-50), which has proved useful in analyzing

$$\alpha = \frac{d\,\Delta G^{\ddagger}}{d\,\Delta G^{0}} = \frac{1}{2}\left(1 + \frac{\Delta G^{0}}{4\Delta G^{\ddagger}_{\text{mean}}}\right) \qquad (7\text{-}50)$$

substituent and isotope effects in proton transfer reactions (79) and in indicating the extent of proton transfer in the transition state, that is, whether the transition state is "early" or "late."

Kurz (74) emphasized the point that Marcus rate theory describes the reaction coordinate in terms of a single displacement variable and that some processes involve the making or breaking of several bonds, such as some of the acid- or base-catalyzed processes encountered herein [e.g., Eq. (7-40)]. The transition state may then be early with respect to one process and late with respect to another. These and other reactions can be analyzed in terms of free-energy surfaces in which displacement vectors parallel and perpendicular to the reaction coordinate are considered. The reader can find extensive accounts of the development (82–86) and applications (50, 54, 76, 78, 80, 87) of such an approach elsewhere.

Another technique that has been used in recent years to help delineate transition state structures for chemical and biochemical reactions involves rate measurements in protiated and deuteriated solvents and in mixtures of the two. This is sometimes called the *proton inventory* technique, leading references to which can be found elsewhere (88–92).

In a few cases general acids appear to function as traps for intermediates formed in a prior step; if the trapping rate is essentially independent of the strength of the general acid, then a Brønsted coefficient near zero will be observed. Such a process has been called *spectator catalysis* by Schowen (93). An example of a process that has some of these characteristics is the measurably slow proton transfer from hydrogen-bonded protonated diamines to a series of phenolate ions, in which a change in the base strength of the latter has a negligible effect on the rate of reaction [Eq. (7-51)]. The slowness of the

$$(7\text{-}51)$$

overall proton transfer has been ascribed partly to the non-hydrogen-bonded intermediate being present in low concentration and partly to proton removal from this intermediate in the trapping step being sterically hindered. The negligible Brønsted coefficient is a result of the trapping step being thermodynamically favorable (*95*).

How closely do most reactions that are subject to general acid or base catalysis conform to the Brønsted relation, and what are the structural factors in the catalyst that are likely to lead to deviations in individual cases? Reactions that take place by nucleophilic catalysis (Section III,C) frequently give poor Brønsted plots, particularly if the catalysts are highly polarizable (e.g., I^- or RS^-) or contain an atom with an unshared electron pair adjacent to the nucleophilic center (the α effect; e.g., HO_2^- or H_2NNH_2). On the other hand, when such species act as general bases, they generally fit fairly well on a Brønsted plot (*96*).

Although a number of reactions obey the Brønsted relation even when a variety of catalyst types are used [e.g., the dehydration of acetaldehyde hydrate (*99*) and to some extent the reaction of Fig. 7-2], in other cases there is a clear separation between the Brønsted lines generated by families of catalysts. This is frequently found to be the case when amines are the catalysts, with primary, secondary, and tertiary amines generating their own Brønsted lines. This is sometimes caused by the amines acting not as simple bases but as reagents that react with the substrate itself. The classic case in this regard is the formation of Schiff base intermediates during a number of condensation or other reactions of carbonyl compounds (*100–104*). An example of such a process is the deuterium exchange of the proton in acetone catalyzed by amines (*101*). The reaction takes place via the protonated enamine (Schiff base) intermediate as shown in Eq. (7-52), which is a lower energy route to

$$H_3C{-}\overset{\overset{\displaystyle O}{\|}}{C}{-}CH_3 + RND_2 \overset{D^+}{\rightleftharpoons} H_3C{-}\overset{\overset{\displaystyle {}^{+}NDR}{\|}}{C}{-}CH_3 \xrightarrow{-H^+} H_3C{-}\overset{\overset{\displaystyle NDR}{|}}{C}{=}CH_2 \xrightarrow{D_2O} H_3C{-}\overset{\overset{\displaystyle O}{\|}}{C}{-}CH_2D \qquad (7\text{-}52)$$

products than that in which the amine acts as a general base [Eq. (7-31)].

Tertiary amines cannot form enamines, and they are less effective than primary or secondary amines for this reason. Nonetheless, when compared to anionic oxygen bases their reactivity is generally much greater than would be expected, considering the differences in base strength. Thus, oxaloacetic acid undergoes facile enolization in the presence of tertiary amines, apparently via the carbinolamine intermediate shown in Eq. (7-53) (*105, 106*). The methylene

$$^{-}O_2C{-}\overset{\overset{\displaystyle O}{\|}}{C}{-}CH_2CO_2^- + R_3N \overset{H^+}{\rightleftharpoons} {}^{-}O_2C{-}\underset{\underset{\displaystyle {}^{+}NR_3}{|}}{\overset{\overset{\displaystyle OH}{|}}{C}}{-}CH_2CO_2^- \xrightarrow{R_3N} {}^{-}O_2C{-}\overset{\overset{\displaystyle OH}{|}}{C}{=}CHCO_2^- \qquad (7\text{-}53)$$

unit in the carbinolamine is activated to a greater extent by the groups attached to the adjacent carbon atom than is the case in the starting material. It undergoes rapid proton loss to a second molecule of amine in an elimination process that regenerates the tertiary amine and forms the enol.

Even when there is no alternative to direct proton transfer, as for example in the ionization of nitroethane, tertiary amines are more effective than oxyanions and, indeed, in this case more effective than their primary and secondary analogues. In accordance with the ideas enunciated earlier in this section, tertiary amines generate a Brønsted line with a lower slope than do oxyanions or other amines (98). That is, higher reactivity tends to be associated with lower Brønsted slopes, though exceptions to this rule of thumb certainly exist (107).

It is clear from this result and from that displayed in Fig. 7-2 that the role of steric effects in proton abstraction processes is not straightforward. Whereas there are many examples of sterically hindered bases behaving as expected and reacting somewhat sluggishly in proton abstraction processes there are several clear-cut cases of such species outperforming their unencumbered analogues in such reactions. In particular, steric retardation of proton transfer is a common occurrence when the base is a hindered pyridine (108–111), and steric acceleration has been observed in several instances when the base is a hindered carboxylate ion (60, 61, 112–114). (As noted earlier, steric retardation tends to be more pronounced in nucleophilic than in proton transfer reactions.)

Finally, it should be noted that the hydroxide ion and, to a lesser extent, the hydronium ion are much less effective in general-acid- and general-base-catalyzed reactions than would be suggested by the formal pK values for proton loss from the relevant forms ($pK_{H_3O^+} = -1.74$, $pK_{H_2O} = 15.74$). They can deviate by as much as several logarithmic units, which is fortunate, since otherwise these species would often obliterate the effect of weaker acids or bases, particularly for reactions with high Brønsted slopes (63, 116).

Anomalous Brønsted Coefficients

It had been an article of faith that Brønsted coefficients (sometimes called exponents because of the exponential nature of the Brønsted relation) could not be outside the range of 0 to 1. It was known, of course, that there were cases of weaker acids transferring protons at faster rates than stronger acids (e.g., phenol and 2,4-pentanedione), but it was assumed that these and countless other similar cases could be ignored because of the requirement for structural similarity in a Brønsted relationship. In 1969 F. G. Bordwell (117) drew attention to the inverse relationship between rate and extent of ionization of the closely related compounds nitromethane, nitroethane, and 2-nitropropane, which had been discovered almost 30 years earlier (118, 119).

Here proton abstraction by hydroxide ion has the relative rates of $113:18:1$, whereas the K_{HA} values change in the opposite manner, $6 \times 10^{-11}, 3 \times 10^{-9}$, and 2×10^{-8}. This corresponds to a Brønsted α value of -0.5 and, since $\alpha + \beta$ must sum to unity for forward and reverse reactions, protonation of the anions has a β value of 1.5. (As before, we use α when the structure of the acid varies and β when the structure of the base varies.) This result, which has been confirmed and refined (120) cannot be dismissed on the grounds that the structural changes in the nitroalkane series are too severe, since Bordwell showed that varying the aryl substituent in the series shown in Eqs. (7-54) and (7-55) also produced anomalous results (121). In these cases the ring substituents change the rate and equilibrium in the same direction, but the effect on the rate is *greater* than on the equilibrium, which produces α coefficients that are greater than unity. For protonation of the anions, of course, the β values will be negative.

$$ArCH_2NO_2 + HO^- \longrightarrow ArCHNO_2^- + H_2O \qquad \alpha = 1.54 \qquad (7\text{-}54)$$

$$\underset{\underset{CH_3}{|}}{ArCH_2CHNO_2} + HO^- \longrightarrow \underset{\underset{CH_3}{|}}{ArCH_2CNO_2^-} + H_2O \qquad \alpha = 1.61 \qquad (7\text{-}55)$$

What is special about the ionization of nitroalkanes that produces such unusual results? There now appears to be a consensus that the transition state and the nitronate anion have quite different geometries and charge distributions. It is believed that in the transition state most of the charge that has been transferred from the hydroxide ion resides on the α-carbon atom, whereas in the nitronate ion, which is planar, it is almost entirely on the oxygen atoms (67, 120, 122–126). Hence, the transition state (4) resembles neither reactant nor product (5) and is affected by substituents in its own way, thereby giving rise to

Brønsted coefficients that are out of the normal range. The importance of solvation in such reactions (127) is illustrated by the reaction shown in Eq. (7-54) giving a "normal" Brønsted coefficient (0.92) when the reaction is conducted in dimethyl sulfoxide (DMSO) solution (122).

When the structure of the nitroalkane is kept fixed and that of the base is varied, Brønsted coefficients in the normal range are obtained (98, 126, 128). The value of β for ionization of nitroethane by tertiary amines is 0.45, for secondary and primary amines 0.6, and for oxyanions 0.71. This is also the order of decreasing reactivity; for example, a tertiary amine whose pK_{BH^+} is 10.0 is 13-fold and one whose pK_{BH^+} is 6.2 is 130-fold as reactive as an oxyanion of the same base strength (98). That is, the more reactive the series, the lower the Brønsted coefficient tends to be, as was found to be the case in most of the examples cited earlier. (In the absence of curvature, of course, the Brønsted lines must cross at some point and the reverse situation obtain.)

E. Normal Acids and Pseudoacids

Eigen classified acids and bases as "normal" when minimal structural and electronic reorganization accompanies proton loss or gain (129). Most oxygen acids and nitrogen bases are of this type, and proton transfer between oxygen atoms or between oxygen and nitrogen atoms is almost always fast. Virtually all carbon acids, on the other hand, have extensive charge delocalization in their anions that requires substantial geometrical and solvation changes to take place. Furthermore, the incipient proton in the un-ionized acid does not have the benefit of hydrogen bonding to a water molecule as does the proton of an oxygen acid. As a result the ionization of most carbon acids is slow, and the term *pseudoacid* has been used for many years for such species.

Normal acids give rise to biphasic Brønsted plots (Fig. 7-3), which are commonly called Eigen curves. The slopes are unity and zero over most of the range of acid and base strength, with there being narrow regions of sharp curvature where the proton is transferred between bases of comparable strength. The regions of zero slope correspond to exergonic (energetically downhill) proton transfer in which the rate of diffusion is the limiting factor.

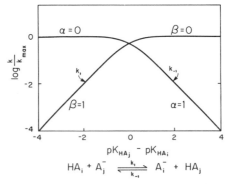

Fig. 7-3 Brønsted plot showing the effect of changes of equilibrium acid strength on the rate of proton transfer for "normal" acids. The maximum rate is that for diffusion control.

There are a few carbon acids that undergo minimal structural change on ionization and thus might be expected to behave more as normal acids than as pseudoacids—for example, hydrogen cyanide, alkynes, and haloforms. Whereas the rate of proton loss from hydrogen cyanide is indeed fast, that from phenylacetylene or chloroform, for example, is slow. However, Kresge *et al.* showed that these two carbon acids, whose pK_{HA} values they have estimated to be 20.0 and 24.1, respectively, give Brønsted plots of essentially unit slope [Eqs. (7-56) and (7-57)]. They argued that these reactions

$$C_6H_5C{\equiv}CH + B\text{:} \longrightarrow C_6H_5C{\equiv}C^- + BH^+ \qquad (7\text{-}56)$$

$$CHCl_3 + B\text{:} \longrightarrow {}^-CCl_3 + BH^+ \qquad (7\text{-}57)$$

correspond to the thermodynamically uphill regions of ionization of normal acids and that proton transfer is rapid and reversible. This would mean that the rate-controlling step of the dissociation is separation of the ion pairs that are the initial products of the proton transfer (*124*).

F. Concerted and Bifunctional Acid–Base Catalysis

In 1925 Lowry and Faulkner suggested that those reactions that are catalyzed by both acids and bases normally take place by simultaneous attack of acid and base on the substrate (*130*). Such a proposal is not inconsistent with the second-order kinetics normally observed in general-acid- or general-base-catalyzed reactions in aqueous solution since the third species could be a water molecule, which can function as an acid or a base as required. Furthermore, in the presence of substantial concentrations of buffer a third-order term has been observed in a number of cases, in the enolization of acetone, for example, a reaction that has been repeatedly examined over a period of more than half a century.

Dawson and Spivey (*131*) in 1930 noted that the rate law for the iodination of acetone in acetate buffer contained a small term of the form $[\text{acetone}][CH_3CO_2H][CH_3CO_2{}^-]$, and this was subsequently confirmed by the work of a number of investigators, including some of the most eminent in the field (*132–134*). The third-order term corresponds to a termolecular mechanism, probably of the type suggested by Hand and Jencks (*135*) and shown in Eq. (7-58). Because of the presence of hydrogen bonding (*136, 137*)

$$\underset{H_3C-\overset{\overset{\displaystyle HA}{\cdots}}{\underset{\|}{C}}-CH_3 + A^-}{} \longrightarrow \underset{H_3C-\overset{\overset{\displaystyle A^-}{\cdots}}{\underset{|}{C}}=CH_2 + HA}{} \qquad (7\text{-}58)$$

the forward and reverse reactions as depicted are not in the strictest sense termolecular, although that term is often used for any mechanism that involves the concerted action of acid and base, as is the term *push–pull mechanism* (*138*).

The question arises as to whether the catalytic term in acetic acid and the catalytic term in acetate ion correspond to concerted proton transfers such as those shown in Eqs. (7-59) and (7-60). It turns out that the contribution of the

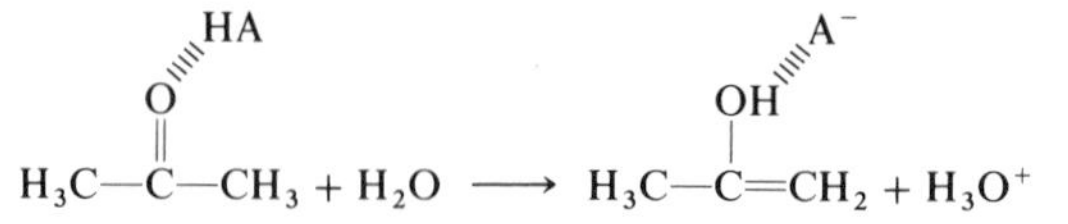

$$H_3C-\overset{\overset{\displaystyle HA}{\underset{\|}{O}}}{C}-CH_3 + H_2O \longrightarrow H_3C-\overset{\overset{\displaystyle A^-}{\underset{|}{OH}}}{C}=CH_2 + H_3O^+ \qquad (7\text{-}59)$$

$$H_3C-\overset{\overset{\displaystyle H_2O}{\underset{\|}{O}}}{C}-CH_3 + A^- \longrightarrow H_3C-\overset{\overset{\displaystyle {}^-OH}{\underset{|}{OH}}}{C}=CH_2 + HA \qquad (7\text{-}60)$$

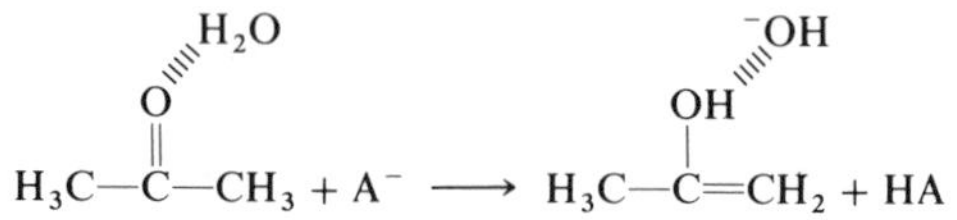

third-order term is insignificant with some substrates and, even in the case of acetone, with some buffer acid–base pairs, and Bell has marshaled a considerable body of evidence to show that concerted routes such as those given in Eqs. (7-59) and (7-60) are *not* major pathways for the general-acid- and general-base-catalyzed routes (*139*). That is, they cannot compete effectively with the A-2 and B-2 mechanisms given earlier (Sections III,A and B), at least in aqueous solution.

With regard to the concerted or termolecular reaction of acetone, which has been depicted here as proceeding through the mechanism of Eq. (7-58), Hegarty and Jencks showed that changes in strength of the buffer acid have only a small effect on the third-order rate constant. This result is not unexpected since changes in the strength of the acid should be largely offset by changes in the strength of the base, this being the one route that has both protons simultaneously in flight (*133, 134*).

The role of bifunctional (tautomeric) catalysts, such as the monoanions of carbonic, phosphonic, and other polyprotic acids, is described in Chapter 4, Section II,A,2. These and other species such as 2-pyridone often can effect proton transfer from one site to another in an organic substrate by adding and removing protons in a concerted fashion (*140*).

An interesting case of bifunctional catalysis that does not involve concerted proton transfers was observed by Hine *et al.* (*141–144*). They showed that diamines containing primary and tertiary amino groups separated by the appropriate number of carbon atoms can catalyze the isotopic exchange of the α-hydrogen atoms of ketones in a number of ways. As outlined earlier, formation of a protonated Schiff base (iminium ion) activates carbonyl groups toward proton loss to external bases, and it is clear that in the present case the primary amine forms the reactive iminium intermediate and the tertiary amine then acts as an internal base. Hine showed that in some cases the α-hydrogens on both sides of the carbonyl group undergo "all-at-once" exchange; this takes place, for example, during exchange of cyclopentanone-d_4 catalyzed by 3-(dimethylamino)propylamine, where the concentration of either the starting material (d_4) or the final product (d_0) is at all times greater than the combined concentrations of the d_1, d_2, and d_3 ketones (143).

With acetone and certain diamines, however, all of the hydrogen atoms on one side of the carbonyl group tend to exchange before an appreciable amount of exchange occurs on the other side. That is, the principal intermediate that can be isolated during the conversion of $(CD_3)_2C{=}O$ to $(CH_3)_2C{=}O$ is CD_3COCH_3. These results can be explained by the effect of structural variations on the rate constants of the several steps of Eq. (7-61) and by the fact

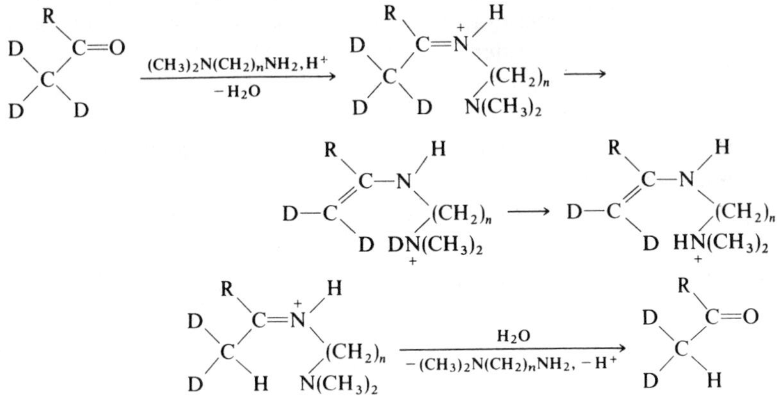

$$\tag{7-61}$$

that the cis–trans nature of the double bond in imines will cause the basic group to be near one set of α-hydrogen atoms and distant from the other. If the imine undergoes cis–trans isomerization faster than it is hydrolyzed back to ketone, rapid exchange will occur on both sides of the carbonyl group within the imine intermediate, and by the time reversion to ketone occurs most of the active hydrogens will have been exchanged. If imine isomerization is relatively slow, however, the tertiary amino group may effect exchange only on the one side before the carbonyl group is regenerated.

G. Rate–Acidity Profiles

The rates of those reactions that are subject to specific acid or specific base catalysis are simple functions of the acidity of the solution, since the rates depend only on the equilibrium concentration of the protonated or deprotonated forms of the substrate. A plot of $\log k$ against pH for a specific-acid-catalyzed reaction of a molecule Z will give a straight line of negative unit slope in the region where the degree of protonation is slight. This will change to a line of zero slope as the degree of protonation becomes high, provided that pK_{BH^+} occurs in the aqueous region. Similarly, a specific-base-catalyzed reaction of a molecule YH will give a straight line of positive unit slope when the extent of formation of Y^- is small and one of zero slope when the degree of ionization is high, provided that pK_{HA} occurs in the aqueous region.

With very weakly acidic or basic substrates a high degree of ionization can be attained only if powerfully acidic or powerfully basic media are used; under such conditions, where activity coefficients depart appreciably from unity, the rate may continue to change as the acidity or basicity is further increased.

Reactions that are subject to both specific acid and specific base catalysis (but not to general catalysis) are rare. Their pH–rate profiles will resemble that shown in Fig. 7-6 for the general-acid- and general-base-catalyzed enolization of acetone in the absence of buffers, since in the latter case the contribution of the water (uncatalyzed) term does not exceed the sum of the H^+ and OH^- terms at any pH. Specific-acid- or specific-base-catalyzed reactions presumably also have "uncatalyzed" routes available. For example, acetal hydrolysis, which normally takes place by unimolecular reaction of the protonated form could also take place by spontaneous decomposition of the neutral molecule to produce an ion pair, $R_2\overset{+}{C}OR$ and RO^-, which could, in turn, be trapped by reaction with water. It is difficult to gauge the degree of activation that prior

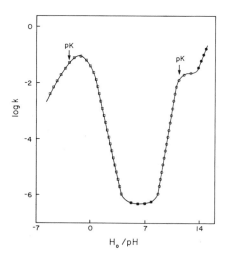

Fig. 7-4 Rate of iodination of the 4-methyl group in **6** [Eq. (7-62)] as a function of the acidity of an unbuffered aqueous solution. The symbols do not represent experimental points but, rather, indicate the predominant mechanism at the particular acidity. ●, Eq. (7-62); ○, Eq. (7-63); ■, Eq. (7-64); □, Eq. (7-65). Data from ref. *145*.

protonation of the substrate produces in such cases, but we shall see that for many general-acid-catalyzed reactions the effect is very large, often greater than 10^7.

Reactions that are subject to general acid or general base catalysis produce a wide variety of rate–acidity plots whose shapes are determined by the rate and equilibrium constants that are pertinent in each particular case. This is only to be expected since in general catalysis a variety of weak acids and bases, usually in the form of buffers, can contribute to the reaction.

Even in unbuffered systems fairly complex curves can result from a change in the acidity of the medium. In Fig. 7-4 is shown a plot of the rate of isotopic exchange of the hydrogens of the 4-methyl group of the imino pyrimidine **6** as the acidity of the solution is altered from 5 M aqueous sodium hydroxide to 75% sulfuric acid. The rate changes by a factor of almost 10^6 over this span, with the minimum appearing near pH 6. (The trough in the neutral region can be filled in to a considerable extent if selected buffers are added to the solution.) The reactions that make the principal contributions to the ionization of the reactive methyl group are the four shown in Eqs. (7-62) to (7-65);

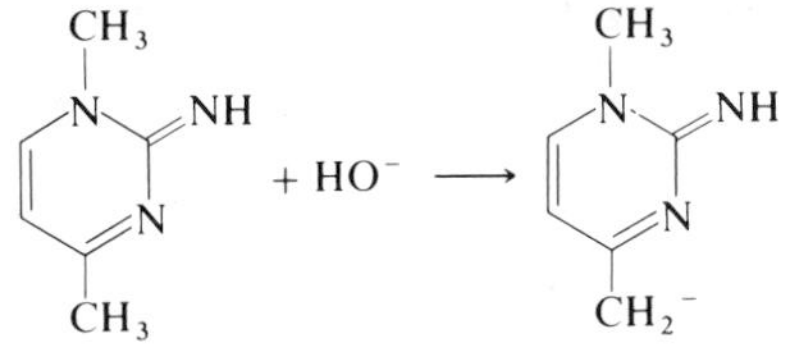

$$\text{(7-62)}$$

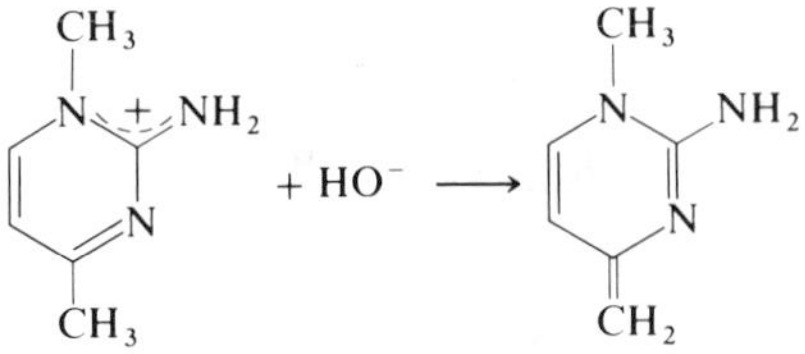

$$\text{(7-63)}$$

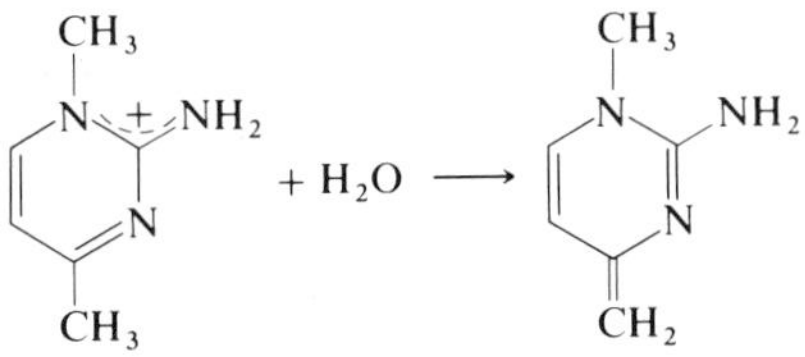

$$\text{(7-64)}$$

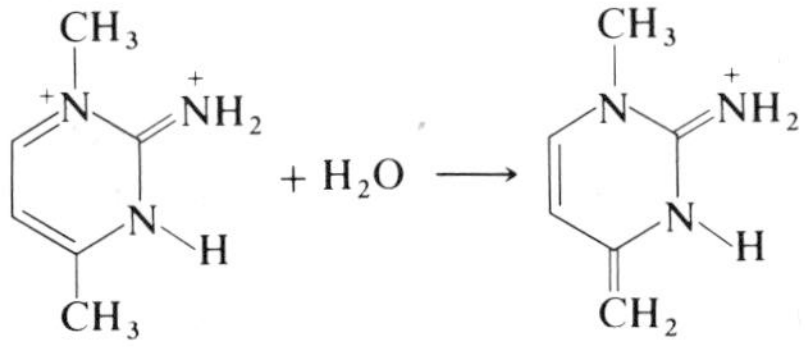

$$\text{(7-65)}$$

for simplicity's sake all hydrogen atoms are shown in the protio form and no indication of delocalization of charge is given in the intermediates that are formed and that lead to isotopic exchange.

The regions of the acid–base spectrum that are dominated by each of these four mechanisms is indicated in the figure. The first and second protonations of **6** occur at the points shown and are consistent with the mechanistic assignments given on the plot. The decrease in rate in strongly acidic solution is caused principally by the decrease in activity of water, which acts as the general base in this region [Eq. (7-65)].

The reaction of Eq. (7-63), which is the dominant route between pH 8 and 13, is kinetically equivalent to reaction between the neutral iminopyrimidine and water, but as will be seen later the latter route is expected to be the less important of the two in cases such as this.

For the protolytic reaction of a monobasic substrate Z in aqueous solution, there are four possible reaction paths [Eqs. (7-66)–(7-69)].

$$ZH^+ + HO^- \longrightarrow \text{reactive intermediate} \tag{7-66}$$

$$ZH^+ + H_2O \longrightarrow \text{reactive intermediate} \tag{7-67}$$

$$Z + HO^- \longrightarrow \text{reactive intermediate} \tag{7-68}$$

$$Z + H_2O \longrightarrow \text{reactive intermediate} \tag{7-69}$$

For purposes of illustrating the contribution each route makes to the reaction, we assign the following relative second-order rate constants (these are arbitrary but not untypical values for prototropic processes):

$$k_{Z+HO^-} = 1$$

$$k_{ZH^+ + H_2O} = 10^{-3}$$

$$k_{ZH^+ + HO^-} = 10^7$$

$$k_{Z+H_2O} = 10^{-10}$$

We can now calculate the contribution that each route will make for substrates of various basicities. The results for three substrates whose pK_{BH^+} values are, respectively, -3.0, 3.0, and 10.0 are shown in Fig. 7-5. There are a number of points of difference in the plots in the three cases, the most obvious being the different shapes. Furthermore, a closer inspection reveals that the kinetically equivalent routes ($Z + H_2O$ and $ZH^+ + HO^-$) switch their order of importance in going from graph (a) to (b), that is, when the basicity of the substrate is increased from -3.0 to 3.0. In the third case, where $pK_{ZH^+} = 10$, the relative contribution of the $Z + H_2O$ route is relatively small, and there is little doubt that this condition also applies to the iminopyrimidine **6** (pK_{BH^+} = 11.3), which was discussed earlier. For the two routes to make equal contributions in this case the rate constant of the $ZH^+ + HO^-$ route would

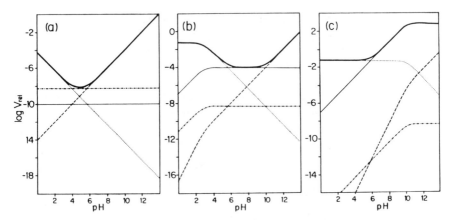

Fig. 7-5 Plots of logarithm of relative rate against pH of each of the pathways of Eqs. (7-66) to (7-69) for a protolytic reaction of substrate Z. (a) $pK_{ZH^+} = -3.0$; (b) $pK_{ZH^+} = 3.0$; (c) $pK_{ZH^+} = 10.0$. The rate constants are those assigned in the text. The heavy curve represents the observed rate, the light solid curve the contribution of $ZH^+ + HO^-$, the dashed curve that of $Z + HO^-$, the dotted curve that of $ZH^+ + H_2O$, and the dashed–dotted curve that of $Z + H_2O$. Reproduced with permission from ref. *160*.

have to be only 10^4 times greater than that of the $Z + H_2O$ route, which would correspond to unusually small degrees of activation of the substrate by protonation (ZH^+ versus Z) and of the base by deprotonation (HO^- versus H_2O). Activating factors such as these, designated paf and dpf, are described in a subsequent section. We shall see there that some weakly basic organic compounds prefer to react by the alternative route, that is, by the reaction of water with the neutral substrate.

For some compounds one can find conditions for which all four routes shown in Eqs. (7-66) to (7-69) make comparable contributions to the rate. Exchange of the methylene protons of creatinine (**7**) in neutral aqueous solution is one such system. The pK_{BH^+} of this compound is 4.7, and between pH 6 and 7, where most of the substrate is in the neutral form, the four routes shown in Eq. (7-70) all make significant contributions to the overall rate (*14*,

$$\text{(7-70)}$$

7

146). Indeed, the base strength of this compound is such that it can itself act as a base with the consequence that a fifth path due to self-catalysis can also be observed in all but very dilute solutions of the substrate.

Returning to the shapes of rate–acidity curves, it is sometimes found that compounds of closely related structure behave quite differently. Thus, the detritiation of 1,2,4-triazole-*t* (**8**) at 80°C studied by Jones *et al.* (*147*) gives a flat-topped bell-shaped curve in aqueous solution, with the maximum rate occurring in the pH region 3–7. This is because the mechanism of Eq. (7-71),

$$
\underset{\textbf{9}}{\text{structure}} \xrightarrow{\text{H}^+} \underset{\textbf{8}}{\text{structure}} \xrightarrow{\text{H}^+} \underset{\textbf{10}}{\text{structure}} \xrightarrow{\text{HO}^-} \underset{\textbf{11}}{\text{structure}} \tag{7-71}
$$

which produces the species **11**, is the only one of importance in purely aqueous solution. The pK_{BH^+} and pK_{HA} values of **8** are 2.0 and 8.9, respectively, at this temperature, and the decline in rate below pH 3 is caused by the hydroxide concentration decreasing while the concentration of the cation **10** approaches its maximum value. Between pH 3 and 8 the concentrations of **10** and hydroxide ion are inversely affected by changes in the acidity of the solution, which means that the rate is independent of pH. At higher pH the anion **9** is formed, which further reduces the concentration of the active intermediate, leading to a decline in the rate of reaction.

By way of contrast, the rate profile for the closely related tetrazole-*t* (**12**) is

$$
\underset{\textbf{12}}{\text{structure}}
$$

the inverse of that just described. Here, the minimum rate is in the pH region 6–11, with the rate rising sharply in both acidic and basic solution. In basic solution deprotonation of the tetrazole anion becomes important, and in acidic solution a term in the neutral substrate and the proton appears. The term in acid could be the result of either rate-controlling carbon protonation or of water acting as a general base on the conjugate acid of the substrate, although it should be pointed out that general catalysis by buffers was not detected in this or analogous systems. A discussion of the mechanism and rate profiles of a number of more complex heterocyclic systems, particularly purines, can be found in the review of Jones and Taylor (*148*).

Bell-shaped curves are a common feature of reactions of the carbonyl group. For example, the reaction of acetone with excess hydroxylamine gives a curve that peaks between pH 4 and 5 and drops off sharply on either side.

Jencks showed that this profile can be quantitatively accounted for on the basis of a change in pH causing a change in the rate-controlling step of the two-step reaction shown in Eq. (7-72) (149).

$$(CH_3)_2C{=}O + H_2NOH \longrightarrow (CH_3)_2C(OH)NHOH \xrightarrow{\ H^+\ } (CH_3)_2C{=}NOH \quad (7\text{-}72)$$

Formation of the addition compound takes place principally by way of attack of neutral hydroxylamine on the carbonyl group. Since hydroxylamine is largely in the protonated form in acidic solution [$pK_{BH^+} = 5.95$ (150)] and since the second step is catalyzed by the proton (little or no buffer catalysis is observed), the first step is rate determining at low pH. As the pH is raised the rate of the first step increases (because of an increase in the concentration of free hydroxylamine) and that of the second step decreases (because of the decrease in the concentration of the acid catalyst), with the result that the second step becomes rate determining. As a consequence, the overall rate passes through a maximum and then declines with increasing pH.

A pH–rate profile that is bell-shaped is often [though not always (151)] a sign that the reaction path is multistepped, with the rate-controlling step changing as the pH changes (153). Rate–acidity plots that are concave upward are much more common and show that there are available two or more reaction paths that have different pH dependencies. There is a close analogy here to Hammett plots, where concave-downward curves also indicate a change in rate-controlling step and concave-upward plots the presence of alternative reaction paths.

Buffer Catalysis

It was pointed out earlier that hydroxide ion and, to a lesser extent, hydronium ion are less effective as general catalysts than would be expected from their equilibrium strengths. Were this not so, catalysis by low concentrations of buffer would frequently be unobservable, particularly for reactions with high Brønsted coefficients, since the hydronium or hydroxide terms would be dominant in all regions. (The isotopic exchange reactions of the ring hydrogens of heterocyclic compounds, discussed earlier, are exceptional in generally showing hydroxide dominance even in buffer solutions.)

The changes in the shape of rate–acidity profiles that are brought about by the presence of buffers are dependent on a number of reaction parameters. For a reaction such as the enolization of a carbonyl compound these would be the catalytic coefficients of the acid and anion of the particular buffer used, the pK_{BH^+} value of the substrate, and the value of the Brønsted coefficient.

Taking acetone as an example of a typical substrate, the curves shown in Fig. 7-6 represent the rate of enolization (or enolate anion formation) under three sets of conditions: in the absence of buffer, in the presence of acetic

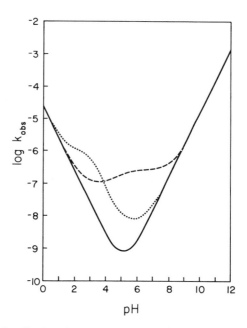

Fig. 7-6 Rate of enolization of acetone in aqueous solution at 25°C. ——, No buffer present; ---, 1 M acetate buffer present; ·····, 1 M chloroacetate buffer present. Curves calculated from data of refs. *154* and *160*. Rates are in units of reciprocal seconds.

acid–acetate buffer, and in the presence of chloroacetic acid–chloroacetate buffer. For the sake of convenience, the concentration of buffer has been taken as 1 M. The difference in the shapes of the two buffer curves is caused by the buffer anion being the more effective catalyst in the case of the acetic acid system and the carboxylic acid being the more effective catalyst in the case of the chloroacetic acid system. The position of the minimum in the uncatalyzed reaction is given by Eq. (7-73), and it can be shown that contributions from

$$[H^+]_{min} = (10^{-14}k_{HO^-}/k_{H^+})^{1/2} \qquad (7\text{-}73)$$

both components of the buffer will be most clearly evident when the pK of the buffer acid falls near this point.

The higher the pK_{HA} of the buffer acid the more effective is the anionic form and the less effective the acidic form, which means that a phenol buffer, for example, will show only phenoxide, not phenol, catalysis of acetone enolization. Conversely, only the contribution of the neutral form of a buffer acid with a very low pK will be detectable, which is almost the situation with the chloroacetic acid buffer (pK_{HA} = 2.87) shown in Fig. 7-6. Only in a narrow region near pH 6 is the dominant catalytic species the chloroacetate ion, and then only when rather high buffer concentrations are used. On the other hand, in acetate buffer the anion is the dominant species in solution over a span of several pH units, even at quite low buffer concentration.

The catalytic coefficients for the acidic and anionic components of acetate and chloroacetate buffers in the enolization of acetone are as follows: acetate, $k_{HA} = 1.01 \times 10^{-7}$ and $k_{A^-} = 2.43 \times 10^{-7}$; chloroacetate, $k_{HA} = 1.27 \times 10^{-6}$ and $k_{A^-} = 5.0 \times 10^{-9}$ [the values are expressed as $M^{-1}\,\text{sec}^{-1}$ at $25°C\,(154)$]. It is clear that the relative effectiveness of the acidic and anionic forms is reversed in going from one buffer system to the other. That is, the anion is the more effective component in the acetate case, while the acid is the more effective component in the chloroacetate case, and this is responsible for the shapes of their curves in Fig. 7-6 being different. It can be readily shown that the strength of the buffer acid whose acidic and basic components have identical coefficients in a prototropic reaction is given by Eq. (7-74), in which K_{ZH^+} is the

$$K_{HA} = K_{ZH^+}/\text{paf} \qquad (7\text{-}74)$$

acidity constant of the conjugate acid of the substrate and paf is a quantity that measures the degree of activation of the substrate that protonation brings about.

The quantity paf [proton activating factor (14)] is the ratio of the rate constants for reaction of a base with the protonated and unprotonated forms of a substrate [Eq. (7-75)]. It is not simply the ratio of the experimental

$$\begin{array}{l} ZH^+ + A^- \xrightarrow{\ K_{ZH^+ + A^-}\ } \\[4pt] H^+ \Big\Updownarrow \qquad\qquad\qquad\qquad \text{products or} \qquad\qquad (7\text{-}75)\\[4pt] Z + A^- \xrightarrow{\ K_{Z + A^-}\ } \qquad \text{reaction intermediates} \end{array}$$

catalytic constants k_{HA} and k_{A^-}, since the equilibrium base strength of both the substrate and the base are also relevant. (The rate constant k_{Z+A^-} and the catalytic constant k_{A^-} are, in fact, identical, whereas the quantities $k_{ZH^+ +A^-}$ and k_{HA} are not.) Equation (7-76) expresses paf for a prototropic or similar

$$\text{paf} = \frac{k_{ZH^+ +A^-}}{k_{Z+A^-}} = \frac{k_{HA}K_{ZH^+}}{k_A - K_{HA}} \qquad (7\text{-}76)$$

reaction of a substrate Z with a base A^- (frequently a buffer anion) in terms of the catalytic constants k_{HA} and k_{A^-} and the dissociation constants of the protonated substrate and the buffer acid.

Values of paf can sometimes be obtained for nitrogen bases by using quaternary salts as surrogates for the conjugate acid. In these cases a simple comparison of rate constants for reaction of a base with the substrate and with the alkylated cation will give paf directly, provided that the effects on the reactive site of alkyl and hydrogen are the same. There are a number of cases in which this condition appears to be nearly met (155), and a number in which it is not (156, 157). Curiously, alkylation adjacent to the reactive site of a prototropic molecule often produces a greater degree of activation than does protonation, which is not what would be expected on simple electronic or steric grounds.

The magnitude of paf for a prototropic reaction can vary considerably, depending on the structure of the substrate and on the extent to which the charge in the product or intermediate is delocalized. For deprotonation at the α position of aliphatic ketones, where the negative charge in the carbanion is spread between an oxygen and a carbon atom, values in the range of 10^7 to 10^8 are commonly found. On the other hand, in a molecule such as 13 (Table 7-5) there is a high degree of delocalization in the anion, and here the paf is only $\sim 10^2$. This is a result of the rate constant for the base-catalyzed route being much larger for this compound than for, say, acetone, whereas the rate constants for the general-acid-catalyzed routes for these two compounds are much closer together.

Table 7-5

COMPARISON OF THE CATALYTIC CONSTANTS[a] AND paf VALUES FOR DEPROTONATION OF ACETONE AND 2,4,6-TRIMETHYLLUMAZINE (13) BY ACETATE ION[b]

Parameter	CH_3—$\overset{O*}{\underset{\|}{C}}$—$CH_3^{\ddagger}$	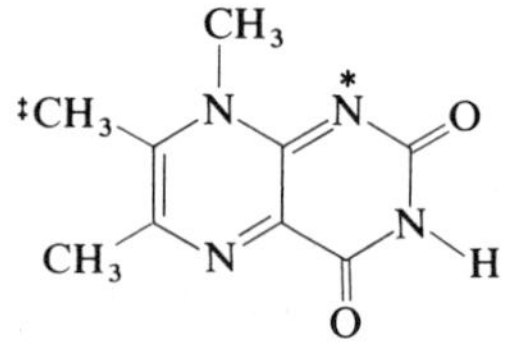 13
$k_{ZH^+ + A^-}$	5.0^c	2.9
$K_{Z + A^-}$	2.4×10^{-7}	0.014
paf	2.0×10^{7c}	210

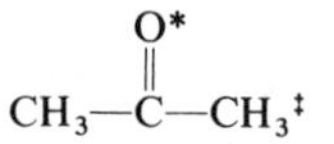

[a] Rate constant in units of M^{-1} sec^{-1} in water at 25°C (14, 154, 158).
[b] Position of protonation marked by asterisk (*); site of proton loss marked by double dagger ($\ddagger$).
[c] Calculated using a pK_{BH^+} of -2.9 for acetone; if the more negative values that have sometimes been proposed for this composed are used, the paf value will be increased by two to three powers of 10 (see Chapter 3, Section V,A).

There are some carbon acids that depend on polar effects rather than resonance to stabilize their anions; as a consequence deprotonation of the conjugate acid produces mesoionic intermediates. 9-Isopropylpurine (**14**) is

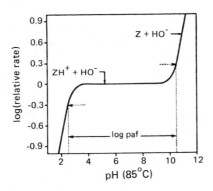

14

one such molecule. Here buffer catalysis in aqueous solution cannot be observed since exchange at the reaction site is due entirely to hydroxide ion, which acts either on the neutral compound or on its conjugate acid (*159*). (In **14** the reaction site is marked by a dagger and the important protonation sites are marked by asterisks.) In this case paf can be determined directly from the rate–pH profile since the two routes, $Z + HO^-$ and $ZH^+ + HO^-$, must make equal contributions when $[H^+] = K_{ZH^+}/\text{paf}$, that is, at a point 0.3 logarithmic unit above the plateau in Fig. 7-7. The hydrogen ion concentration at this point is 3×10^{-11}, and since K_{ZH^+} for 9-isopropylpurine at the temperature of the exchange studies (85°C) is 3×10^{-3}, a value of 10^8 for paf is obtained. This quantity can also be read from the graph, as shown in Fig. 7-7.

Whereas activation of the substrate in a prototropic reaction is caused by protonation, activation of the base is caused by deprotonation. The relevant term here is dpf (deprotonating factor). Almost all the information available on dpf comes from the H_2O/HO^- pair, which gives dpf values ranging from $\sim 10^8$ for 2,4-pentanedione to $\sim 10^{15}$ for thiazole (*14*). In general, dpf can be expected to be larger than paf since the proton is removed directly from the

Fig. 7-7 Rate–acidity profile for the exchange of the 8-hydrogen in 9-isopropylpurine (**14**). Reproduced with permission from ref. *14*. Copyright 1978 American Chemical Society.

atom that is the basic site in the former and added to an atom that is, at best, adjacent to the atom that is the acidic site in the latter.

It is clear that the mechanism operative at the plateau in Fig. 7-7, that is, between pH 3 and 10, is the reaction of ZH^+ with HO^- and not the kinetically equivalent reaction of Z with H_2O. Were the latter the mechanism, the pH–rate profile would not turn down as the pH approached the pK of the substrate, as is observed, but would turn up as the contribution of the $ZH^+ + H_2O$ route came into play.

With simple ketones, on the other hand, there are good reasons for believing that the more important of the kinetically equivalent routes is that between neutral ketone and water, not that between the ketone conjugate acid and hydroxide ion. It has been frequently noted that replacing a hydrogen atom of a ketone by halogen increases both the water- and hydroxide-catalyzed rate of proton loss, whereas the hydrogen ion–catalyzed reaction is almost unaffected. Were the water-catalyzed reaction due to the reaction of hydroxide ion with the conjugate acid of the ketone, the effects of halogen on the basicity of the carbonyl oxygen and on the rate of the proton abstraction would be expected to largely offset one another, as is the case with hydrogen ion catalysis.

For those reactions for which the catalytic constants for H^+, HO^-, and H_2O are all known the relative velocities V of the two kinetically equivalent routes, though not which is the greater, can be estimated using Eq. (7-77);

$$\log \frac{V_{Z+H_2O}}{V_{ZH^+ + HO^-}} = \pm \log \frac{K_w k_{H^+} k_{HO^-}}{k_{H_2O}^2} \tag{7-77}$$

here k_{H_2O} is the first-order catalytic constant, that is, the rate constant for which the concentration of water is not considered. Equation (7-77) is derived on the assumption that paf is independent of the base (H_2O or HO^-) and that dpf is independent of the ionic state of the substrate (Z or ZH^+). For aliphatic ketones the ratio of rates given by Eq. (7-77) is one or two orders of magnitude, which, for reasons already given, can be presumed to be the extent to which the $Z + H_2O$ route is favored over the $ZH^+ + HO^-$ route (160).

IV. Reactions Conducted under Strongly Acidic or Strongly Basic Conditions

It was noted by Zucker and Hammett in the 1930s that the response of acid-catalyzed reactions to the use of strongly acidic conditions, say, $> 10\%$ aqueous sulfuric acid, depends on the kind of reaction being studied (160a). They observed that sucrose hydrolysis, which is subject to specific acid catalysis (an A-1 reaction involving unimolecular decomposition of the conjugate acid of the substrate) was affected more than was ketone enolization (an A-2 reaction between the conjugate acid and a molecule of base) by an increase in the amount of sulfuric acid in the solution. Since water is the predominant base in such media and since its activity declines as the concentration of acid is increased, it is entirely reasonable that there should be a smaller rate increase exhibited by the A-2 than by the A-1 reaction. Furthermore, in the sucrose case a plot of $\log k$ against the acidity function H_0 was found to be linear and of unit slope, which must be a result of the ratio of the activity coefficients of substrate and transition state being equal to that of the indicator pairs used to construct the H_0 scale. In the ketone enolization case an analogous relationship between $\log k$ and acid molarity was observed, and for a time in the 1940s and 1950s it appeared as though all A-1 and A-2 reactions might follow this pattern (the Zucker–Hammett hypothesis) (161). It subsequently became clear that there is a gradation of responses of rate to increases in the acid concentration of the medium.

In the mid-1960s J. F. Bunnett developed a general method of treating reactions conducted in aqueous sulfuric acid and similar media that helps delineate the role of water in such reactions (162). He showed that plots of $\log k + H_0$ against the logarithm of the activity of water were in most cases linear but that the slopes of the plots, designated w values, varied according to the process occurring in the rate-controlling step. Considering only those processes in which protonation occurs at a nitrogen or oxygen atom, he found that A-1 reactions give slopes that are close to zero or are negative (as with sucrose). On the other hand, A-2 reactions give positive slopes, with a further division between those in which water acts as a base and those in which it acts as a nucleophile being made on the basis of the magnitude of the positive slope; the more positive values of w, which correspond to smaller increases in rate with increasing concentration of acid, are indicative of water acting as a base rather than a nucleophile. Although the division into three categories according to the value of w is not hard and fast, it frequently provides a useful indication of reaction mechanism.

It is obvious that treatments of this type must take into consideration the extent of protonation of the substrate. When the substrate is largely protonated, further changes in the observed rate will depend on the effect of the changing medium on the relevant activity coefficients. For reactions involving water (most A-2 processes) the rate frequently decreases at higher acidities as the activity of water declines. This occurs with amide hydrolysis, where benzamide, for example, has a rate maximum in 29% aqueous sulfuric acid (164). In concentrated acid the hydrolysis is so slow that it is often possible to convert nitriles to amides quantitatively, whereas in dilute acid the reaction proceeds all the way to the carboxylic acid level (166).

The generally accepted mechanism for amide hydrolysis [Eq. (7-78)] is consistent with these results. However, when R′ is a *tert*-butyl group, an A-1 route becomes available and, in this case, the reaction rate continues to rise as the amount of sulfuric acid in the mixture is increased [Eq. (7-79)] (*167*).

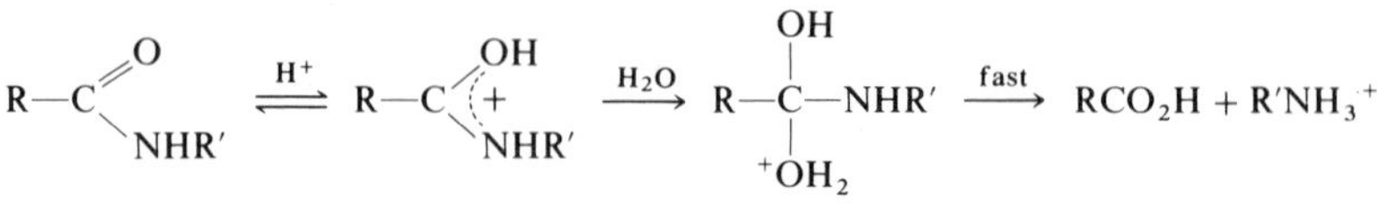

$$R-C\overset{O}{\underset{NHR'}{\diagup}} \xrightarrow{H^+} R-C\overset{OH}{\underset{NHR'}{\diagup}}+ \xrightarrow{H_2O} \overset{OH}{\underset{^+OH_2}{R-\overset{|}{\underset{|}{C}}-NHR'}} \xrightarrow{fast} RCO_2H + R'NH_3^+$$

$$(7\text{-}78)$$

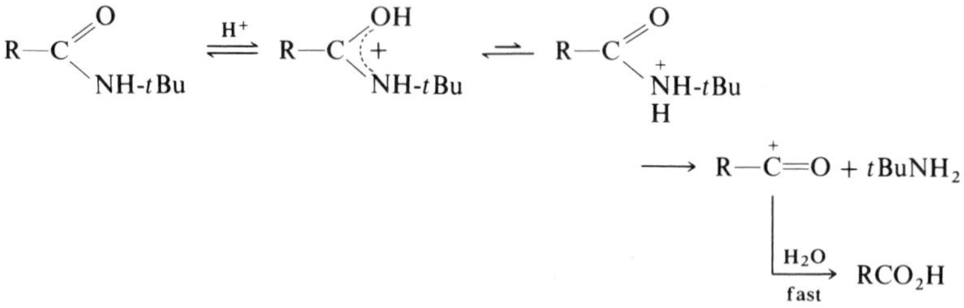

$$R-C\overset{O}{\underset{NH\text{-}t\,Bu}{\diagup}} \xrightarrow{H^+} R-C\overset{OH}{\underset{NH\text{-}t\,Bu}{\diagup}}+ \rightleftharpoons R-C\overset{O}{\underset{\underset{H}{\overset{+}{N}H\text{-}t\,Bu}}{\diagup}}$$

$$\longrightarrow R-\overset{+}{C}{=}O + t\text{BuNH}_2$$

$$\quad\quad\quad \underset{fast}{\overset{H_2O}{\longrightarrow}} RCO_2H \quad (7\text{-}79)$$

Other compounds that exhibit rate maxima are thion esters; ethyl thionbenzoates, for example, have fairly sharp rate maxima in the region 70–80% sulfuric acid. Again an A-2 reaction (with acyl–oxygen or –sulfur cleavage) is implicated [Eq. (7-80)]. By way of contrast, the rate of hydrolysis of ethyl thiolbenzoates increases monotonically as the concentration of sulfuric acid is increased (*168, 169*), and it is clear that acylium ions are being formed, probably as a result of rate-controlling protonation at sulfur [Eq. (7-81)] (*170*).

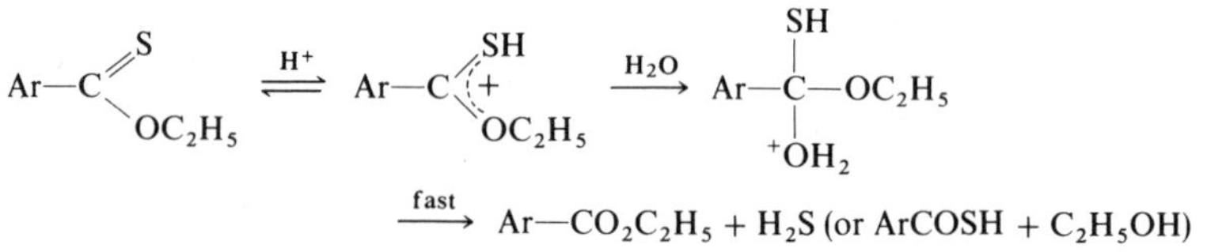

$$Ar-C\overset{S}{\underset{OC_2H_5}{\diagup}} \xrightarrow{H^+} Ar-C\overset{SH}{\underset{OC_2H_5}{\diagup}}+ \xrightarrow{H_2O} \overset{SH}{\underset{^+OH_2}{Ar-\overset{|}{\underset{|}{C}}-OC_2H_5}}$$

$$\xrightarrow{fast} Ar-CO_2C_2H_5 + H_2S \text{ (or ArCOSH} + C_2H_5OH) \quad (7\text{-}80)$$

$$Ar-C\overset{OH}{\underset{SC_2H_5}{\diagup}}+ \rightleftharpoons Ar-C\overset{O}{\underset{SC_2H_5}{\diagup}} \xrightarrow{H^+} Ar-\overset{+}{C}{=}O + C_2H_5SH$$

$$\quad\quad\quad \underset{fast}{\overset{H_2O}{\longrightarrow}} ArCO_2H \quad\quad\quad (7\text{-}81)$$

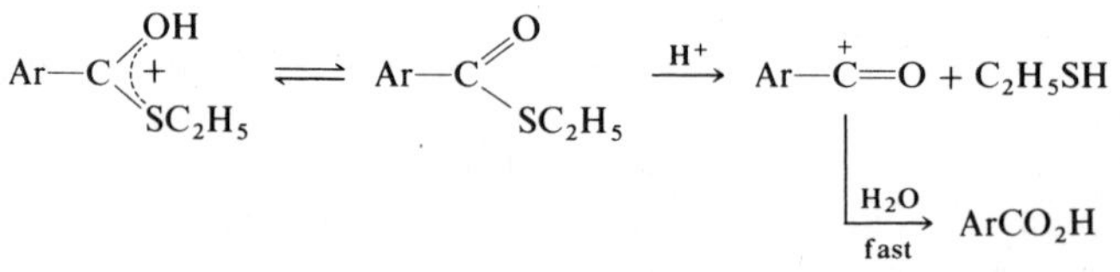

Ordinary esters, such as ethyl benzoate, hydrolyze by the A-2 route with acyl–oxygen cleavage in solutions more dilute than 80% sulfuric acid and by the A-1 route giving acylium ions in more concentrated acid (*170, 171*).

In addition to acidity function methods (*161, 172*), other treatments involving excess acidity (*173*), the *r* parameter (which designates the number of water molecules involved in the rate-controlling step) (*174*), and transition state activity coefficients (*175*) have been used to study the effect of changing acid concentration on reaction rate, particularly with regard to mechanism. The reader is referred to Liler's 1971 monograph for a comprehensive account of the subject (*166*) and to papers by Yates and others for details of more recent advances (*170–175*).

From a preparative point of view undoubtedly the most extensively studied reactions conducted in concentrated acid are electrophilic aromatic substitution and related aromatic isomerization reactions (*176*). Alkane protolysis, which leads to isomerization or formation of cracked products (*179*), is a subject of considerable current interest, as is the conversion of methanol and other heterosubstituted methanes to ethylene and higher hydrocarbons under conditions of high temperature and high acidity (*180*). The use of strongly acidic conditions in homogeneous photolytic conversions is considered in Chapter 6, Section III,C.

Turning to strongly basic conditions, the means of producing such systems and how they are used to determine equilibrium acidities of feeble organic acids are described in Chapter 2, Section IV,C. We shall confine ourselves here to a brief discussion of some of the other uses of strongly basic systems in organic chemistry. [For a comprehensive account of the base-catalyzed reactions of hydrocarbons and related compounds see the monograph by Pines and Stalick (*181*).]

Apart from the Cannizzaro reaction and some hydride transfer reactions of alkoxide ions (Chapter 4, Sections II,B and IV) most organic reactions that take place in strongly basic solution involve removal of a proton from an activated position in an organic molecule. They include oxidation, racemization, isotope exchange, isomerization, and halogenation reactions, and myriad condensation processes and alkylation processes. The chief activating groups in these cases are carbonyl, cyano, nitro, aryl, alkynyl, alkenyl, sulfonyl, and various heterocyclic units.

Fluorene (**15**) is an example of a weakly acidic hydrocarbon that readily undergoes autoxidation in strongly basic solution. This reaction has been thoroughly studied (*182*) and begins with proton removal from the methylene group of the central ring [Eq. (7-82)], followed by the initiation step [Eq. (7-83)] and the propagation steps [Eqs. (7-84) and (7-85)] of a chain reaction, and finally a nonradical step [Eq. (7-86)].

15

$$R_2CH_2 + B: \longrightarrow R_2CH^- + BH^+ \qquad (7\text{-}82)$$

$$R_2CH^- + O_2 \longrightarrow R_2CH\cdot + O_2\cdot^- \qquad (7\text{-}83)$$

$$R_2CH\cdot + O_2 \longrightarrow R_2CHOO\cdot \qquad (7\text{-}84)$$

$$R_2CHOO\cdot + R_2CH^- \longrightarrow R_2CHOO^- + R_2CH\cdot \qquad (7\text{-}85)$$

$$R_2CHOO^- \longrightarrow R_2C{=}O + HO^- \qquad (7\text{-}86)$$

Nitroaromatics accelerate the reaction significantly by providing a rapid route [Eq. (7-87)] for converting carbanion to the chain-carrying radical $R_2CH\cdot$

$$R_2CH^- + ArNO_2 \longrightarrow R_2CH\cdot + ArNO_2\cdot^- \qquad (7\text{-}87)$$

Fluorene has a pK_{HA} of 22.2 (Table 2-21), and so rather strongly basic solutions are required to produce significant concentrations of the carbanion. Both tBuOH/K^{+-}O-tBu and DMSO/tBuOH/K^{+-}O-tBu systems have been used for this purpose, with the former producing high yields of fluorenone. The DMSO mixture is sufficiently basic to produce high concentrations of the carbanion, as is revealed by the red color of the reaction mixture. The chief product in this case is not the ketone but rather 9-hydroxy-9-(methylsulfinylmethyl)fluorene. Other weakly acidic organic compounds are subject to autoxidation in strongly basic solution by mechanisms that also involve radical-anion intermediates (*182*).

Most carbon acids are much weaker than fluorene, and a number of techniques have been developed to produce significant quantities of carbanion so that autoxidation can proceed. In addition to the use of polar aprotic solvents such as DMSO, and dimethylformamide, phase-transfer catalysis has been successfully employed, with cryptands, crown ethers, polyethylene glycols, and quaternary ammonium salts being the favored catalysts (*183, 184*).

Powerfully basic media have also been used to introduce deuterium and tritium labels into organic molecules. This subject has been reviewed by Jones (*185, 186*).

A reaction of considerable importance in synthesis is alkylation at positions α to carbonyl and other activating groups, and strongly basic conditions are usually required for the purpose (*187, 188*). Although kinetic measurements are often used to obtain estimates of equilibrium acid strengths (Chapter 2, Section IV,C,3), it is not uncommon to find that the orders of kinetic and equilibrium basicities of hydrogen atoms in an organic compound are not the same. This is seldom the major source of error in using reaction rates to estimate pK values, since even a 10-fold shift in relative rates between a pair of compounds of the same equilibrium acidity would produce a discrepancy of only one pK unit, which is often within the uncertainty range of other methods of pK determination. However, a 10-fold shift between rate and equilibrium acidities of a pair of protons in the same molecule can have a significant effect on the distribution of products in a subsequent reaction. It is known, for example, that in 2-pentanone the less hindered proton is preferentially abstracted, and under nonequilibrium conditions (use of a strong, hindered base, e.g., lithium diisopropylamide, at low temperature in an aprotic medium, e.g., tetrahydrofuran) the terminal enolate anion is the chief product [Eq. (7-88)].

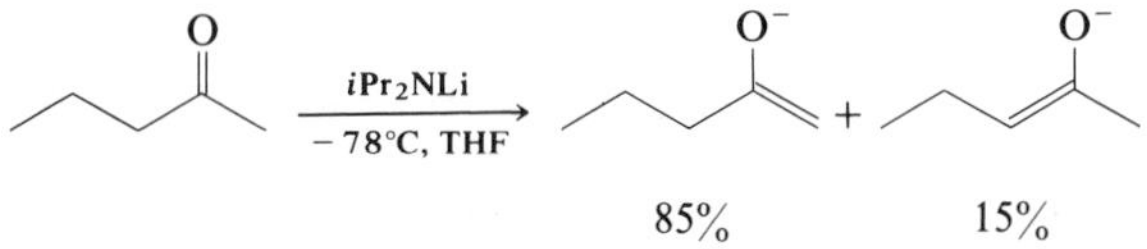

$$\text{(7-88)}$$

$$85\% \qquad 15\%$$

On the other hand, under equilibrium conditions (room temperature or above and use of a strong base, often in a protic solvent) the internal enolate anion becomes the major product, the degree of preference being cation and solvent dependent (*189, 190*).

α,β-Unsaturated ketones also give the less stable carbanion (i.e., the less conjugated ion) under kinetic conditions, with equilibration producing the more stable, more conjugated product [Eq. (7-89)].

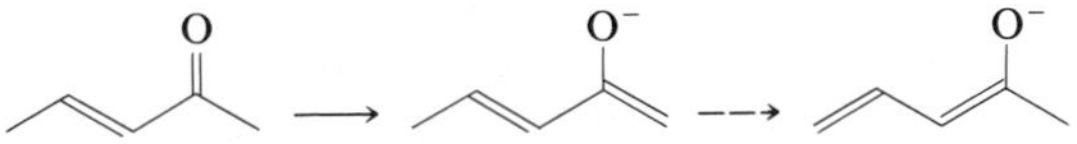

$$\text{(7-89)}$$

Even α-vinyl protons may be removed kinetically in preference to those in a conjugated position. When **16** is treated with lithium diisopropylamide in tetrahydrofuran at $-80°C$, the hydrogen atom marked with the asterisk is removed in preference to the unmarked hydrogen, although abstraction of the latter would lead to a resonance-stabilized anion (*191*).

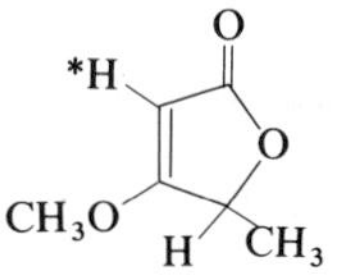

16

Sulfonium and ketimino groups are similar to carbonyl groups in activating protons on adjacent carbon atoms, and here, too, the regiospecificity depends on whether kinetic or equilibrium conditions are used (*192, 193*).

Dianions can often be generated from carbonyl compounds, particularly those containing more than one carbonyl group. The less stable carbanion, that corresponding to loss of the second proton, tends to be the more reactive in these cases, making it possible for alkylation or condensation reactions to take place at what would not normally be the preferred reaction site (*187, 194*).

The matters of reactivity, regiospecificity, and diastereoselectivity in these and other condensation and alkylation reactions that proceed through deprotonation processes are the subject of a vast literature, and the reader is referred elsewhere for leading references (*187, 189, 195–199*).

References

1. M. L. Bender, "Mechanism of Homogeneous Catalysis from Protons to Proteins," pp. 51–53. Wiley (Interscience), New York, 1971.

2. C. A. Bunton and J. L. Wood, *J. Chem. Soc.* p. 1522 (1955).

3. P. Salomaa, *Acta Chem. Scand.* **11**, 125, 132, and 141 (1957).

4. R. H. Smith, C. L. Denlinger, R. Kupper, S. R. Koepke, and C. J. Michejda, *J. Am. Chem. Soc.* **106**, 1056 (1984).

5. T. A. Modro and K. Yates, *J. Am. Chem. Soc.* **98**, 4247 (1976).

6. C. C. Chmiel and F. A. Long, *J. Am. Chem. Soc.* **78**, 3326 (1956).

7. C. K. Ingold, "Structure and Mechanism in Organic Chemistry," Chap. VII. Cornell Univ. Press, Ithaca, New York, 1953.

8. See, for example, J. P. Guthrie and P. A. Cullimore, *Can. J. Chem.* **58**, 1291 (1980).

9. For examples of protonation activating an organic compound toward oxidation, see refs. *10* and *11*.

10. F. Banoo and R. Stewart, *Can. J. Chem.* **47**, 3199 (1969).

11. D. G. Lee and M. van den Engh, *Can. J. Chem.* **50**, 2000, 3129 (1972).

12. R. Stewart and R. van der Linden, *Discuss Faraday Soc.* p. 211 (1960).

13. R. Stewart, *in* "Isotopes in Organic Chemistry" (E. Buncel and C. C. Lee, eds.), Vol. 2, pp. 289–291. Elsevier, Amsterdam, 1976.

14. R. Stewart and R. Srinivasan, *Acc. Chem. Res.* **11**, 271 (1978).

15. Based on analogy with the reaction described in ref. *16*.

16. R. Stewart and M. M. Mocek, *Can. J. Chem.* **41**, 1160 (1963).

17. R. P. Bell, "The Proton in Chemistry," 2nd ed., Chap. 8. Cornell Univ. Press, Ithaca, New York, 1973.

18. J. Toullec, *in* "Advances in Physical Organic Chemistry" (V. Gold and D. Bethell, eds.), Vol. 18, p. 1. Academic Press, London, 1972.

19. R. Taylor, *in* "Comprehensive Chemical Kinetics" (C. H. Bamford and C. F. H. Tipper, eds.), Vol. 13, Chap. 1. Elsevier, Amsterdam, 1972.

20. R. A. Cox and K. Yates, *Can. J. Chem.* **57**, 2944 (1979).

21. A. J. Kresge, G. Mylonakis, and L. E. Hakka, *J. Am. Chem. Soc.* **94**, 4197 (1972).

22. A. J. Kresge and T. S. Straub, *J. Am. Chem. Soc.* **105**, 3957 (1983).

23. See, for example, ref. *17*, p. 293 and ref. *18*, p. 61.

24. B. Capon and C. Zucco, *J. Am. Chem. Soc.* **104**, 7567 (1982). These authors suggest that the loss of the enolic proton is concerted with the protonation step in acid catalysis.

25. R. A. Burt, Y. Chiang, A. J. Kresge, and S. Szilagyi, *Can. J. Chem.* **62**, 74 (1984) and refs. therein.

26. J. M. Williams and M. M. Kreevoy, *in* "Advances in Physical Organic Chemistry" (V. Gold and D. Bethell, eds.), Vol. 6, p. 63. Academic Press, London, 1968.

27. H. Dahn and M. Ballenegger, *Helv. Chim. Acta* **52**, 2417 (1952).

28. Ref. *17*, p. 171.

29. T. H. Fife, *Acc. Chem. Res.* **5**, 264 (1972).

30. Y. Chiang, A. J. Kresge, M. O. Lahti, and D. P. Weeks, *J. Am. Chem. Soc.* **105**, 6852 (1983) and refs. therein.

31. C. A. Bunton, F. Davoudazedeh, and W. E. Watts, *J. Am. Chem. Soc.* **103**, 3855 (1981).

32. C. D. Ritchie and H. Fleischhauer, *J. Am. Chem. Soc.* **94**, 3481 (1972).

33. J. F. Kirsch and W. P. Jencks, *J. Am. Chem. Soc.* **86**, 837 (1964).

34. T. C. Bruice and G. L. Schmir, *J. Am. Chem. Soc.* **79**, 1663 (1957).

35. See ref. *33*.

36. The evidence for there being a switch in mechanism is described in ref. *1*, pp. 177–179 and in ref. *37*, pp. 483–487.

37. W. P. Jencks, "Catalysis in Chemistry and Enzymology." McGraw-Hill, New York, 1969.

38. W. P. Jencks and J. Carriuolo, *J. Biol. Chem.* **234**, 1272 and 1280 (1959).

39. See refs. *40* and *41* for other examples of activation of water by general bases.

40. R. M. Pollack and T. C. Dumsha, *J. Am. Chem. Soc.* **97**, 377 (1975).

41. J. Suh and I. M. Klotz, *J. Am. Chem. Soc.* **106**, 2373 (1984).

42. A. L. Green, G. L. Sainsbury, D. Saville, and M. Stansfield, *J. Chem. Soc.* p. 1583 (1958).

43. J. O. Edwards and R. G. Pearson, *J. Am. Chem. Soc.* **84**, 16 (1962).

44. K. Taira and D. G. Gorenstein, *J. Am. Chem. Soc.* **106**, 7825 (1984).

45. Ref. *37*, pp. 107–110.

46. J. A. Feather and V. Gold, *J. Chem. Soc.* p. 1752 (1965); see also ref. *1*, p. 105, ref. *37*, p. 96, and ref. *46a*.

46a. C. Roussel, A. T. Balaban, U. Berg, M. Chanon, R. Gallo, G. Klatte, J. A. Memiaghe, J. Metzger, D. Oniciu, and J. Pierrot-Sanders, *Tetrahedron* **39**, 4209 (1983).

47. A. J. Kirby, *in* "Advances in Physical Organic Chemistry" (V. Gold and D. Bethell, eds.), Vol. 17, p. 183. Academic Press, London, 1980.

48. R. P. Bell, *in* "Advances in Physical Organic Chemistry" (V. Gold, ed.), Vol. 4, p. 1. Academic Press, London, 1966.

49. W. P. Jencks *in* "Progress in Physical Organic Chemistry" (S. G. Cohen, A. Streitwieser, and R. W. Taft, eds.), Vol. 2, p. 63. Wiley (Interscience), New York, 1964.

50. J. P. Guthrie, *J. Am. Chem. Soc.* **102**, 5286 (1980).

51. W. P. Jencks, *Acc. Chem. Res.* **9**, 425 (1976).

52. L. H. Funderburk, L. Aldwin, and W. P. Jencks, *J. Am. Chem. Soc.* **100**, 5444 (1978).

53. R. Eliason and M. M. Kreevoy, *J. Am. Chem. Soc.* **100**, 7037 (1978).

54. R. A. McClelland and M. Coe, *J. Am. Chem. Soc.* **105**, 2718 (1983).

55. R. Stewart, *Can. J. Chem.* **62**, 907 (1984).

56. J. E. Critchlow, *J. Chem. Soc., Faraday Trans. 1* **68**, 1774 (1972).

57. J. P. Richard and W. P. Jencks, *J. Am. Chem. Soc.* **106**, 1396 (1984).

58. See, for example, R. L. Schowen, *in* "Progress in Physical Organic Chemistry" (A. Streitwieser and R. W. Taft, eds.), Vol. 9, pp. 309–310. Wiley (Interscience), New York, 1972.

59. V. Gold and D. C. A. Waterman, *J. Chem. Soc. B* p. 839 (1968).

60. R. Srinivasan and R. Stewart, *J. Am. Chem. Soc.* **98**, 7648 (1976).

61. R. Srinivasan and R. Stewart, *J. Chem. Soc., Perkin Trans. 2* p. 674 (1976).

62. F. Hibbert, *Acc. Chem. Res.* **17**, 115 (1984).

63. A. J. Kresge, *Chem. Soc. Rev.* **2**, 475 (1973); see also N.-Å. Bergman, Y. Chiang, and A. J. Kresge, *J. Am. Chem. Soc.* **100**, 5954 (1978).

64. Ref. *17*, p. 203.

65. F. G. Bordwell and D. L. Hughes, *J. Org. Chem.* **47**, 3224 (1982).

66. A. J. Kresge, S. G. Mylonakis, Y. Sato, and V. P Vitullo, *J. Am. Chem. Soc.* **93**, 6181 (1971); see also ref. *67*.

67. A. J. Kresge, *in* "Proton Transfer Reactions" (E. F. Caldin and V. Gold, eds.), Chap. 7. Chapman and Hall, London, 1975.

68. M. L. Ahrens, M. Eigen, W. Kruse, and G. Maass, *Ber. Bunsenges. Phys. Chem.* **74**, 380 (1970).

69. R. P. Bell and O. M. Lidwell, *Proc. R. Soc. London, Ser. A* **176**, 88 (1940).

70. A. Pross and S. S. Shaik, *J. Am. Chem. Soc.* **104**, 1129 (1982) and refs. therein.

71. See, for example, refs. *72* and *73*; see also D. J. Hupe and E. R. Pohl, *J. Am. Chem. Soc.* **106**, 5634 (1984).

72. J. R. Gandler and W. P. Jencks, *J. Am. Chem. Soc.* **104**, 1937 (1982).

73. W. P. Jencks, S. R. Brant, J. R. Gandler, G. Fendrich, and C. Nakamura, *J. Am. Chem. Soc.* **104**, 7045 (1982).

74. J. L. Kurz, *J. Org. Chem.* **48**, 5117 (1983).

75. A. Pross, *J. Org. Chem.* **49**, 1811 (1984).

76. E. Buncel, H. Wilson, and C. Chuaqui, *J. Am. Chem. Soc.* **104**, 4896 (1982).

77. C. D. Johnson, *Tetrahedron* **36**, 1461 (1980).

78. M. M. Kreevoy and I.-S. H. Lee, *J. Am. Chem. Soc.* **106**, 2550 (1984).

79. See, for example, refs. *63, 80, 81,* and *81a.*

80. J. R. Keeffe and A. J. Kresge, *in* "Techniques of Chemistry" (C. F. Bernasconi, ed.), 4th ed., Vol. 6, Chap. 10. Wiley (Interscience), New York, 1985.

81. C. J. Schlesener, C. Amatore, and J. K. Kochi, *J. Am. Chem. Soc.* **106**, 3567 (1984).

81a. M. M. Kreevoy, *in* "Isotopes in Organic Chemistry" (E. Buncel and C. C. Lee, eds.), Vol. 2, Elsevier, Amsterdam, 1976.

82. E. R. Thornton, *J. Am. Chem. Soc.* **89**, 2915 (1967).

83. W. J. Albery, *Prog. React. Kinet.* **4**, 355 (1967).

84. R. A. More O'Ferrall, *J. Chem. Soc. B* p. 274 (1970).

85. D. A. Jencks and W. P. Jencks, *J. Am. Chem. Soc.* **99**, 7948 (1977).

86. T. H. Lowry and K. S. Richardson, "Mechanism and Theory in Organic Chemistry," 2nd ed., Chap. 2 and Chap. 4. Harper and Row, New York, 1981.

87. C. F. Bernasconi and J. R. Gandler, *J. Am. Chem. Soc.* **100**, 8117 (1978).

88. W. J. Albery, *in* "Proton Transfer Reactions" (E. F. Caldin and V. Gold, eds.), Chap. 9. Chapman and Hall, London, 1975.

89. R. D. Gandour and R. L. Schowen, "Transition States of Biochemical Processes." Plenum, New York, 1978.

90. K. S. Venkatasubban, K. R. Davis, and J. L. Hogg, *J. Am. Chem. Soc.* **100**, 6125 (1978).

91. G. Gopalakrishnan and J. L. Hogg, *J. Org. Chem.* **49**, 1191 (1984).

92. D. Gerritzen and H.-H. Limbach, *J. Am. Chem. Soc.* **106**, 869 (1984).

93. K. S. Venkatasubban and R. L. Schowen, *J. Org. Chem.* **49**, 653 (1984), and refs. therein; see also ref. *94.*

94. W. P. Jencks, *Acc. Chem. Res.* **13**, 161 (1980).

95. G. H. Barnett and F. Hibbert, *J. Am. Chem. Soc.* **106**, 2080 (1984).

96. See, for example, refs. *97* and *98.*

97. R. F. Pratt and T. C. Bruice, *J. Org. Chem.* **37**, 3563 (1972).

98. P. Y. Bruice, *J. Am. Chem. Soc.* **106**, 5959 (1984).

99. R. P. Bell and W. C. E. Higginson, *Proc. R. Soc. London, Ser. A* **197**, 141 (1949).

100. Ref. *49*, p. 69.

101. M. L. Bender and A. Williams, *J. Am. Chem. Soc.* **88**, 2502 (1966).

102. R. D. Roberts, H. E. Ferran, M. J. Gula, and T. A. Spencer, *J. Am. Chem. Soc.* **102**, 7054 (1980).

103. T. A. Spencer, "Bioorganic Chemistry," Vol. 1. Academic Press, New York, 1977.

104. J. Hine, M. S. Cholod, and R. A. King, *J. Am. Chem. Soc.* **96**, 835 (1974).

105. P. Y. Bruice, *J. Am. Chem. Soc.* **105**, 4982 (1983).

106. P. Y. Bruice and T. C. Bruice, *J. Am. Chem. Soc.* **100,** 4793 (1978).

107. See, for example, ref. *108.*

108. J. A. Feather and V. Gold, *J. Chem. Soc.* p. 1752 (1965).

109. E. S. Lewis and L. H. Funderburk, *J. Am. Chem. Soc.* **89,** 2322 (1967).

110. J. Hine, J. G. Houston, J. H. Jensen, and J. Mulders, *J. Am. Chem. Soc.* **87,** 5050 (1965).

111. C. D. Gutsche, D. Redmore, R. S. Buriks, K. Nowotny, H. Grassner, and C. W. Armbruster, *J. Am. Chem. Soc.* **89,** 1235 (1967).

112. R. P. Bell, E. Gelles, and E. Möller, *Proc. R. Soc. London, Ser. A* **198,** 308 (1949).

113. R. Stewart, T. W. S. Lee, R. R. Perkins, K. Nagarajan, S. T. Tomic, and K. P. Shelly, unpublished results.

114. See also ref. *115.*

115. R. Stewart and T. W. S. Lee, *J. Org. Chem.* **47,** 2075 (1982).

116. C. F. Bernasconi, S. A. Hibdon, and S. E. McMurray, *J. Am. Chem. Soc.* **104,** 3459 (1982).

117. F. G. Bordwell, W. J. Boyle, J. A. Hautala, and K. C. Yee, *J. Am. Chem. Soc.* **91,** 4002 (1969).

118. D. Turnbull and S. Maron, *J. Am. Chem. Soc.* **65,** 212 (1943).

119. G. W. Wheland and J. Farr, *J. Am. Chem. Soc.* **65,** 1433 (1943).

120. A. J. Kresge, *Can. J. Chem.* **52,** 1897 (1974).

121. F. G. Bordwell and W. J. Boyle, *J. Am. Chem. Soc.* **94,** 3907 (1972); F. G. Bordwell, W. J. Boyle, and K. C. Yee, *J. Am. Chem. Soc.* **92,** 5926 (1970).

122. J. R. Keeffe, J. Morey, C. A. Palmer, and J. C. Lee, *J. Am. Chem. Soc.* **101,** 1295 (1979).

123. F. G. Bordwell, J. E. Bartmess, and J. A. Hautala, *J. Org. Chem.* **43,** 3107 (1978).

124. A. C. Lin, Y. Chiang, D. B. Dahlberg, and A. J. Kresge, *J. Am. Chem. Soc.* **105,** 5380 (1983).

125. R. A. More O'Ferrall, *in* "Proton Transfer Reactions" (E. F. Caldin and V. Gold, eds.), Chap. 8. Chapman and Hall, London, 1975.

126. D. B. Dahlberg, M. A. Kuzemko, Y. Chiang, A. J. Kresge, and M. F. Powell, *J. Am. Chem. Soc.* **105,** 5387 (1983).

127. C. D. Ritchie, *J. Am. Chem. Soc.* **91,** 6749 (1969).

128. R. F. Pearson and F. V. Williams, *J. Am. Chem. Soc.* **76,** 258 (1954).

129. M. Eigen, *Angew. Chem., Int. Ed. Engl.* **3,** 1 (1964).

130. T. M. Lowry and I. J. Faulkner, *J. Chem. Soc.* p. 2883 (1925).

131. H. M. Dawson and E. Spivey, *J. Chem. Soc.* p. 2180 (1930).

132. R. P. Bell and P. Jones, *J. Chem. Soc.* p. 88 (1953).

133. A. F. Hegarty and W. P. Jencks, *J. Am. Chem. Soc.* **97,** 7188 (1975).

134. W. J. Albery and J. S. Gelles, *J. Chem. Soc., Faraday Trans. 1* **78,** 1569 (1982).

135. E. S. Hand and W. P. Jencks, *J. Am. Chem. Soc.* **97,** 6221 (1975).

136. C. G. Swain, D. A. Kuhn, and R. L. Schowen, *J. Am. Chem. Soc.* **87,** 1553 (1965).

137. M. M. Kreevoy, T. Liang, and K.-C. Chang, *J. Am. Chem. Soc.* **99,** 5207 (1977).

138. C. G. Swain, *J. Am. Chem. Soc.* **72,** 4578 (1950).

139. Ref. *17,* pp. 148–156; see also ref. *18.*

140. See, for example, K.-A. Engdahl, H. Bivehed, P. Ahlberg, and W. H. Saunders, *J. Am. Chem. Soc.* **105,** 4767 (1983) and refs. therein.

141. J. Hine, *Acc. Chem. Res.* **11,** 1 (1978).

142. J. Hine and H.-M. Tsay, *J. Org. Chem.* **48,** 3797 (1983).

142a. J. Hine and A. Sinha, *J. Org. Chem.* **49,** 2186 (1984).

143. J. Hine, D. E. Miles, and J. P. Zeigler, *J. Am. Chem. Soc.* **105,** 4374 (1983).

144. J. Hine and W.-S. Li, *J. Am. Chem. Soc.* **98,** 3287 (1976).

145. R. Stewart, S. J. Gumbley, and R. Srinivasan, *Can. J. Chem.* **57,** 2783 (1979).

146. R. Srinivasan and R. Stewart, *Can. J. Chem.* **53,** 224 (1975).

147. J. A. Elvidge, J. R. Jones, R. Salih, M. Shandala, and S. E. Taylor, *J. Chem. Soc., Perkin Trans. 2* p. 447 (1980); *J. Chem. Res. (S)* p. 172 (1980).

148. J. R. Jones and S. E. Taylor, *Chem. Soc. Rev.* **10**, 329 (1981).

149. W. P. Jencks, *J. Am. Chem. Soc.* **81**, 475 (1959).

150. P. Lumme, P. Lahermo, and J. Tummavuori, *Acta Chem. Scand.* **19**, 2175 (1965).

151. See, for example, Eq. (7-71) and ref. *152.*

152. C. A. Bunton, D. R. Llewellyn, K. G. Oldham, and C. A. Vernon, *J. Chem. Soc.* p. 3574 (1958).

153. W. P. Jencks, *in* "Progress in Physical Organic Chemistry" (S. G. Cohen, A. Streitwieser, and R. W. Taft, eds.), Vol. 2, p. 73. Wiley (Interscience). New York, 1964.

154. R. P. Bell and O. M. Lidwell, *Proc. R. Soc. London, Ser. A* **176**, 88 (1940).

155. R. Srinivasan, S. J. Gumbley, and R. Stewart, *Tetrahedron* **35**, 1257 (1979).

156. T. W. S. Lee, S. J. Rettig, R. Stewart, and J. Trotter, *Can. J. Chem.* **62**, 1194 (1984).

157. R. Stewart and R. Srinivasan, *Can. J. Chem.* **53**, 2906 (1975).

158. R. Stewart and J. M. McAndless, *J. Chem. Soc., Perkin Trans. 2* p. 376 (1972).

159. J. A. Elvidge, J. R. Jones, C. O'Brien, E. A. Evans, and H. C. Sheppard, *J. Chem. Soc., Perkin Trans. 2* p. 1889 (1973).

160. R. Stewart and R. Srinivasan, *Can. J. Chem.* **59**, 957 (1981).

160a. L. Zucker and L. P. Hammett, *J. Am. Chem. Soc.* **61**, 2791 (1939).

161. F. A. Long and M. A. Paul, *Chem. Rev.* **57**, 935 (1957).

162. J. F. Bunnett, *J. Am. Chem. Soc.* **83**, 4956, 4968, and 4978 (1961); see also ref. *163.*

163. J. F. Bunnett and F. P. Olsen, *Can. J. Chem.* **44**, 1899 and 1917 (1966).

164. J. T. Edward and S. C. R. Meacock, *J. Chem. Soc.* p. 2000 (1957); see also ref. *165.*

165. A. Cipiciano, P. Linda, G. Savelli, and C. A. Bunton, *J. Org. Chem.* **48**, 1349 (1983).

166. M. Liler, "Reaction Mechanisms in Sulphuric Acid," p. 207. Academic Press, London, 1971.

167. L. M. Druet and K. Yates, *Can. J. Chem.* **62**, 2401 (1984).

168. J. T. Edward, S. C. Wong, and G. Welch, *Can. J. Chem.* **56**, 931 (1978).

169. J. T. Edward and S. C. Wong, *J. Am. Chem. Soc.* **99**, 7224 (1977).

170. R. A. Cox and K. Yates, *Can. J. Chem.* **60**, 3061 (1982).

171. R. A. Cox, M. F. Goldman, and K. Yates, *Can. J. Chem.* **57**, 2960 (1979).

172. C. C. Greig, C. D. Johnson, S. Rose, and P. G. Taylor, *J. Org. Chem.* **44**, 745 (1979).

173. R. A. Cox and K. Yates, *Can. J. Chem.* **59**, 2853 (1981) and refs. therein.

174. K. Yates, *Acc. Chem. Res.* **4**, 136 (1971).

175. K. Yates and T. A. Modro, *Acc. Chem. Res.* **11**, 190 (1978).

176. See, for example, refs. *166, 177,* and *178.*

177. R. M. G. Roberts, *J. Org. Chem.* **47**, 4050 (1982).

178. G. A. Olah, *Angew. Chem., Int. Ed. Engl.* **12**, 173 (1973).

179. P.-L. Fabre, J. Devynck, and B. Trémillon, *Chem. Rev.* **82**, 591 (1982) and refs. therein.

180. G. A. Olah, H. Doggweiler, J. D. Felberg, S. Frohlich, M. J. Grdina, R. Karpeles, T. Keumi, S. Inaba, W. M. Ip, K. Lammertsma, G. Salem, and D. C. Tabor, *J. Am. Chem. Soc.* **106**, 2143 (1984) and refs. therein.

181. H. Pines and W. M. Stalick, "Base-Catalyzed Reactions of Hydrocarbons and Related Compounds." Academic Press, New York, 1977.

182. G. A. Russell, E. G. Janzen, A. G. Bemis, E. J. Geels, A. J. Moye, S. Mak, and E. T. Strom, *Adv. Chem. Ser.* **51**, 112 (1965).

183. M. Halpern, Y. Sasson, and M. Rabinovitz, *J. Org. Chem.* **49**, 2011 (1984).

184. R. Neumann and Y. Sasson, *J. Org. Chem.* **49**, 1282 (1984) and refs. therein.

185. J. R. Jones, "Ionization of Carbon Acids," Chap. 11. Academic Press, London, 1973.

186. J. R. Jones, *in* "Progress in Physical Organic Chemistry" (A. Streitwieser and R. W. Taft, eds.), Vol. 9, p. 241. Wiley (Interscience), New York, 1972.

187. H. O. House, "Modern Synthetic Reactions," pp. 546–570. Benjamin, Menlo Park, 1972.

188. A. A. Solav'yanov, J. P. Beletskaya, and O. A. Reutov, *J. Org. Chem. USSR (Engl. Transl.)* **19**, 1592 (1983).

189. J. d'Angelo, *Tetrahedron* **32,** 2979 (1976).

190. A. S. Narala, *Tetrahedron Lett.* **22,** 4119 (1981).

191. O. Miyata and R. R. Schmidt, *Angew. Chem., Int. Ed. Engl.* **21,** 637 (1982).

192. V. Cerè, C. Paolucci, S. Pollicino, E. Sandri, and A. Fava, *Tetrahedron Lett.* **24,** 839 (1983).

193. J. K. Smith, D. E. Bergbreiter, and M. Newcomb, *J. Am. Chem. Soc.* **105,** 4396 (1983).

194. M. Alderdice, C. Spino, and L. Weiler, *Tetrahedron Lett.* **25,** 1643 (1984) and refs. therein.

195. S. Masamune, J. W. Ellingboe, and W. Choy, *J. Am. Chem. Soc.* **104,** 5526 (1982).

196. H. O. House, W. V. Phillips, and D. Van Derveer, *J. Org. Chem.* **44,** 2400 (1979).

197. I. Reichelt and H.-U. Reissig, *Justus Liegigs Ann. Chem.* p. 531 (1984).

198. A. Krief, *Tetrahedron* **36,** 2531 (1980).

199. J. S. Sawyer, T. L. Macdonald, and G. J. McGarvey, *J. Am. Chem. Soc.* **106,** 3376 (1984).

Index

307

ORGANIC CHEMISTRY

A SERIES OF MONOGRAPHS

EDITOR

HARRY H. WASSERMAN
Department of Chemistry
Yale University
New Haven, Connecticut

1. Wolfgang Kirmse. CARBENE CHEMISTRY, 1964; 2nd Edition, 1971

2. Brandes H. Smith. BRIDGED AROMATIC COMPOUNDS, 1964

3. Michael Hanack. CONFORMATION THEORY, 1965

4. Donald J. Cram. FUNDAMENTALS OF CARBANION CHEMISTRY, 1965

5. Kenneth B. Wiberg (Editor). OXIDATION IN ORGANIC CHEMISTRY, PART A, 1965; Walter S. Trahanovsky (Editor). OXIDATION IN ORGANIC CHEMISTRY, PART B, 1973; PART C, 1978; PART D, 1982

6. R. F. Hudson. STRUCTURE AND MECHANISM IN ORGANO-PHOSPHORUS CHEMISTRY, 1965

7. A. William Johnson. YLID CHEMISTRY, 1966

8. Jan Hamer (Editor). 1,4-CYCLOADDITION REACTIONS, 1967

9. Henri Ulrich. CYCLOADDITION REACTIONS OF HETEROCUMULENES, 1967

10. M. P. Cava and M. J. Mitchell. CYCLOBUTADIENE AND RELATED COMPOUNDS, 1967

11. Reinhard W. Hoffmann. DEHYDROBENZENE AND CYCLOALKYNES, 1967

12. Stanley R. Sandler and Wolf Karo. ORGANIC FUNCTIONAL GROUP PREPARATIONS, VOLUME I, 1968; 2nd Edition, 1983; VOLUME II, 1971; VOLUME III, 1972

13. Robert J. Cotter and Markus Matzner. RING-FORMING POLYMERIZATIONS, PART A, 1969; PART B, VOLUMES 1 AND 2, 1972

14. R. H. DeWolfe. CARBOXYLIC ORTHO ACID DERIVATIVES, 1970

15. R. Foster. ORGANIC CHARGE-TRANSFER COMPLEXES, 1969

16. James P. Snyder (Editor). NONBENZENOID AROMATICS, VOLUME I, 1969; VOLUME II, 1971

17. C. H. Rochester. ACIDITY FUNCTIONS, 1970

18. Richard J. Sundberg. THE CHEMISTRY OF INDOLES, 1970

19. A. R. Katritzky and J. M. Lagowski. CHEMISTRY OF THE HETEROCYCLIC N-OXIDES, 1970

20. Ivar Ugi (Editor). ISONITRILE CHEMISTRY, 1971

21. G. Chiurdoglu (Editor). CONFORMATIONAL ANALYSIS, 1971

22. Gottfried Schill. CATENANES, ROTAXANES, AND KNOTS, 1971

23. M. Liler. REACTION MECHANISMS IN SULPHURIC ACID AND OTHER STRONG ACID SOLUTIONS, 1971

24. J. B. Stothers. CARBON-13 NMR SPECTROSCOPY, 1972

25. Maurice Shamma. THE ISOQUINOLINE ALKALOIDS: CHEMISTRY AND PHARMACOLOGY, 1972

26. Samuel P. McManus (Editor). ORGANIC REACTIVE INTERMEDIATES, 1973

27. H. C. Van der Plas. RING TRANSFORMATIONS OF HETEROCYCLES, VOLUMES 1 AND 2, 1973